DON'T TRY THIS AT HOME!

THE PHYSICS OF HOLLYWOOD MOVIES

BY ADAM WEINER

PUBLISHING
New York

Vice President and Publisher: Maureen McMahon
Editorial Director: Jennifer Farthing
Acquisitions Editor: Megan Gilbert
Development Editor: Cynthia Ierardo
Production Editor: Karina Cueto
Production Designer: Todd Bowman
Cover Designer: Carly Schnur
Illustrator: Dusty Deyo

Published by Kaplan Publishing, a division of Kaplan, Inc.
1 Liberty Plaza, 24th Floor
New York, NY 10006

Printed in the United States of America

September 2007
10 9 8 7 6 5 4 3 2 1

ISBN-13: 978-1-4195-9406-9
ISBN-10: 1-4195-9406-0

Advance Praise for
Don't Try This At Home! The Physics of Hollywood Movies

"Holy Newtonian mechanics, Batman! A book that makes physics fun?! Yes, it's true! Can you really blow up an asteroid the size of Texas? Can you drill down to the center of the Earth? Boldly go where no movie has ever gone before; real science!"

PHIL PLAIT
NASA Education Resource Director (NERD), Sonoma State University E/PO Group and author of *Bad Astronomy*

"The dramatic conclusion of the movie *Speed* shows Keanu Reaves and Sandra Bullock hurtling along in a city bus toward a 50-foot gap in an LA freeway bridge. Can the bus really fly over the gap, or is this another stunt where Hollywood cheats on the laws of physics? In a fun new approach to physics, Adam Weiner engages students in the physics analysis of famous movie scenes.

Combining a careful and systematic development of the physics principles with a jaunty and entertaining description of the movies, Weiner repeatedly shows students how simple approximations can reduce a complicated movie scene to its physics essentials. This reduction of a complicated problem to essential elements is central to the practice of physics and is a widely applicable tool of critical thinking that physics can teach the general student. Weiner cleverly disguises lots of practice in such thinking with the fun of Hollywood movies."

THOMAS O'NEIL
Professor of Physics, University of California, San Diego

"There is sound educational research that suggests the best way to teach students new concepts in ways they will remember them is to confront their own misconceptions. There is perhaps no place better to turn to for built-in misconceptions about physics than Hollywood Movies. There's lots of fodder here for students to have fun with, and to use as a springboard to learn about the real world, which, as I often say, is actually far more fascinating than the science fiction universe."

LAWRENCE M. KRAUSS
Ambrose Swasey Professor and Director of Center for Education and Research in Cosmology and Astrophysics, Case Western Reserve University, and author of *The Physics of Star Trek* and *Hiding in the Mirror*

"With cinematic effects advancing at light speed, for many of us, the boundary between the possible and the impossible has become indistinct. Teaching a 'Hollywood Physics' course myself, this ambiguity has proven an exceptional tool. Adam Weiner's approach not only untangles the real from the computer generated, but also through common-sense estimation and arguments, shows us that we are all physicists by nature. His writing is clear, engaging, funny, and just as accessible as the popular movies he discusses."

WESSYL KELLY
Ph.D. candidate who teaches a course on Hollywood Movie Physics, Brown University

Table of Contents

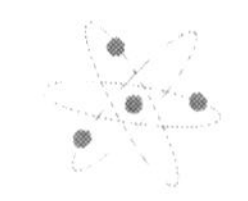

About the Author

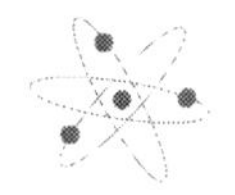

ADAM WEINER teaches physics and advanced placement physics at The Bishops School, a private college preparatory school in La Jolla, California. He has also done physics content editing for Kaplan review books for the SAT exam, the MCAT, and its British equivalent, the BMAT. With a Masters in Geophysics from the University of Hawaii and a second Masters in the Performing Arts from Binghamton University, science education through entertaining and engaging means became his teaching trademark. In addition to teaching science (and watching movies), he is an avid runner, surfer, and reader. This book is dedicated to his wife Rebecca, without whom this book would never have been written, his son Jacob, and his 47-year-old Amazon parrot Ralph.

Introduction

HOW MANY TIMES HAVE YOU asked yourself while watching some phenomenal stunt in an action movie, or some amazing technological achievement in a science fiction movie, "Could that really happen?" Have you ever wondered if it's possible to drive a vehicle on an asteroid, jump over a house while riding a motorcycle, drill to the center of the Earth, or travel to distant planets? What kinds of forces are involved in car collisions and high-speed chases? Could the ocean really freeze in a matter of seconds if it got cold enough? If the Earth's magnetic field disappeared what would happen to the planet? Well, with a rudimentary knowledge of some basic physics and a sense of humor, you can answer many of these questions yourself. Why not use film scenes as a springboard into learning fundamental principles of physics? In this book we look at the physics represented in Hollywood movies, analyze the stunts, and assess the scientific principles.

I teach physics in a private high school, and the idea for this book actually came out of a class project that we do in my classes every year. This project consists of critiquing scenes from Hollywood movies in terms of the accuracy of the physical principles involved. How realistic are they? When have we crossed over from the realm of the maybe possible, to the completely and insultingly ridiculous? The results are sometimes so entertaining, and provide such an engaging teaching tool, that I thought, why not do an entire book that covers introductory physics material through the analysis of movie physics? Obviously, action and science fiction movies lend themselves best to this kind of project, but sometimes you can find real gems where you might not expect them—in children's movies, slapstick comedies, and even true-to-life dramas.

This book is intended not just for students taking a first year physics course; it's for anyone interested in physics and movies, or movies and physics! The level of physics is fundamental, the math level is algebra based (similar to that in a high school AP Physics B course, or a non-calculus based introductory college course).* But unlike many "popular science books" we will analyze scenes quantitatively. Each chapter includes a review of the physical principles we will need to understand

* For students and teachers: This is obviously not a typical Kaplan review book, but you can use it pretty effectively to cover most of the major physics principles that you'll find on the SAT Subject Test Physics, and it contains much (but not all) of the material covered in the AP Physics B course. I think you'll find it an effective and engaging way to learn physics.

to analyze the movie scenes, supplemented with activities and additional problems to solidify the concepts (although we have to leave out derivations of equations to keep things concise).

Keep in mind that as we assess the physics of these movie scenes we will often use the art of estimation to get to the gist of a problem. Many of the solutions will not be as exact as those you'll find in the end of the chapter textbook problems. For that reason we will be a bit more free and loose with things like significant figures. So I'm telling you ahead of time, don't worry if we round something off here or there, or if the numbers of "sig figs" are inconsistent from one problem to another. We are going for the big picture here: Can we apply fundamental principles of physics in a meaningful way in order to understand what is happening in a scene, and to assess whether or not it is possible? Can we use the action in a movie scene to learn physics, and to better understand how the physical universe works? The answer is yes!

Oh, and one more thing. I also think of this book as a *public service!* It's time to take a stand against the insidious proliferation of bad movie physics, and expose it to the light of the Sun (which is generated by nuclear fusion of high energy hydrogen nuclei!) It is important for the younger generation (and everyone else) to understand that what they see in the movies—the stunts, the supposed scientific jargon, etc.—may or may not correspond to reality as we know it. Just because it's in a movie doesn't mean it makes any sense or that it's true. Even James Bond can't collide with a wall at 50 miles per hour and emerge undamaged. Is it really possible to jump off of a bridge and catch a helicopter? The odds of accomplishing that is so low, it shouldn't constitute part of any action hero strategy. Just because some "scientist" in a movie tells you that microwave radiation from the Sun can burn through a bridge, don't necessarily believe it! Learn some physics, and then we can discuss it like reasonable citizens of an enlightened society!

Disclaimer: Remember, many of the scenes and stunts that we are going to analyze were done with "professional drivers on closed courses," camera sleight of hand, or computer graphics. They may range from the potentially realistic, to the highly dangerous, to the absolutely impossible, including the completely incomprehensible. Never forget the warning plastered across the cover of this book. And when in doubt . . . DON'T TRY THIS AT HOME!

Kinematics

ONE OF THE MOST EXCITING and heart-stopping movie standbys is the high-speed car chase. Cars careen around corners, smash through department store windows, and crash into oncoming street signs on their way to the final climatic moment. Sometimes they jump over obstacles or drive on two wheels in order to fit through an alley. How realistic are these scenes? When do they cross the line from the possible to the ludicrous? We can analyze many of these situations with a basic and fundamental understanding of Newtonian mechanics.

XXX AND THE DIFFERENCE BETWEEN DISPLACEMENT AND DISTANCE

Action movies are not particularly known for their scintillating dialogue, sophisticated plots, or Academy Award–winning acting. They are cherished for their car crashes, their explosions, and the almost superhuman abilities and amazing good luck of their heroes. Extreme situations and improbable or impossible stunts are on the menu when you rent last summer's action-packed blockbuster. What's not usually on the menu is an accurate presentation of physical principles as we know them.

We begin our excursion into the physics of movies, with a scene from the quintessential action genre blockbuster, *XXX*. This movie epitomizes the Hollywood action movie, with its bad acting, bad writing, clichéd plot devices, and some pretty outlandish action sequences. We are going to discuss several scenes from *XXX* at varying levels of detail because it is full of outrageous physics (and at times affronts common sense).

The Movie

In *XXX,* Xander Cage (how's that for a name?), played by Vin Diesel (how's that for a name?), is an extreme sports counter-culture rebel looking for adrenaline thrills while sticking it to "the man" whenever he can. He's tattooed, tough, and fearless. The National Security Agency is having big problems keeping their agents alive in the Czech Republic, where the criminal group called Anarchy 99, run by a homicidal Russian lunatic named Yorgi (Marton Csokas) is up to something particularly nasty. In fact, in addition to the usual drugs and prostitution, Yorgi and his cohorts have

Photo: Rounding a sharp turn in the movie *Speed.*

decided to unleash deadly poison gas on major cities throughout the world. This will, according to Yorgi's logic, unleash a chain reaction of political disintegration worldwide, culminating in the demise of civilization. In the subsequent anarchy Yorgi feels that he will finally be able to do what he wants (even though it would appear he already does whatever he wants).

Now Yorgi is pretty savvy about these CIA operatives, thus their short shelf life, so the plan is to find some real hardened tough-guy types that the NSA can send over there. These guys should fit in better with the Russian criminals without giving themselves away—and they're more expendable than the real agents are. It's a win-win situation! They recruit Vin Diesel for the job, and threaten him with federal prison if he doesn't comply. Once in Europe he infiltrates Yorgi's operations, falls in love with his girlfriend Yelena (Asia Argento) who predictably turns out to be one of the good guys, gets discovered, but finally, through a series of incredibly heroic stunts, foils Yorgi's plan and saves the world!

The Scene. In *XXX's* climactic scenes 25 and 27, Yorgi has just released the automated boat containing the toxic gas containers onto the Danube. It is traveling "80 mph at least" according to Vin Diesel, as he and Yelena frantically try to stay parallel with it while driving on the road adjacent to the river in their specially outfitted GTO. They must find a way to board the boat and disarm the containers before it's too late. The situation becomes even more challenging when the road veers away from the waterfront, travels a couple of miles inland, and passes through a charming country village before making its way back to the river. Fortunately the car has been equipped with rocket launchers that the two heroes use to blast wooden carts and bales of hay out of their way so they don't have to slow down much. They do have to make a few pretty sharp turns however, and with the extra distance traveled, we might be surprised to see that they return to the river at the exact moment the boat arrives there. Let's examine these scenes to illustrate some fundamental kinematics concepts.

KINEMATICS

Kinematics is the study that describes *how* an object moves. It answers questions like: "Where is the object?" and: "How fast is it going?" without worrying about the *causes* of motion. To describe motion we need to review some fundamental definitions.

Distance and Displacement

Distance and *displacement* are similar ideas in that they both represent an amount that is measurable in feet, meters, miles, or any other length unit. *Distance* is the total length traveled during some motion. Distance does not depend on which direction you are traveling, or whether you change direction during the motion. A quantity that does not depend on any particular direction is called a *scalar* quantity. For example, mass is an example of a scalar quantity, as the property of mass has no direction.

Unlike distance, *displacement* does have a direction associated with it. A quantity that is specified by both an amount and a direction is called a *vector* quantity.

If an object moves from one point to another its displacement is the straight-line distance from the beginning to the end of a motion. Displacement represents the overall change in the objects position rather than the total distance traveled.

$$\text{Displacement} = \text{final position} - \text{initial position} = x_f - x_i = \Delta x.$$

Example 1: A driver being pursued in a high-speed car chase enters the highway at mile marker 6 heading north. The highway is straight. At mile marker 8 he skids to a stop and reverses direction due to a roadblock ahead. He drives back to mile 7 where he is forced to stop. What is the total distance of the chase? What is the total displacement?

Solution: For the trip he covers a total distance of 3 miles because he drives 2 miles north, turns around, and travels another mile south, but his displacement is one mile to the north. In other words he ended up one mile north of where he entered the highway.

Speed and Velocity

Average *speed* (a scalar) is defined as the total distance traveled during a motion divided by the time it took to complete the motion.

$$s = \frac{d}{\Delta t}$$

Average *velocity* (a vector) is defined as the displacement from the beginning to the end of a motion divided by the time it took to complete that motion.

$$v = \frac{\Delta x}{\Delta t} = \frac{x - x_0}{t - t_0} \text{ (in one dimension)}$$

Velocity is represented by the preceding formula, where x_0 and t_0 represent the initial position and beginning time, respectively, for the motion, and x and t represent, in order, the final position and time. Usually we will set t_0 equal to 0.

Example 2: In the car chase mentioned in example 1, it takes the driver 70 seconds to get from mile marker 6 to mile marker 8, and another 30 seconds to get back to mile marker 7.

a) What is his average speed for the entire chase?

b) What is his average velocity for the entire chase?

Solution:

a) $s = \frac{d}{\Delta t} = \frac{3 \text{ mi}}{100 \text{ s}} \times \frac{3{,}600 \text{ s}}{1 \text{ hr}} = 108 \text{ mi/hr}$

b) $v = \frac{\Delta x}{\Delta t} = \frac{1 \text{ mi}}{100 \text{ s}} \times \frac{3{,}600 \text{ s}}{1 \text{ hr}} = 36 \text{ mi/hr due north}$

THINKING PHYSICS

You throw a motorcycle straight up into the air. (You're very strong, similar to Superman, the Incredible Hulk, or Vin Diesel in *XXX*.) Refer to the lesson on acceleration below to answer the following questions. What is the direction of the acceleration (a) on the way up, (b) at the top of its flight, and (c) on the way down?

Answer: The acceleration is downward in each case. Remember that acceleration refers to the rate of *change* of velocity. On the way up the motorcycle is slowing down, and therefore the acceleration is downward. The motorcycle never spends any time at the top. As soon as it finishes slowing down upward, it starts speeding up downward. The acceleration is continuously in the downward direction.

Another way to look at it is for the first 70 seconds v = 2 mi/70s × 3,600 s/hr = 103mi/hr due north followed by v = 1 mi/30s = 120 mi/hr due south. Because 70 percent of the time he is traveling at a velocity of 103 mi/hr due north, and 30 percent of the time he is traveling 120mi/hr due south, we can calculate the average velocity as follows:

$$108 \text{ mi/hr } (0.70) + (-120 \text{ mi/hr})(0.30) = 36 \text{ mi/hr}$$

Acceleration

If the velocity of an object is changing then the object is accelerating. *Acceleration* (a vector) is the rate at which velocity is changing.

$$a = \frac{\Delta v}{\Delta t} = \frac{v - v_0}{t - t_0}$$

If an object has an *acceleration* of 5 m/s^2, this means that each second it will change its *velocity* by an amount equal to 5 m/s.

A special case of acceleration occurs for freely falling objects near the surface of the Earth. Neglecting significant air resistance, all objects will accelerate towards the earth at a rate of 9.8 m/s^2.

Example 3: If you throw the motorcycle upward at an initial upward velocity of 25 m/s, what is its velocity 1 second, 2 seconds, and 3 seconds later?

Solution: Each second the velocity changes by an amount equal to 9.8 m/s. On the way up the motorcycle is slowing down, therefore after one second v = 25 m/s – 9.8 m/s = 15.2 m/s. At t = 2 s, v = 15.2 – 9.8 = 5.4 m/s. At t = 3 s, v = 5.4 – 9.8 = –4.4 m/s. Therefore, the motorcycle reaches the top of its flight somewhere between 2 and 3 s.

Graphs of Motion

It is often convenient to represent the motion of objects as a function of time visually on a graph.

A position time (x-t) graph represents the *location* of an object at any time. Because the rate of change of position with time is the velocity, the *slope* of any section of an x-t graph represents the *average velocity* of the object during that time. The instantaneous velocity can be determined from the slope of a tangent line at a specific point on the *x-t* graph.

A velocity time (*v-t*) graph directly tells you the velocity of an object at any time. The *slope* of the *v-t* graph gives you the *acceleration* of the object.

The following *x-t, v-t*, and *a-t* graphs all represent the same motion of an object speeding up at a constant rate without changing direction.

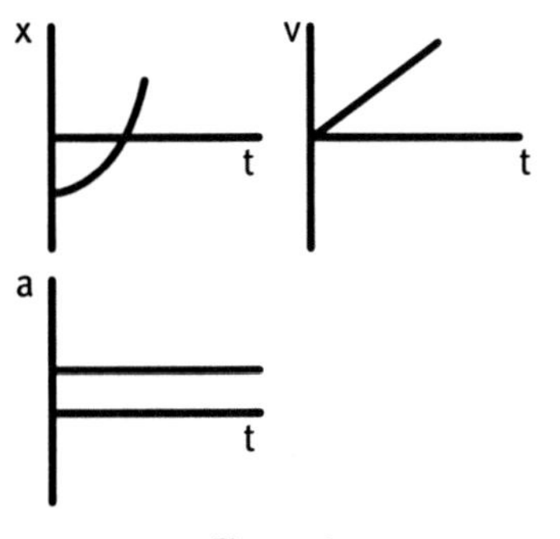

Figure 1

Motion with Constant Acceleration

In situations where the rate of acceleration of an object is *constant* (unchanging) over some period of time, we can obtain some useful equations that quantify the relationship between the position, velocity, acceleration, and time of motion of the object. We define the following variables:

x_0 = initial position
x = final position
v_0 = initial velocity
v = final velocity
a = acceleration
t = elapsed time (assuming motion starts at $t = 0$)
$\Delta x = x - x_0$

The following *kinematic equations* can be derived:

$$1)\ v = v_0 + at$$

$$2)\ x - x_0 = v_0 t + \frac{1}{2}at^2$$

$$3)\ x - x_0 = \frac{v + v_0}{2}t$$

$$4)\ v - v_0 = 2a(x - x_0)$$

Example 4: Getting back to the movie, let's say Vin Diesel's stunt double needs to speed up as fast as he can go on his motorcycle before attempting an extreme jump. If he has an initial speed of just above 50 miles per hour (22 m/s), and is able to reach a final speed of 70 miles per hour (31 m/s) over a 50-meter distance just before his jump, what average acceleration must the motorcycle be capable of to do this?

Solution: Because we know the initial and final velocity, and the distance over which this velocity change occurs, we can see that equation 4 from the previous list will yield the acceleration of the motorcycle directly:

$$a = \frac{v^2 - v_0^2}{2(x - x_0)} = \frac{(31\ \text{m/s})^2 - (22\text{m/s})^2}{2(50\ \text{m})} = 4.8\ \text{m/s}^2$$

This means that the motorcycle has to increase its speed by 4.8 m/s every second. How much time does it take to increase its speed by this much?

$$t = \frac{v - v_0}{a} = \frac{(31\ \text{m/s} - 22\ \text{m/s})}{4.8\ \text{m/s}^2} = 1.9\ \text{s}$$

The Physics—Catching a Boat Before It Destroys Prague. Now what about the boat chase? What does the car have to do to catch up with the boat if it takes the detour? Clearly, because the car must travel farther than the boat, it has to achieve a higher average speed for that part of the chase. Fortunately we are given some pretty specific information when the scene cuts to the GTO on-board GPS. The screen plots the location of both the car and the boat at that moment on a map:

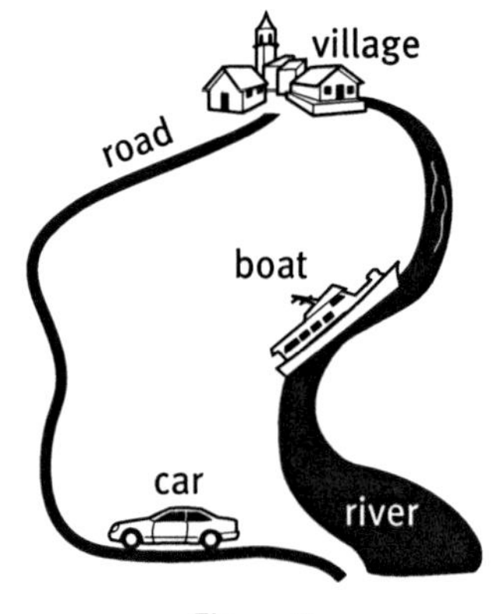

Figure 2

Well, from the instant we see its location on the GPS map, it looks like the car has at least twice as far to travel as the boat before the two routes reconnect. Yelena says it's 15 miles back to the river at that moment, so according to the map it looks like the distance the boat travels to the same point can't be more than 7 miles. Actually, the GTO has to travel more than twice as fast as the boat does to catch up with it.

With *XXX* we are dealing with the stereotypical action blockbuster and as a result the movie incorporates many of the obligatory clichés of the genre. For example, every time a car, boat, or snowmobile crashes or is hit by a bullet in *XXX*, it explodes. As you know, automobile accidents happen every day, in every city, state, and country. Unfortunately, some of them involve high-speed impacts resulting in major damage and fatalities. Yet how often do these cars actually explode? It's really quite rare, and quite unlikely. Gasoline will only ignite with the right mixture of gasoline vapor and air. The ratio is confined to a very small range, and the odds that a collision will release sufficient vapor in the right quantity to mix with air and also be exposed to a spark at that time are extremely small. Moreover, it is by no means certain that bullets produce sufficient sparking when they impact metal to generate enough heat to ignite gas vapor. Yes, in action movies, this exploding car phenomenon is a beloved staple. It is done so often, and has become such a cliché, that every time a car crashes we expect it to blow up. *XXX* supports this tradition with admirable and ridiculous consistency.

Which vehicle has the greater average velocity between the time that the car leaves the river and when it returns? (Note: Here we are talking about the complete detour from point A to point B on the map.)

In fact, they both have the exact same average velocity. If you thought that the car had the greater velocity, go back and review the definitions of velocity and speed. Average velocity depends on the displacement between the beginning and the end of the motion. It is a vector quantity. For this part of the trip, the displacements are identical for both vehicles, as is the total time of travel.

Although both the car and boat have the same total displacement, the car definitely travels a greater distance. Therefore the average speeds of the car and the boat are not the same. Vin Diesel says that the (average) speed of the boat is 80 mi/hr. Again, assessing the detour between the moment when we see the GPS and the moment they rejoin the river, we get a travel time for the boat:

$$\Delta t = \frac{d}{v} = \frac{7 \text{ mi}}{80 \text{ mi/hr}} = 0.088 \text{ hr}$$

Because the car travels for the same amount of time, we can calculate its average speed.

$$v = \frac{d}{\Delta t} = \frac{15 \text{ mi}}{0.088 \text{ hr}} = 170 \text{ mi/hr}$$

Traveling at 170 miles per hour is pretty fast—as fast as an Indy 500 car. The only way the specially outfitted GTO could maintain that speed would be if it were speeding down a straight stretch of highway. In our scene, however, the car is navigating though narrow village streets and around sharp turns. A car that careens through an unbanked curve at this speed would (in most cases) skid, roll, and crash, although probably not explode like hundreds of action movies would have you believe! (See the discussion of exploding cars above.) In Chapter 4 we will go through some calculations that set limits on how fast a car can go and still successfully make a sharp turn without an accident. It's sufficient to say that the car would have to slow down significantly on the curves or it would skid off the road. This means that they would have to make up the lost time on the straightaways. If they have to average 170 then they might have to exceed 200 miles per hour on the straights. That's hard to do! It is an amazing coincidence coupled with some good luck that Vin Diesel and Yelena arrive at back at the river at the exact moment when the boat is passing.

Conclusion. Is it possible for Vin and Yelena to stay up with the boat while on their inland excursion? If they hadn't shown us the map, it would have been left to our imaginations how much out of the way the car had to go, and we might have hypothesized that they were only a few hundred yards away from the river. In that case the scene would be merely in the realm of the highly improbable. However, knowing what we know about the route, the highly improbable has degenerated completely into the realm of the absolutely impossible.

The Physics—More *XXX* Kinematics. In scene 22, Vin Diesel purposely starts an avalanche to delay pursuers while he races down the mountain on a snowboard. He stays about 10 yards ahead of the avalanche for at least a mile before he leaps with his board into the air and grabs a power station antenna about 15 feet above the ground. The avalanche then sweeps over him as it continues on its way down the mountain. A few seconds later our hero pops through the top of the new layer of powder, unscathed (never mind that a few hundred yards up the slope, the tumbling snow completely obliterates a wood frame house).

Powder avalanches can travel at speeds ranging from 60 to 90 m/s. That's 130 to almost 200 miles per hour. According to the Guinness Book of World Records, the fastest recorded snowboard run, which took place on a slick bobsled track, not loose powder, is just less than 50 miles per hour.

Let's say that Vin Diesel, motivated by the XXX-treme conditions, and the XXX-treme pressure under which he finds himself, completely obliterates the world snowboarding speed record. Let's say he gets to a speed of almost 80 miles per hour (36 m/s), and let's assume the avalanche is traveling at the slower end of the velocity range. How long until the avalanche, moving at 130 mi/hr, catches up to him, considering his 10-meter head start?

It isn't very long! Because the avalanche should be closing the gap with Diesel at a relative speed of 50 miles per hour (22 m/s), the snow will catch him in 0.45 seconds.

$$\Delta t = \frac{d}{v} = \frac{10 \text{ m}}{22 \text{ m/s}} = 0.45 \text{ s}$$

How far would he make it down the mountain before being caught? It's not very far!

$$d = v\Delta t = 36 \text{ m/s}(0.45 \text{ s}) = 16 \text{ m}$$

Conclusion. Question: Can a person on a snowboard outrun an avalanche? Answer: Never.

Our hero executes about a half-dozen astonishing and highly improbable motorcycle leaps in various scenes in *XXX*. It may be the world record for cinematic motorcycle jumping outside of a motocross movie. We are going to introduce and analyze the motion of projectiles next but it will be much more gratifying for us to wait to analyze these particularly astonishing motorcycle jumping scenes in Chapter 2 after we are fully armed with Newton's laws of motion—so be patient!

One of the more common action movie staples is the heart-arresting car jump. Sometimes the only escape for the protagonist is to risk a fatal crash by projecting him or herself over a seemingly insurmountable gap in the road. (It's amazing how often they're successful!) How difficult is it to pull off such a jump in reality? Let's investigate the motion of these cinematic projectiles through that famous leaping bus immortalized in the movie *Speed* (1994).

SPEED AND PROJECTILES

The entire basis for the movie *Speed* is a variation of the high-speed chase. In this case, it's a bus we're talking about, and it's not exactly being chased.

The Movie

A crazed ex-police demolitions expert (played by that ultimate portrayer of the psychotic and disgruntled—Dennis Hopper) is out to get revenge on his old colleagues, and extort a few million out of the municipal budget by terrorizing the city with deadly hidden bombs. It's up to heroic, emotionally stable, and uncommonly good-looking current police demolitions expert Keanu Reeves and unsuspecting, innocent, perky-yet-vulnerable city-bus-passenger-turned-driver Sandra Bullock to save the day.

The Scene. In scene 15, a bomb has been planted on a city bus. In a sadistic twist, once the bomb is armed (when the speed first *exceeds* 50 miles per hour) it will not detonate unless the speed of the bus drops below 50 miles per hour. Keanu and Sandra have to keep the bus moving above the requisite speed through a series of difficult obstacles, including traffic-clogged city streets, tight 90-degree turns, and a 50-foot gap in a highway bridge.

Do Keanu and Sandra have a chance to save the city? Can a transit bus actually leap over a 50-foot chasm? Our problem can be assessed using ***kinematics*** in two dimensions, in particular, the kinematics of projectile motion.

ACTIVITY

a) Write down any information that we can explicitly obtain from the scene in *Speed* that you think will help in determining whether or not the bus can successfully negotiate the jump.

b) Write down information that is not given, shown, or stated explicitly, but that would be useful in solving the problem. Could you approximate or assume any of this information by carefully observing the scene? (This step is often crucial in the analysis of movie physics, but it is also used in science more often than you might think. This is where the creativity and leaps of intuition come in—and where real discoveries are made. The art of approximation and reasoned assumption is an extremely valuable skill!)

Projectile Motion

Any object that is given an initial velocity and then released into the air, such that only gravity (and air resistance) is acting on the object after the initial push, is called a *projectile*. Projectiles can move in two or three dimensions. However, to analyze motion in more than one dimension all we need to do is to break it up into separate one-dimensional motions. Anticipating Newton's laws, if we neglect air resistance, then the only force acting on the projectile is gravity, and therefore the projectile will always be accelerating straight downwards towards the ground. It will have no acceleration horizontally, and will simply maintain a constant velocity in that dimension. Therefore, the projectile has two simultaneous motions: a constant velocity horizontally superimposed on a constant downward vertical acceleration. When these two motions occur simultaneously the path traced out by the projectile is that of a parabola.

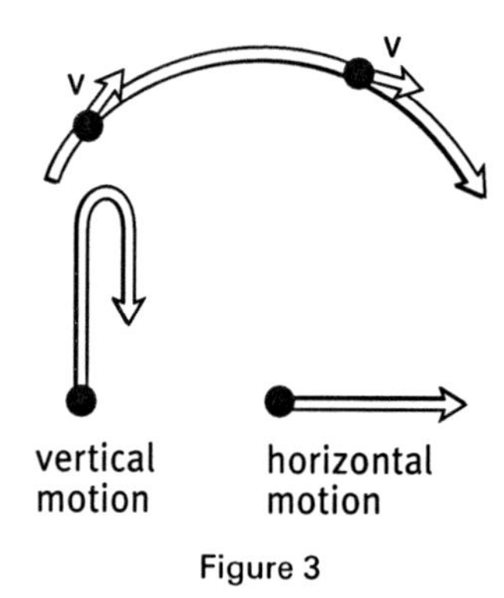

Figure 3

Example 5: Having analyzed the physics in some of the scenes from *Speed* you throw the DVD out the second-story window of your house in a fit of anger and outrage. Your window is 3 meters above the ground, and the DVD lands 4 m horizontally from a spot just underneath the window. If you threw it horizontally, at what speed (no pun intended) did you throw the DVD?

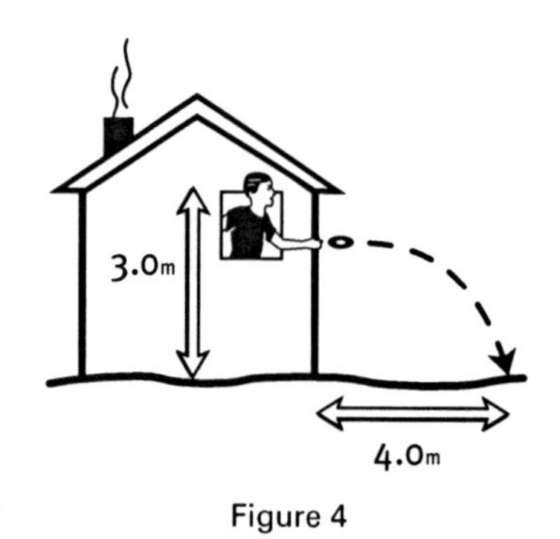

Figure 4

Solution: Let's reason our way through the solution to this problem. In the *horizontal direction* the velocity is constant and equal to the initial velocity of the DVD. Therefore, the initial velocity is

PROBLEM

If instead you throw the DVD with the same initial velocity but at an initial direction 30 degrees above horizontal, how far away from the house will it land? What is the maximum height above the ground that it will achieve during its flight?

given by $v = v_x = \Delta x/\Delta t$. We know $\Delta x = 4$ m, but we don't know Δt. To find this we need to look at the vertical motion.

To help put this in perspective consider how your friend would describe the motion of the DVD if she were running underneath it with the same horizontal velocity as the DVD.

Vertically we have a constant downward acceleration, due to gravity, of 9.8 m/s^2. In addition, we know that the initial vertical velocity = 0 because it was thrown horizontally. However, from your friend's point of view, running underneath the DVD, it is not moving horizontally at all—it is simply falling. The vertical motion is exactly the same as simply dropping the DVD from a height of 3 m. Therefore we can find the falling time as follows:

$$\Delta y = v_{0y}t + \frac{1}{2}a_y t^2 .$$

Taking downward to be the negative direction we get:

$$-3.0m = (0)t + \frac{1}{2}(-9.8\ \text{m/s}^2)t^2 \text{, which gives:}$$

$$t = \sqrt{\frac{-6\ \text{m}}{-9.8\ \text{m/s}^2}} = 0.78\ \text{s},$$

Now that we know t we can solve for v_x, which equals the initial velocity of the DVD:

$$v = \frac{4\ \text{m}}{0.78\ \text{s}} = 5.1\ \text{m/s}$$

ACTIVITY

Find a tennis ball, or any other kind of ball. Run along smoothly at a constant speed and drop the ball. What is the path of the ball according to you? Do this with a partner who stands by while you run past. What is the path of the ball according to your friend?

The Physics—One Giant Leap. Now let's apply what we now know to the jumping bus. As she heads toward the jumping point, Sandra Bullock pushes the accelerator to the floor, and the speed of the bus gradually increases to 70 miles per hour—this is the velocity that the bus has as it hits the end of the road. We are told that the gap in the highway bridge is "50 feet, at least!" The next crucial piece of information we need to obtain is the angle of the road relative to horizontal. The greater the angle of launch is relative to horizontal—up to 45 degrees—the greater the horizontal distance of the jump assuming no significant air resistance. The horizontal distance that a projectile travels is called the *range* of the projectile. We will also assume that the launch height is the same as the landing height.

Because we are not given the angle of the highway in the film, we have to employ the art of estimation and approximation to our problem. A visual inspection of the scene can give us an estimate of the launch angle of the bus. Looking carefully at the film clip, however, presents us with a potentially disastrous problem. It looks as though the highway is practically flat on each side of the gap. What would happen if this were true? Without doing any calculations at all, we know that no matter how fast the bus is traveling, the instant it leaves the ramp it will start to fall due to gravity, and by the time it gets to the other side it will have lost height. The faster it is moving horizontally the less time it will take to get to the other side, and the less distance it will fall. Nevertheless, whether it's 1 foot or 10 feet below the level of the bridge doesn't really matter—the bus will crash and the onboard bomb will explode.

ACTIVITY

Prove mathematically that a 45-degree launch gives the maximum horizontal range. Then verify the angle experimentally. Using a rubber band, a marble (or a very small bus), and a protractor, launch the marble at a variety of angles and see what happens. Make sure that you pull the rubber band back the same amount each time. You can also do the experiment with more sophisticated equipment such as spring loaded launchers that you may find in some physics classrooms.

For example, if the bus is traveling at 70 miles per hour when it reaches the gap, how far below the level of the bridge will the front of the bus be when it reaches the other side?

(Note: Convert all units to SI units for consistency.)

The time it takes to cross the gap is $t = \dfrac{\Delta x}{v}$.

$v = 70 \text{ mi/hr} = 31 \text{ m/s}$ and $\Delta x = 50 \text{ ft} = 15 \text{ m}$.

Therefore, $t = \dfrac{15 \text{ m}}{31 \text{ m/s}} = 0.48 \text{ s}$.

How far will the bus fall in about half of a second?

$$\Delta y = v_{0y}t + \frac{1}{2}a_y t^2 = 0 + \frac{1}{2}(-9.8 \text{ m/s}^2)(0.48 \text{ s})^2 = -1.1 \text{ m},$$

or about 3½ feet below the level of the bridge. This is not good.

In the movie there may well be a slight launch angle of a degree or two; it's hard to tell. (A 5-degree angle is actually rather steep, and would be quite obvious.) Nevertheless, let's optimistically estimate that the ramp is 2 degrees, and see how we do.

The Physics—Hitting the Gap. We now have the same initial velocity of 31 m/s, but we need to break this velocity vector into vertical and horizontal components. Recall that two-dimensional motion can be broken into two separate one-dimensional motions. In the case of a projectile, neglecting significant air resistance, the motion in the horizontal direction maintains constant velocity, while the motion in the vertical direction has a constant downward acceleration due to gravity.

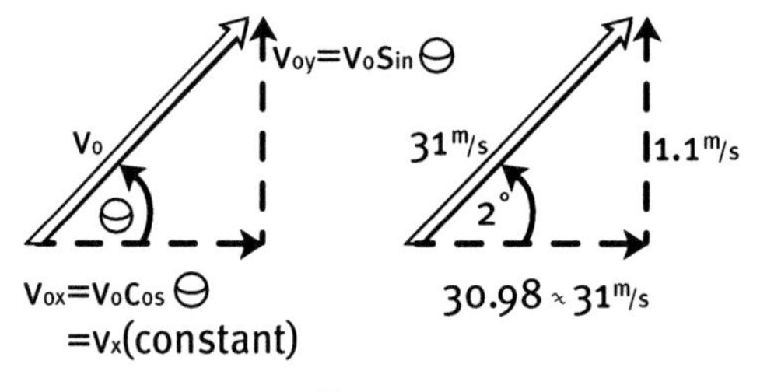

Figure 5

Initial conditions:

Horizontal: $v_{0x} = v_x$ (because the horizontal v is constant) = 31 m/s cos 2 = 30.98 = 31 m/s.

Vertical: v_{0y} = 31 m/s sin 2 = 1.1 m/s, $a_y = -9.8$ m/s^2 and, because we want to find out how far the bus has gone horizontally when it falls back to the same height, $y - y_0$ or $\Delta y = 0$.

We can use the vertical motion to find out how much time it takes the bus to go up and back down to the same height:

$$\Delta y = v_{0y}t + \frac{1}{2}a_y t^2,$$

which makes $t = 0.22$ seconds.

How far does the bus go horizontally in 0.22 s?

$$\Delta x = v_x t = (31 \text{ m/s})(0.22\text{s}) = 6.8 \text{ m (about 22 feet)},$$

which is less than half the distance of the gap.

This is a best-case scenario because we assumed that (1) the angle was 2 degrees (which is probably an overestimate) and (2) there was no air resistance. Air resistance reduces the horizontal velocity and would make this situation even worse, causing the bus to cover less distance during its flight. There is no way to successfully make the jump under these conditions and the only thing we can do when watching this scene is either suspend our disbelief or consider an alternate—albeit abbreviated and less-positive—ending to the movie.

Another troubling aspect in watching this particular scene is that for some reason just before the bus hits the end of the road and propels itself into the air, it does a wheelie, and the wheelie somehow helps the bus alter its trajectory to a steeper launch angle. While this perhaps adds drama to the moment, how this occurs is a mystery. For the front end of the bus to rear up like a frightened horse there has to be some cause—in other words, some force to push it up that way. There is no evidence of anything within the vicinity of the bus that could possibly account for this unusual and unexpected behavior.

THINKING PHYSICS

Do you think air resistance would have a large or a small effect on the motion of the bus? Why do you think so? In what type of situations would air resistance be only a minor problem, and when would it be a major problem? (Hint: Stick your hand out of the car window when driving and note the effects of air resistance at a variety of speeds.)

Conclusion. Despite all the bad news, it might still be possible to launch a bus over a 50-foot gap if the bus either is going faster or launched at a steeper angle than in the movie. (What might happen to the bus when it lands is another matter.) In fact, it turns out that in a stunt show a bus actually successfully leaped a 71-foot gap from ramp to ramp. This was reported to be the bus-jumping world record. The ramp angle was a lot greater than the 0 to 2 degrees that we see in the *Speed* jump, however, where they were clearly in no position to be challenging world records.

ADDITIONAL QUESTIONS

1) This time we really want that bus to make the "*Speed* jump" successfully. Either we have to increase the takeoff speed significantly or we're going to have to find a steeper ramp somewhere. Let's try increasing the speed first. At what speed must the bus be moving when it hits a 2-degree incline to make it over the 50-foot gap?

2) Now let's assume that the bus is limited to a maximum speed of 70 miles per hour. The police scribble down a few calculations and realize that the bus can't make it with the incline as it is, so they are able to arrive ahead of time and quickly install a makeshift ramp. What is the minimum angle at which they must launch the bus at this speed in order to make it to the other side successfully?

3) Watch some other movies that feature vehicles making huge leaps similar to the one in *Speed,* such as *Road Trip* or *Gone In 60 Seconds.* What information can you glean from the movies themselves and what can you infer from your knowledge of physics to determine if any of these jumps could actually happen?

4) In the movie *2 Fast 2 Furious,* two cars are racing toward a raised bridge at high speeds. They need to make it over a gap in the road, and they are lucky enough to have a steep ramp to take off. The car in front hits the end of the bridge at around 130 miles per hour. The car in back speeds up to 160 miles per hour before the leap, and arrives at the end of the bridge a second or two later. Which car should hit the road on the other side first? What happens in the movie? Using projectile kinematics, explain why this is a problem.

Newton's Laws

SPIDER–MAN AND OTHER STICKY THINGS

The superhero/fantasy genre has generated some of the most entertaining movies in recent years. Who can forget such memorable scenes as Superman reversing the direction of the Earth's rotation by flying around it really fast in the opposite direction, allowing him to turn back time and rescue Lois Lane before she is buried alive? These kinds of movies can be a lot of fun, and the continuing improvement in special-effects technology has allowed your favorite comic book heroes to come to life on the big screen in a way that wasn't possible a few decades ago. Superman, Batman, and The Incredible Hulk have all had their 15 minutes to two and a half hours of fame (not including sequels). Some of these superheroes have superpowers, others (like Batman) only determination, grit, ingenuity, and a lot of money, but in each case, their exploits and escapades give us the opportunity to delve into the physics of Hollywood from a fresh perspective.

The Spider–Man movies are a lot of fun in particular, and despite the completely impossible premise that being bitten by a genetically enhanced spider changes your own genetic structure, it is a pleasant fantasy. We live vicariously through the former nerdy nice guy Peter Parker's transformation. He truly combines the best of man and spider. However, despite Spider–Man's superpowers, his incredible strength, his resilience to pounding blows, his formidable acrobatic skills, and his web-slinging abilities, we still may suspect that his heroics violate fundamental physical principles. To prove our hypothesis, let's look at two or three scenes from *Spider–Man,* and analyze them in terms of Newton's laws of motion.

The Movie

Peter Parker (Tobey McGuire) has been bitten by a genetically modified spider (what happened to no GMOs?), and has developed a plethora of formidable spider-like superpowers. At a pivotal point in the early part of the movie, even though he could have prevented it, Parker allows a criminal to escape from justice. Ironically this leads directly to the murder of his kindhearted uncle Ben who has been the only father he has ever known. As a result, the guilt-ridden Parker becomes "Spider–Man"

Photo: The Green Goblin hovering on his "glider" in *Spider–Man.*

and dedicates his life to fighting crime in the mean streets of New York City. Meanwhile Parker's extremely good-looking, monumentally wealthy, but sadly underachieving best friend Harry, played by James Franco, is dating sexy, but troubled, MJ, played by Kirsten Dunst, with whom Parker has always secretly been in love. To complicate things further, Harry's father, the brilliant entrepreneur scientist Norman Osborne (Willem Dafoe), loses a huge government contract and is then fired as CEO of his own company. In an act of desperation, Osborne performs an ill-fated experiment on himself in which he gains superpowers of his own, but at the expense of his sanity. He becomes the vicious Green Goblin bent on wreaking havoc in the city and taking revenge on everyone who did him wrong or stands in his way. (Although he still has a pretty good sense of humor.) Only Spider–Man has the ability to stop him! Let's see if he can.

The Scenes

Super Powerful Punches and Sticky Fingers. In scene 8, Spider–Man is still only mild-mannered Peter Parker, just discovering his powers. After accidentally knocking some food on the school bully, Flash, Spider–Man is forced to defend himself from being beaten up. At this moment we get a glimpse into the "Zen of the Spider" as Peter Parker's awareness goes into overdrive. Everything moves in slow motion. He can see the individual wing beats of a passing fly. He easily evades Flash's attacks, and in a very satisfying turnabout he punches the frustrated Flash so hard he knocks him across the room. How would a blow that hard affect not only the one receiving the punch, but the puncher as well?

Overstimulated and confused, Peter runs out into an alley in a state of breathless agitation, where suddenly he has the epiphany that he can climb up sheer walls! The shot cuts to a magnified close up of tiny serrated insect-like hairs popping out of his fingers. They're a little gross, but these hairs are presumably the key to the "sticking power" necessary to climb up the wall. How strong are these hairs going to have to be? That is, how much force must each hair exert on the wall to get Peter away from the bad guys?

An Object In Motion. Later in scene 8, Spider–Man is firing on all cylinders. After climbing up to the rooftops, he jumps from building to building. He can run extremely fast and leaps 50 or 60 feet at a pop with exuberant ease. After a particularly energetic leap he comes to an abrupt halt as he lands right at the edge of the roof.

How fast must Spider–Man run in order to leap a horizontal distance of 50 feet and what kind of forces must his legs be capable of exerting? Considering how fast he must have been going when he landed, what kind of force would have to be exerted on him to stop so quickly? How much friction should there be between his feet and the roof if he is to be successful?

How Strong Is the Green Goblin? In scene 26, as we approach the climax of the movie, the Green Goblin, as a way of getting revenge on Spider–Man, has kidnapped MJ and taken her to the top of some sort of urban edifice. It looks like a beam on the top of a really high suspension bridge. Whatever it is, it's extremely narrow and scary. As Spider–Man swings in for the rescue/confrontation the

Goblin has prepared a diabolical choice. In one outstretched hand the Goblin holds MJ suspended over the void. In the other, he holds a thick cable wound over a pulley connected to a cable car filled with innocent children that he has dislodged from the track. The car also hangs suspended over the void. We are suspended on the edge of our seats. How strong does the Green Goblin have to be to hold on to MJ and the cable car?

In order to assess the physics in each of these scenes a basic understanding of Newton's three laws of motion is in order.

NEWTON'S LAWS

In Chapter 1, we looked at scenes that could be analyzed using kinematics, which is the *description* of motion. Now we will discuss *causes* of motion or, more specifically, what causes an object to *change* its state of motion. The study of the causes of changing motion is called *dynamics*.

Forces

We can think of a *force* as a push or pull acting on an object. The push or pull is due to another object interacting with it, applying that force. Force is a vector quantity having both magnitude and direction. The SI units of force are the Newton.

$$1\text{N} = 1 \text{ kg m/s}^2$$

An object may have multiple forces acting on it due to its interaction with multiple other bodies. Sometimes these forces cancel or balance each other out, and we say that there is no net force acting on the object. If the forces do not all cancel out, we say there is an *unbalanced*, or *net force, acting on the object.* Now let's look at what Newton has to say specifically about how forces affect the motion of objects.

Newton's First Law of Motion and Inertia

Newton's first law of motion states that "An object in motion will stay in motion at a constant velocity (or an object at rest will stay at rest) unless acted on by a net, or unbalanced force." Newton's first law is also called the law of inertia. *Inertia* is a term that refers to an object's resistance to a change in motion. The *mass* of an object is a quantitative measure of its inertia. The more inertia an object has, the harder it is to change its motion, and the more force is required to do so. For example, it is much harder to accelerate or change the velocity of a bowling ball than it is a tennis ball. The important point of Newton's first law is that for an object to stay at a constant velocity the *absence* of force is

ACTIVITY

Sit on a skateboard holding on to a tennis ball. Have someone push you along at a constant speed. Toss the ball straight up into the air. Where will it land? Compare this to a situation in which you are driving in a car at a constant speed with the windows closed and you toss a ball straight up into the air. Where will this ball land? Would it be different with the windows open? Explain what happens in each case in terms of Newton's first law.

required, specifically the absence of a net force. It is easy to draw the erroneous conclusion that a constant force is required to maintain constant velocity. Consider sliding a box across the floor. The force one would need to apply to keep it moving is not the only force acting on the box. Friction is acting against the box's motion. Therefore, to keep it moving at a constant velocity we must apply a force exactly balancing the friction force, resulting in a net force of zero. If the applied force is greater than the friction force the object will speed up. If it is less, or if no push is applied at all, it will slow down until it stops. Once the box stops there is no longer a friction force so the net force is zero, and the box stays at rest.

Newton's Second Law of Motion

According to the first law, if an object does not experience a net force, it will maintain a constant velocity; however, if it does experience a net force then the object will change its velocity and accelerate. Newton's second law of motion states that if an object experiences a net force then it will accelerate in the direction of the net force. The acceleration will be directly proportional to the magnitude of the net force and inversely proportional to the mass of the object.

Mathematically:

$$F_{net} = ma$$

The second law can also be written in terms of momentum, as follows.

$$F_{net} = \frac{\Delta p}{\Delta t}$$

Momentum (p) is defined as the product of the mass and velocity of an object (mv). Like force and acceleration, momentum is a vector quantity. Because:

$$a = \frac{\Delta v}{\Delta t}$$

$$F_{net} = m\frac{\Delta v}{\Delta t} = \frac{\Delta p}{\Delta t}$$

Example 1: Spider–Man has netted a 75-kg criminal in the act of attempting a bank heist and is dragging him along the ground with his web.The pulling force = 250 N, and there is a 100-N friction force on the ground opposing the motion. Draw a "force diagram" showing the forces that are acting on the criminal, and determine his acceleration.

Solution: The forces are shown in the following diagram. The forces in the vertical direction cancel, and the net force in the horizontal direction is 250 N – 100 N = 150 N. Therefore, $a = \frac{F}{m} = \frac{150\text{ N}}{75\text{ kg}} = 2\text{ m/s}^2$ to the right.

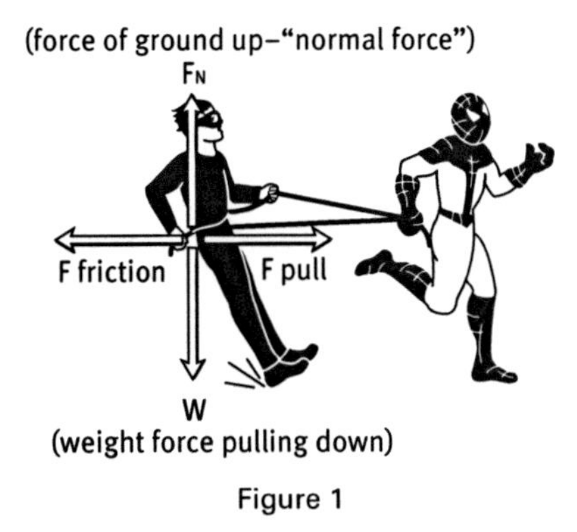

Figure 1

Example 2: A 2-kg mass experiences the forces shown in figure 2a below.

a) What is the magnitude and direction of the net force acting on the mass?
b) What is the acceleration of the mass?

Solution:

a) Add all of the horizontal and vertical forces separately. The 15-N force can be broken down into a vertical and a horizontal component. Using trigonometry the horizontal component = 15 cos 30 = 13 N, and the vertical component = 15 sin 30 = 7.5 N as shown in figure 2b.

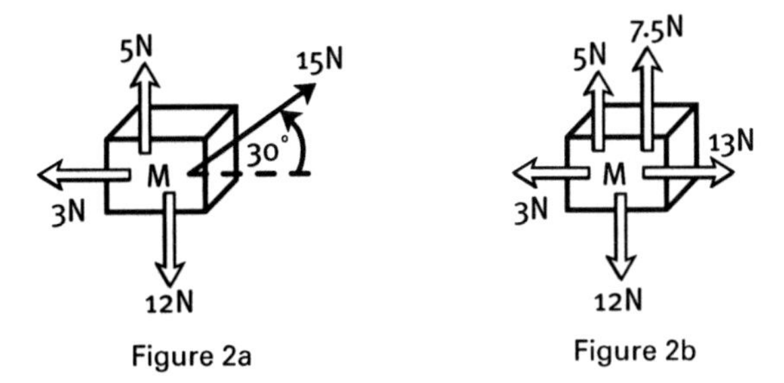

Figure 2a Figure 2b

The net horizontal force is then 13 N – 3 N = 10 N, and the net vertical force is 5 N + 7.5 N – 12.5 N = 0. Therefore, the net force is in the horizontal direction to the right with a magnitude of 10 N.

b) $a = \frac{F}{m} = \frac{10\text{ N}}{2\text{ kg}} = 5\text{ m/s}^2$ in the same direction as the net force.

Contact versus Noncontact Forces. Forces can be divided between *contact* and *noncontact* forces. These categories are exactly as they sound. Contact forces require direct contact between the interacting objects. For example, pushing a box or pulling a mass with a string is a contact force. Noncontact forces are also called *fundamental* forces. These forces act at a distance, across empty space, and include gravity, the electromagnetic force, and nuclear forces. In reality all forces are ultimately due to these fundamental forces. Contact forces are actually due to the electromagnetic force at the atomic level. When you push on an object it is electrostatic repulsion between the electrons in your hand and those in the object that is ultimately responsible for the force that is exerted between you and the object.

The Weight Force. The force of gravity is a fundamental force which is always attractive (it is the best-looking of the fundamental forces), that acts between objects. It is dependent on the mass of each object. Newton's universal law of gravitation gives the magnitude of the gravitational force between any two masses:

$$F = \frac{Gm_1 m_2}{r^2}$$

ACTIVITY

Take a balloon and fill it with air. Without doing anything more to the balloon, place it against the wall. What happens? Now rub the balloon vigorously against your hair (giving the balloon an electrostatic charge), and place it against the wall again. What happens this time? Which is stronger: the gravitational force or the electrostatic force?

Where F is the force between the two masses m_1 and m_2, r is the distance between the centers of each mass, and G is called the universal gravitation constant and is equal to 6.67×10^{-11} Nm^2/kg^2 (notice the miniscule value of the constant).

The law of gravitation is very similar in form to Coulomb's law which describes the electrostatic force between charges (Chapter 5 discusses Coulomb's law in greater detail). Like Coulomb's law, the law of gravitation is an inverse square law, and diminishes rapidly with distance between the interacting objects. Because the constant G is extremely small, the force of gravity is actually extremely weak compared to the other fundamental forces. It requires at least one of the objects that are being attracted to each other to be very large for the force to be noticeable in everyday life. Therefore, due to the Earth's relatively large mass, objects at the Earth's surface experience a noticeable pull directed towards the center of the Earth. This force of gravity on an object is called the *weight force*. From the law of gravitation we can see that the more massive the object, the greater the weight force acting on it.

However, it can be shown experimentally that if objects of different masses are dropped, as long as air resistance is not a major factor, they will all experience the same acceleration due to gravity of $9.8\ m/s^2$. It's easy to confirm this by dropping two objects of different weights simultaneously from the same height. This acceleration is often represented by the constant g (lowercase). It is important to note that the value of g given above only pertains at the Earth's surface or close to it. Once an object moves a significant distance from the surface, the value of g will decrease. G is a universal constant of nature, while g simply refers to a specific acceleration under the specific conditions just described. If an object is dropped just above the surface, and air resistance is negligible, we can show that the value of g is the same for objects of different mass from the universal law of gravitation.

$$F = \frac{Gm_{earth} m_{object}}{r_{earth}^2} = m_{object} a_{object}$$

Therefore, $a_{object} = \frac{Gm_{earth}}{r_{earth}^2}$, which is independent of the mass of the object being accelerated towards the earth. If the values for G, the mass, and the radius of the Earth are plugged in, the result gives:

$$a = 9.8 \text{ m/s}^2 = g,$$

and the weight force on an object at or near the Earth's surface reduces to $W = mg$.

Example 3: The Green Goblin has just pulled off a bank job, and he accelerates upward on his nifty glider at a rate of 4 m/s^2, hauling a 30-kg sack of unmarked twenties with a rope. If the rope can withstand a maximum tension of 400 N, will it be strong enough to hold the sack?

Solution:

$$F_{net} = ma$$

Therefore:

$$F_{net} = 30 \text{ kg}(4 \text{ m/s}^2) = 120 \text{ N}$$

Because:

$$F_{net} = F_{rope} - W,$$

$$F_{rope} = F_{net} + W = F_{net} + mg$$

$$120 \text{ N} + (30 \text{ kg})(9.8 \text{ m/s}^2) = 414 \text{ N}$$

Therefore, the Green Goblin's rope isn't strong enough. It will break, making him angrier than he already is, if you can imagine that.

Mass versus Weight. The concepts of mass and weight are sometimes confused. For example, it is common to hear a statement like "the rock *weighs* 3 kg." However, kilograms are units of mass not of weight. Mass is a statement of how much *matter* an object contains. Because the amount of matter relates to the object's inertia, its mass can also be thought of as its resistance to a change in motion. An object's mass is independent of where it is. If a rock is floating around in deep space it has no weight, but it still has the same mass it would have on Earth. Weight is a measure of the *force* that gravity exerts on a mass. In fact, as seen previously $W = mg$ on the Earth's surface. Weight does not exist if there is no gravity. It is sometimes misleading to see a scale that measures in kilograms. The scale is actually measuring weight but because $m = W/g$, and g is a constant at the Earth's surface, the scale is calibrated to convert the weight into the amount of mass that would have that weight on Earth. If you took the scale to the moon, it would not record the correct mass because g

on the moon is 1/6 that on Earth. However, if you had a scale with units of pounds, then this scale would measure your correct weight on the moon, which would be 1/6 your weight on Earth.

Air Resistance and Terminal Velocity. Because Earth has an atmosphere, objects in free-fall may experience a significant friction force due to collisions with air molecules. This friction is commonly called *air resistance*. The force of air resistance on an object increases with its velocity (and is also dependent on the shape of the object). Because this is true, an object in free-fall will have an acceleration that decreases over time. Consider a skydiver at three different moments after he has jumped out of an airplane: first the instant after he jumps out, then a short while later, and finally some time after jumping.

ACTIVITY

Drop a piece of paper and a marble from a meter or two above the ground. How long does it take the paper to reach terminal velocity? Explain. Does the marble reach terminal velocity before it hits the ground? Explain how you know. Just for fun, crumple the paper into a ball and do the experiment again. What's different?

At first, the only force acting on the skydiver is the weight force, and therefore the diver accelerates at rate g. However, as the skydiver's velocity increases the force of air resistance builds up. Therefore, the *net* force on the skydiver decreases. Because $F_{net} = ma$ the acceleration will now be less than g. The skydiver will still be speeding up, but at a lesser rate. Eventually as the velocity and air resistance increase, the air resistance force will equal the force of gravity. At that point the net force will equal zero and acceleration will stop. The velocity at which this occurs is called the *terminal velocity*. Because the weight force on a feather is so small, it takes very little air resistance to balance the weight, and therefore it gains very little speed before it reaches a terminal velocity. Because a skydiver has considerably more weight than a feather, the skydiver's terminal velocity is much greater, and that is why he also carries a parachute.

Sliding Friction. *Friction* is a force that opposes motion or potential motion. Whenever two surfaces are in contact, and one of the surfaces is sliding along the other, there is a force of ***kinetic friction*** (F_k) that exists between the surfaces, acting against the motion of each object. Unlike the force of air resistance, the force of kinetic friction is independent of the relative speed of the objects. It does depend on how "sticky" the two surfaces are when in contact with each other (measured by the coefficient of kinetic friction, μ_k), and how hard the surfaces are being pushed together (measured by the normal force, F_N). The normal force is defined as a force that a surface exerts on another surface.

$$F_k = \mu_k F_N.$$

For example, dry ice sliding along a Formica tabletop experiences much less friction than a brick sliding along a concrete surface. A book on a table will experience a much greater friction force if it is being pushed down into the table.

Static friction may exist when an object is not in motion relative to the surface on which it rests. For example, a book is at rest on a table and a small force is applied in an attempt to push it along the table, yet the book does not move. This is because a static friction force (F_s) is acting against the pushing force, resulting in a net force of zero on the book. Static friction opposes *potential* sliding between two surfaces. Like kinetic friction, it is also dependent on the material properties of each surface (the coefficient of static friction, μ_s), and how hard the surfaces are being pushed together (F_N). Static friction will only be as strong as it needs to be to balance the forces that would cause an object to slide. Therefore, it will increase as it balances these forces. However, it has a certain maximum value and once the pushing forces exceed that value the object will start sliding.

$$F_s \leq \mu_s F_N$$

Note: Coefficients of friction between surfaces vary from zero (no friction at all) to maximum values around one.

Example 4: A crate is in the back of a flatbed truck when the truck begins to accelerate. If the crate accelerates along with the truck, what force is responsible for its acceleration? If the truck accelerates at a rate of 2 m/s^2, and the crate does not slip, what is the amount of friction force acting on the crate? If the crate has a mass of 5 kg, what is the minimum coefficient of friction that will allow the crate to stay with the truck while accelerating?

Solution: The force responsible for accelerating the crate is *static* friction. There is no sliding between the bed of the truck and the crate, although there is the potential for sliding. If there was no friction at all, the truck would just drive out from underneath the crate (Newton's first law). The crate would be sliding relative to the truck in that case. If there was some friction, but not enough to keep the crate accelerating with the truck, then there would be a kinetic friction force acting on the crate. However, in this case, the static friction force is able to "grip" the crate and pull it along with the truck.

Because static friction is the only force acting on the crate in the direction of motion:

$$F_s = ma = (5 \text{ kg})(2 \text{ m/s}^2) = 10 \text{ N}.$$

The minimum coefficient of static friction that is required will be able to provide a maximum F_s = 10 N = $\mu_s F_N$. In this example, F_N must exactly equal the weight force mg because there is no acceleration in the vertical direction. Therefore:

$$\mu_s = F/mg = 10 \text{ N}/(5 \text{ kg})(9.8 \text{ m/s}^2) = 0.2.$$

Newton's Third Law of Motion

Newton's third law is often quoted "for every action there is an equal and opposite reaction." To be more specific: *If object A exerts a force on object B, then object B exerts a force of equal magnitude back on object A in the opposite direction to the force that A exerts on B.* While the statement is simple, it is often misunderstood. First of all the third law says that all forces occur in pairs, that is, all forces are due to the interaction *between* objects. In addition, the action-reaction force pairs are always equal to each other, no matter what the masses of the respective objects are. For example, if a train were to collide head-on with a fly moving in the opposite direction, during the collision the two objects would exert *equal* forces on each other. However, (as the fly would tell you) the *effect* of the forces will be different on the train and the fly. Because $a = F/m$, for a certain force, a fly will experience a much greater acceleration than a train experiencing the same force.

It is important to note that when applying Newton's second law to the motion of an object, only the forces acting *on* that object count. The reaction forces that the object exerts back on the other objects only affect the motion of those other objects.

ACTIVITY

Find two skateboards, a friend, and a nice smooth flat surface. Sit (don't stand) on your respective skateboards. Push your friend backwards. What happens to you? Explain. Try to push your friend without moving, and without touching anything else. Can you do it?

Example 5: A 10-kg cannonball traveling 200 m/s at impact hits Superman, "the man of steel." Superman is, of course, unhurt (perhaps he sustains a small bruise), but if the cannonball is in contact with Superman for 0.01 seconds (a typical collision time between hard objects), what force did Superman exert on the ball to bring it to a stop? What force did the cannonball exert on Superman? What effect should this have on Superman's motion? (By the way, in the original *Superman* movie he says he weighs about 225 pounds, which gives him a mass of 100 kg.)

Solution: This problem requires you to combine several principles. First, determine the acceleration of the cannonball.

$$a = \frac{\Delta v}{\Delta t} = \frac{200 \text{ m/s} - 0 \text{ m/s}}{0.01 \text{ s}} = 20{,}000 \text{ m/s}^2$$

$$\text{And } F_{net} = ma = (10 \text{ kg})(20{,}000 \text{ m/s}^2) = 200{,}000 \text{ N}$$

According to Newton's third law, the force that Superman exerts on the cannonball must be equal to the force that the cannonball exerts back on Superman. Therefore,

$$a_{sprmn} = \frac{F}{m} = \frac{200{,}000 \text{ N}}{100 \text{ kg}} = 2{,}000 \text{ m/s}^2$$

$$\text{and } \Delta v = a\Delta t = (2{,}000 \text{ m/s}^2)(0.01 \text{ s}) = 20 \text{ m/s}$$

Therefore, even though Superman has an extremely tough exterior, he should still be knocked backwards with an initial velocity of 20 m/s (45 mi/hr).

Note: If the cannonball was made of "kryptonite" Superman would probably also get nauseous.

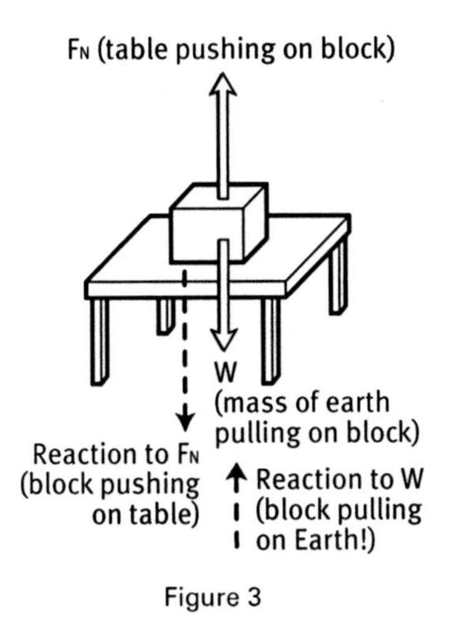

Figure 3

Example 6: A block sits on a table. What forces are acting directly on the block? What are the reaction forces to the forces acting on the block?

Solution: There are two forces acting on the block (as shown in Figure 3): the force of gravity pulling the block towards the Earth, and the force of the table pushing up on the block (the *normal* force). Because the block has no acceleration the net force must be zero. Therefore the force of the table pushing up must be equal to the weight force pulling down.

The weight force is due to the Earth pulling on the block. Therefore, the reaction force to it must be the block pulling back on the Earth. The table pushes up on the block and the reaction to this is the block pushing back down on the table. Contrary to a common misconception the weight force and the force of the table on the block are *not* action-reaction pairs. Although it is true in this example that the force of the table on the block would not be there if gravity wasn't pulling the block down, these forces act on the same object and cannot be action-reaction pairs.

Now, let's get back to our hero.

SUPER PUNCHES

It's time to apply what we know about Newton's laws to Spider-Man's exploits. Of course superheroes are super strong and we have every right to expect amazing feats of physical strength (except when it comes to superhero wannabes like Ironman, who's only claim to superhero status is his fancy iron suit).

Superheroes can obviously punch with great force, and they can brush off blows that would send normal humans into rehab for a couple of decades. Their bones are not easily broken. Their spirits are unflinching in the face of adversity. We can allow them some leeway with physiology, and even turn away with a wink and a nod in the face of the highly improbable, but we have a right to demand a reasonable adherence to fundamental physical principles like Newton's laws in something as familiar as a straight-out fist fight.

The Scene

Do you remember back in scene 8 when Peter Parker punches Flash so hard that he flies across a hallway with his feet flying out from underneath him? You can see in the scene that Flash's trajectory is essentially straight for about five to eight feet before he falls to the ground and slides several more feet. This basic scenario is repeated in a more extreme case later in the movie, at the "Unity Day Festival" (scene 18), when the Green Goblin kicks Spider-Man so hard he is projected at least *50 meters* in a straight line before finally coming to a stop when he smashes into a metal light post (which incidentally breaks in half on impact). It is important that we address the physics involved here without flinching. In each case, the puncher stays firmly rooted to the ground while the "punchee" flies through the air a great horizontal distance.

We reviewed projectile motion in Chapter 1, and we know that from the instant a projectile is launched, it has a downward acceleration equal to g. Therefore, it is clear that anybody or anything that is launched will accelerate towards the ground and eventually hit it (unless they are moving fast enough to go into orbit or escape the Earth's gravity all together). The faster something is moving horizontally, the less far the object will fall as it moves a particular distance. For example, a bullet fired into a wall 10 feet away will fall an almost undetectable distance in the extremely short time it takes to hit the wall. A ball thrown at the same wall will fall an appreciable distance because it takes much longer to get to the wall. Nonetheless, the bullet and the ball *still fall at the same rate downward*. If you launched them both horizontally at the same time in an open field (again neglecting effects of air resistance) they would hit the ground at the same time. The bullet would just land a lot further away horizontally.

ACTIVITY

Roll a ball off the edge of a table such that it has an initial horizontal velocity as it goes over the edge. At the instant it goes over the edge, drop another ball, and note when each hits the ground. Try this with different initial horizontal speeds for the first ball. How does this relate to the example with the bullet?

The Physics—The "Punchees". Maybe the guys taking the punches, (the punchees,) in all of these scenes are punched so hard and are moving so fast that they just don't fall far enough downward for us to notice. Let's examine this. In scene 18 it takes Spider-Man about 3 seconds to fly into the light post. We can easily calculate how far he should fall in this time:

$$\Delta y = v_{0y}t + \frac{1}{2}at^2 = (0)(2\ \text{s}) + \frac{1}{2}(-9.8\ \text{m/s}^2)(3\ \text{s})^2 = 44\ \text{m (almost 100 feet!)}$$

Spider-Man should hit the ground long before he ever hits the post. Either superheroes are able to nullify the effects of gravity temporarily (see the next section on Superman) or the filmmakers are playing fast and loose again.

Conclusion. No matter how hard someone is punched, in the case of a horizontally directed blow, *he or she will not remain airborne*. They will always fall to the ground (and slide across the room if there's not too much friction). The "horizontal trajectory resulting from the really powerful punch" misrepresentation is one of the oldest tricks in the book, typically found in the fantasy action genre. Don't be fooled.

The Physics—The Punchers. We know what should happen to the punchees, but what happens to the punchers (and super-punchers) like Spider-Man and the Green Goblin? What about Newton's third law; what does it tell us?

Newton's third law of motion tells us that when an object (such as the puncher's fist) exerts a force on another object (such as the punchee's face), the other object will exert an equal force back. Therefore, however hard a punchee is punched, a puncher will feel an equal force.

ACTIVITY

Push on the wall lightly with both hands, and feel how hard the wall pushes back on you. Now push really hard on the wall as if you're trying to knock it over. What does the wall do to you or, to put another way: how much force does it exert back on you? For a more visceral experience slap the wall lightly, and then slap it harder. Which hurts your hand more? Why is that?

Given this, you might argue that a puncher doesn't feel the same *effect* as a punchee; that the puncher isn't as hurt and seemingly does not recoil as much. This is true. The puncher usually strikes with a closed hard fist, is generally impacting a less-rigid target, and is often well-planted on his or her feet. Even so, people in fights often hurt their hands pretty badly. In any case, both combatants must experience an equal force. The forces involved in most punches are not hard enough to produce accelerations large enough to fling the combatants backward at significant speeds and therefore, the action-reaction phenomenon is masked. However, in the scenes we are talking about, the force is obviously enough to give the punchee a substantial acceleration, resulting in an obvious and significant change in velocity. The punches would have to produce really big forces to make this happen. We won't stop to consider how forces of this magnitude might affect the body (except to say that, apparently, along with superpowers come proportionally super-strong teeth); we simply want to address the concept of the reaction force.

The impact time of a punch is somewhere in the neighborhood of 0.1 s. In scene 18, Spider–Man therefore accelerates from rest to a velocity of about 17 m/s (based on the fact that subsequently he flies 50 meters in about 3 seconds) in this time. Spidey's acceleration is then:

$$a = \frac{\Delta v}{\Delta t} = \frac{17 \text{ m/s}}{0.10 \text{ s}} = 170 \text{ m/s}^2$$

$$F_{net} = F_{punch} = ma = (75 \text{ kg})(170 \text{ m/s}^2) \approx 13{,}000 \text{ N}$$

This force must be equal to the force that Spider–Man exerts back on the Green Goblin as a reaction. How will this force affect him? Well, because he is experiencing a net push backwards, he will accelerate in that direction. Because the Goblin's mass is pretty close to that of Spider–Man his acceleration backward should be about the same as that of Spider–Man's forward motion. It does not matter how strong he is. There are no other forces to anchor him, except a static friction force with the ground. You can calculate that the amount of this force is less than one-tenth that of the kicking force, and therefore relatively negligible. The Green Goblin *must* recoil, *will* accelerate backward, and, might exclaim (as Mr. Scott from the beloved *Star Trek* series would), "There's nothing I can do. I cannot change the laws of physics!" The same principle applies in scene 8 when Peter Parker punches Flash. There is no way out of it. Unlike what we see in the movie, in reality the amazing Spider–Man would be ignominiously thrust backward into the lockers.

ACTIVITY

Look for other *Spider–Man* scenes that violate Newton's third law. For one of these cases determine/calculate what would really happen. There is at least one additional example in scene 18.

This "absence of the reaction force" mistake, by the way, is very common in superhero movies and TV shows. In an episode of *The Six Million Dollar Man,* Lee Majors throws some extremely heavy-looking equipment (maybe ten times his weight) across the room without being pushed back an inch. He would need more than bionic arms to do that! The general principle of forces for action movies can be summarized as follows: Reaction forces disappear when super strength is called into play. (In the next chapter, we will see that the same issues can be explained in terms of ***conservation of momentum***, which is really Newton's third law in different clothing, yielding the same conclusions.)

STICKY FINGERS

Now let's look at some wall-crawler physiology in detail. Spidey is presumably able to scale walls because those little hairs on his fingertips give him a better grip on the brick than normal fingers would. Notice that soon after Peter Parker's initial discovery of this skill he is in full Spider–Man regalia. His costume covers him completely, including his hands. (We can only assume, and hope, that those millimeter-long hairs are able to poke through the patented Spider–Man gloves.) If you look at the close-up of Spidey's fingers in scene 8 and make an approximate count of his fingertip hairs, you might estimate that he has somewhere between 200 to 400 hairs on each finger. For our purposes, let's estimate he has about 300, which means that on all ten fingers (including thumbs) he has a total of about 3,000 hairs. At the very minimum, these hairs must be able to support Spider–Man's weight. We say at the minimum because to scale the wall he must have brief bursts of acceleration every time he hauls himself up another step. This upward acceleration would require a net upward force (Newton's second law) and therefore the "hair force" would need to be greater than Spider–Man's weight force during the acceleration. On the other hand (no pun intended), he might be able to get some small support from the friction the wall exerts on his feet (there's no way any bug-like hairs are poking through those shoes), so with these two counteracting effects we'll say Spider–Man's finger hairs must exactly support his weight.

Our calculation therefore is really quite simple:

$$F_{hairs} - mg = 0.$$

Assume Spider–Man has a mass of about 75 kg. Note that although the original Peter Parker may have been a little on the scrawny side in keeping with his nerdy tendencies, he did grow some pretty impressive muscles during his transformation (see scenes 4 and 5). Therefore:

$$F = (75\ \text{kg})(9.8\ \text{m/s}^2) = 735\ \text{N}.$$

With 3,000 hairs to work with, each hair need only support a weight of 735/3,000 = 0.25 N (equivalent to the weight force acting on a 25-g mass), which is equal to about 0.05 pounds. (A 25-g mass isn't that heavy.) However, a 25-g mass is about the same as that of a large marble. Point your finger out horizontally. Now imagine a hair sticking out horizontally from the end of your index finger. Now imagine attaching a large marble to the end of it. Can the hair hold up the marble? No. Not even a really rigid super gnarly Spider–Man hair.

Maybe the shot of the sprouting hairs is just intended to be symbolic, however, and actually the hair density is much greater than what's shown on-screen. Maybe it could work. It's not completely far-fetched. Recent advances in nanotechnology have produced microscopic fibers that are dense enough and strong enough to support a person's weight. Nevertheless, assuming what we see is what we've got, these need to be some pretty strong—and really stiff—hairs to be able to support Spidey. If they are, some pretty scratchy fingers might surprise MJ if she ever takes Peter's hand in a moment of overwhelming romantic impulse. It could diminish the magic of the moment. They might even draw blood!

AN OBJECT IN MOTION . . .

Spider–Man is a very good jumper. From our knowledge of projectiles, the first thing we can do is calculate Spider–Man's launch velocity when he leaps into the air. If we look at the scene and compare Spider–Man's body length to the horizontal distance he covers, it looks like he can jump at least 50 feet (18 m) horizontally (almost twice as far as the long jump world record), and his initial angle of launch looks to be pretty close to 45 degrees.

To find his time of flight (assuming no significant air resistance), look at the vertical part of the motion: $\Delta y = v_{0y}t + \frac{1}{2}at^2$. For a complete leap, $\Delta y = 0$, and $v_{0y} = v_0 \sin 45$, so, with $a = -9.8\ m/s^2$:

$$t = \frac{v_0 \sin 45}{4.9\ m/s^2} = \frac{0.707\ v_0}{4.9\ m/s^2}$$

$$\text{Horizontally: } \Delta x = v_x t = v_0 \cos 45 \frac{0.707\ v_0}{4.9\ m/s^2} = 18\ m,$$

which yields $v = 13\ m/s$ (almost 30 mi/hr).

This is faster than world-record sprint pace, so if Spider–Man was not interested in being a crime-fighting superhero, he could easily have a career as a world-class track and field athlete.

Assuming Spider–Man is already moving in the horizontal direction, we can approximate the force Spider–Man has to exert on the ground to generate the additional vertical velocity ($v_y = v_0 \sin 45 = 13\ m/s\ (0.707) = 9.2\ m/s$). The force depends on the time of contact his foot has with the ground during his leap. A contact time of around 0.25 s would be fairly typical for a bent-knee jump. We can then use the second law in the form $F = \frac{\Delta p}{\Delta t}$.

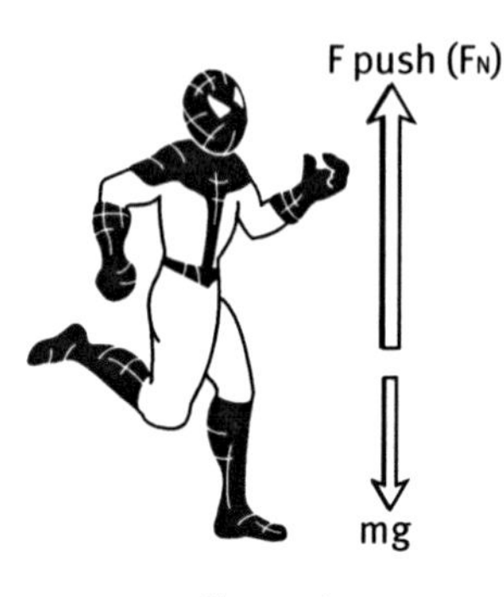

Figure 4

$$F_{net} = F_{push} - mg = \frac{\Delta p}{\Delta t} = m\frac{\Delta v}{\Delta t}$$

$$\text{so } F_{push} = m\frac{\Delta v}{\Delta t} + mg = 75 \text{ kg } \left(\frac{9.2 \text{ m/s}}{0.25 \text{ s}} + 9.8 \text{ m/s}^2\right) = 3{,}500 \text{ N}$$

This is about 800 pounds off of one foot—pretty impressive—but he is Spider-Man, after all!

The last element of this scene that we need to address involves Spider-Man sticking a very tricky landing. How does he stop on a dime, after he lands within a foot or two of the edge of the building? Remember that his horizontal velocity remains unchanged throughout the flight. In this case, $v_x = v_0 \cos 45 = 13$ m/s (0.707) = 9.2 m/s. Therefore, he lands moving horizontally at 9.2 m/s, or a little over 20 miles per hour.

PROBLEM

Calculate Spider-Man's maximum vertical displacement during one of his trans-building leaps and explain why he could also have a very bright future as a basketball player. (Hint: the answer is over 4 meters, and later in the movie we see that he is able to jump two or three times that high.)

According to Newton's first law, a Spider-Man in motion should remain in motion *at a constant velocity* unless acted upon by a net force. Therefore, we need a net force capable of bringing Spider-Man from 20 miles per hour to a stop in less than 2 or 3 feet. In addition, the only force available to us, the only force that could possibly act in the *horizontal* direction (other than a small effect of air resistance), is something that comes in *contact* with Spider-Man (the only relevant noncontact force, gravity, acts *vertically*). The only point of contact with Spider-Man that we can see is between his feet and the building, therefore the only force acting to bring Spider-Man to a stop is the force of kinetic friction. Is it enough?

If Spider-Man slows to a stop in a distance of 1 meter then we can determine his horizontal acceleration:

$$a_x = \frac{v^2 - v_0^2}{2\Delta x} = \frac{0^2 - (9.2 \text{ m/s})^2}{2(1 \text{ m})} = 43 \text{ m/s}^2$$

The force needed to accelerate Spider–Man at this rate would then be:

$$F = ma = (75 \text{ kg})(43 \text{ m/s}^2) = 3{,}200 \text{ N}.$$

This is the amount of the kinetic friction force necessary to bring Spider–Man to a stop:

$$F_k = \mu_k F_{N.}$$

Because the maximum coefficient of friction between surfaces is 1, we will check what coefficient of friction is required to provide the necessary force to arrest Spider–Man's forward motion in a distance of 1 meter. The only additional piece of information we need is the normal force acting on Spider–Man's feet from the roof. If he were just standing there, F_N would simply equal mg, but as he lands the normal force must accelerate Spider–Man in the upward direction to halt his downward vertical velocity. His landing acceleration vertically would be about the same as when he jumps upward, therefore, the vertical force on his feet during the time he slows to a downward stop would also be about the same (at 3,500 N). This helps increase the friction force a lot.

Therefore the minimum coefficient of kinetic friction we need to bring Spider–Man to a stop in time is, $\mu_k = \dfrac{3{,}200 \text{ N}}{3{,}500 \text{ N}} = 0.9$, which puts us right on the borderline. Spider–Man being Spider–Man may stick the landing (although he lands on what looks like very smooth concrete which can be pretty slippery and should give a μ_k significantly less than 1). However, since the friction force acts at the soles of his feet the inertia of the upper half of his body may make it hard for him not to lurch forwards and over the precipice. This normally might not be a serious problem for Spider–Man, however at this point in the movie, he still hasn't figured out how to use his web!

HOW STRONG *IS* THE GREEN GOBLIN (OR MORE ISSUES WITH FRICTION)

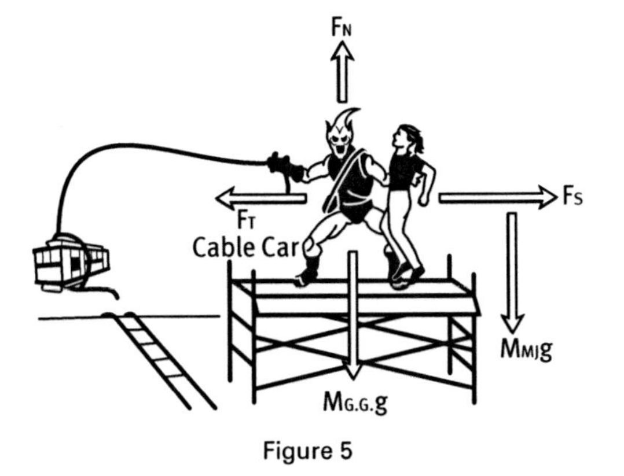

Figure 5

The Scene

We left off our last scene with the Green Goblin holding MJ in one hand and a cable car in the other, with both suspended over a precipice. He's being extremely mean and intimidating, but whether or not he can hold everything depends not so much on his attitude and how strong he is. We assume his strength is "super," and his muscles are physiological wonders, but even so, what is the weak link in the physics here? What forces are acting on the Goblin?

From the force diagram, we see that there is a horizontal force acting on the Green Goblin due to the cable that he holds. The cable is connected to the cable car over a pulley, and the tension force that the cable exerts on the car is essentially the same as the force that the cable exerts on the Goblin. The effect of this force would be to pull him over the edge of the narrow platform. The only force acting to oppose this tension force is the force of static friction between the Goblin's feet and the surface on which he stands. It does not matter how strong he is. It does not matter how strong his muscles must be so that his arms aren't ripped off. The only force acting on the Green Goblin opposing the tension force is static friction—but is it enough?

We know that in the vertical direction the only forces acting are F_N and the weight force (mg) and they must be equal to each other. Because the Goblin holds MJ in one hand, the total weight will include both of them. The G.G. is fully decked out in a costume and a mask, and he's pretty built, so let's give him a mass of 100 kg. MJ looks to be a svelte 120 pounds, so her mass is about 55 kg. Using the maximum possible $\mu_s = 1$, the maximum possible $F_s = \mu_s (m_{gg} + m_{mj})g = 1(55 \text{ kg} + 100 \text{ kg})(9.8 \text{ m/s}^2) = 1{,}520 \text{ N}$.

Now, how big is the tension force acting from the cable? If the cable car was stationary, then $F_T = m_{car}g$. How much mass does a cable car (filled with people) have? Let's make a conservative estimate at 2,000 kg. If the cable car is held without moving then $F_T = (2{,}000 \text{ kg})(9.8 \text{ m/s}^2) \approx 20{,}000 \text{ N}$, which would be the same F_T acting on the Green Goblin. We shouldn't be surprised that the friction force is almost 20 times too small to prevent the G.G. and MJ from accelerating. Therefore, the system—the cable car, the Green Goblin, and MJ—all must accelerate together, and over the edge they will go!

At what rate would they accelerate? Because the car is accelerating, it must have a net force in the downward direction and T will be less than the 20,000 N (as previously calculated). Because everyone accelerates together, we can treat the Goblin, the car, and MJ as one system. We will add all the forces together from each part of the system along the line of motion and set that net force equal to the total mass times the acceleration of the system.

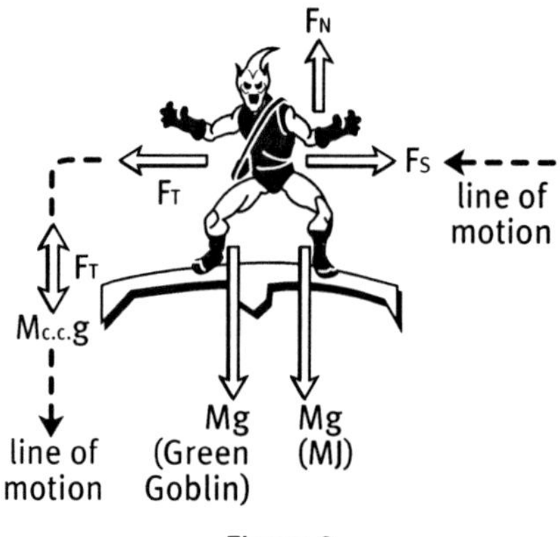

Figure 6

F_{net} (on the system along the line of motion) $= m_{cc}g - F_T + F_T - F_s = (m_{gg} + m_{mj} + m_{cc})\, a$.

Therefore, $a = \dfrac{19{,}000 \text{ N}}{2{,}155 \text{ kg}} = 8.8 \text{ m/s}^2$, which is nearly as much as the free-fall acceleration due to gravity. They would all be flung over the edge pretty quickly.

Conclusion. Clearly there are a fair number of physics no-no's here, but not enough to get us overly agitated or upset considering the genre. Of course, we have to allow some slack when dealing with superpowers and the like, and as we said at the beginning, *Spider–Man* is a fun movie to watch. Therefore, as long as we keep our wits about us, and as long as we maintain a healthy physics perspective at the most exciting moments, we should come through it all right.

HOW *DOES* SUPERMAN FLY?

Superman is an icon among superheroes. Not only was he one of the first ones brought to life on the big screen, he possesses the pinnacle of superpowers, being essentially invulnerable. He is impervious to pain, extremes of hot and cold, and penetrating projectiles. He has x-ray vision, his hearing sets the standard for man and beast, he doesn't seem to need air, he never gets tired, and he can fly. Superman's only weak point is his susceptibility to fragments of kryptonite, which are the remnants of his exploded home planet Krypton. For some reason these fragments are deadly to Superman, but fortunately they are very rare.

We could spend a long time critiquing Superman's impossible physiological attributes, and how they violate both physics and biological principles, but it wouldn't be entirely fair. After all, we know that we are required to suspend our disbelief going into this one. Nevertheless, before we get to the main point of this section, we will discuss a few oddities and make fun of them a little (just to get it out of our systems). In doing so, we will use the original film *Superman* (1978) starring the late Christopher Reeve as our main reference.

Superman's Density

The reason for Superman's incredible powers on Earth, as explained by his father (played by the late Marlon Brando), are due to the "dense molecular structure" possessed by the inhabitants of Krypton. (He says this before sending baby Superman off to Earth in a nifty crystalline spaceship just before Krypton starts disintegrating.) While this alien chemistry/physics affords no obvious advantage on Krypton (as we see clearly see when the planet starts exploding and the Kryptonians are helpless in the face of falling beams, chunks of ceiling, pots and pans, and broken dishes), on Earth it is a completely different matter. Mass on Krypton and mass on Earth must have entirely different properties. Everything seems to be proportionately more massive on Krypton. However, if so, why does Superman only weigh 225 pounds as he claims in scene 26? He seems to be telling the truth (he always does), because he certainly isn't breaking any furniture by sitting on it.

Superman's Anatomy

The Kryptonians look exactly the same as humans. Their mouths, noses, ears, and presumably other parts of their external anatomy are identical to Homo sapiens. Superman eats, and apparently he breathes just like we all do, and yet clearly, despite the amazing similarity with humankind, his body functions must be quite different (an amazing example of extraterrestrial convergent evolution for the biologists in the house). For example, Superman can go into space and exist quite comfortably there without air, television, or other physical comforts. Maybe his lungs are merely vestigial, a *decorative internal ornament* (don't forget about x-ray vision) that activates only in the mating season. We may never know.

Super Vision

Speaking of eyes, does Superman literally have x-ray vision? If he emits x-rays from his eyes it is unclear how the image actually gets back to his eyes, because x-rays are going to penetrate the object that he is looking at and *come out the other side*. Perhaps "x-ray" is just a convenient metaphor for some other type of probing radiation that is, in fact, reflected back off of the interior of the objects of interest. Maybe his eyes emit alpha particles, like in Rutherford's famous gold foil experiment, and Superman's version of optic nerves are able to assemble those alpha particles that bounce back into his eyes into an accurate picture.

Superman's First Law

In the climax of the movie Superman goes out into space and causes the Earth to spin backwards somehow by flying around it really fast opposite to the normal spin direction. How does he effect the Earth's rotation when he is not actually in contact with it? In fact we are going to develop a hypothesis in the upcoming pages to try to explain this strange phenomenon. (Hint: Is there some fundamental force that we don't know about?)

When Superman reverses the Earth's direction there are no inertial effects whatsoever. Considering Newton's first law of motion imagine what might happen if the Earth stopped spinning relatively suddenly! The only effect in the movie is that the backward spin reverses time. Now that's just silly! A lot of things might get broken, but time should not depend on the rotation direction of the Earth. It's also a flagrant violation of the second law of thermodynamics in case you were wondering.

Fantasy has its prerogatives, so let's get to the real reason we are here: to make a heroic attempt of our own. With all of the implausibility and impossibilities encapsulated within the concept of Superman, we are going to find something to hang some physics on. We are going to construct our own theory of how Superman can (or cannot) fly.

The Flight of the Superman

Because we are attempting to come up with a hypothesis explaining how Superman is able to fly, we can do some preliminary work here by considering and answering the following questions. What do birds and airplanes have or what are they able to produce (that Superman seemingly lacks) that makes flight possible? How do birds lift into the air? How do airplanes get off of the ground? How might Superman make up for the apparent deficiency?

If you consider these questions, you might have reasoned that all things that fly must be experiencing a force or forces counter to the force of gravity acting on them. How are these forces generated? In the case of a bird, it flaps its wings, pushing down on the air underneath it. This results in the reaction force of the air pushing back up on the bird's wings (Newton's third law). If the force of air on the wings is greater than the force of gravity down on the bird, then it will accelerate upward (Newton's first and second laws), and if there is any horizontal component of this force it can also accelerate forward.

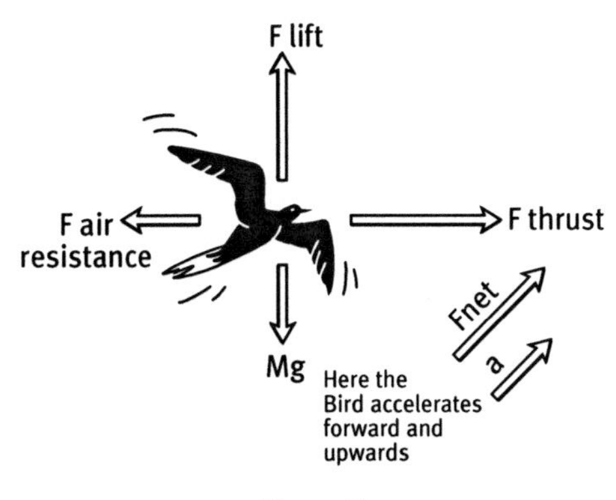

Figure 7

Airplanes don't flap their wings, so they must generate an upward force in a slightly different way. In fact, an airplane can only lift off of the ground (and maintain altitude) if it has a significant horizontal velocity. This is because the shape of the wings directs the air flow more rapidly over the top of the wing than past the bottom. A wing shaped this way is sometimes called an *air foil.* Because fluid pressure decreases with the velocity of the fluid (according to Bernoulli's Principle and Conservation of Energy, see Chapter 5), the pressure underneath the wing is greater than the pressure on top, generating an upward force on the wing called *lift*. As the velocity of the plane increases, so will this upward force, and when the plane reaches sufficient speed, the lift force will equal then exceed the weight force and the plane will accelerate upward. Therefore, an airplane needs something that will provide it with a significant force acting on it in the horizontal direction. In the case of jet planes, the engines suck in air and push it out the back of the engine. By pushing the air out, the air pushes back (third law again), and a forwardly directed force acts on the plane.

PROBLEM

Redraw the previous force diagram to show what would happen if the bird flies at a constant velocity. How would the magnitude of the forces be different?

What we're arriving at is that the only way for Superman to get off the ground, and the only way for him to maintain any kind of flight, would be for him to create a situation like a bird or an airplane where a force or forces are acting on him counter to gravity (and air resistance). Otherwise, "it *is* a bird or a plane!" We must apply the third law for this to work. The only forces Superman experiences while flying must be generated by him pushing on something that pushes back. It is the only way.

Now it's time to construct some hypotheses, see if they make sense, and determine how they might work.

The Physics—Superman the Air Rocket. Let's hypothesize that the forces acting on Superman are similar to those that act on birds and airplanes. Birds and airplanes push on air, and the air pushes back. This is the essence of the "flight forces" exerted on them. If Superman is doing something similar, he must be exerting a force on the air in contact with him. How might he do this? He doesn't have wings, propellers, or jet engines, and he is not shaped like an air foil either.

Let's try this hypothesis: Superman emits thin streams of high-velocity air (through pores in his skin, or other unknown orifices). He is able to control, adjust, and vary which part(s) of his body expel air in order to control, adjust, and vary his flight direction and speed. He can simply replenish the air that he expels by breathing (perhaps that is the real purpose for his lungs, rather than respiration). Therefore, Superman functions much like a rocket, or a missile. He expels fuel (air) by pushing it out of his body, and the reaction force to that, the expelled air pushing back on him, results in a force acting on him.

This hypothesis has some appeal. Superman has an inexhaustible fuel source. He is able to generate the appropriate forces, and we can explain all manner of motion including hovering, horizontal and vertical motion, and accelerated motion via this mechanism.

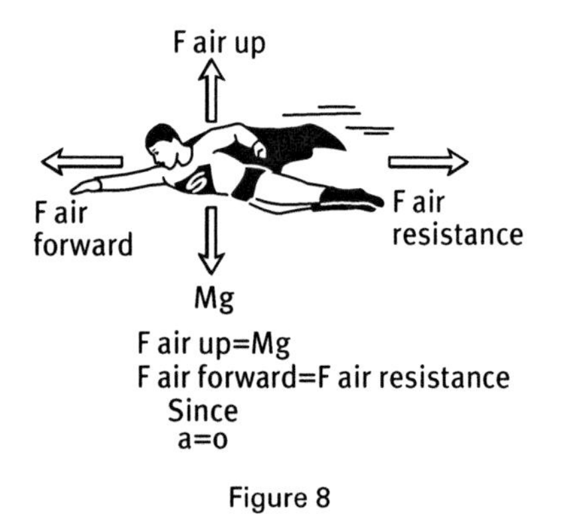

Figure 8

Example 7: Superman is flying horizontally, at a constant velocity over the city of Cleveland. Draw a force diagram showing all of forces acting on Superman. From what general regions of his body should he expel air to maintain this flight?

Solution: Because Superman is flying at a constant velocity, the net force acting on him must equal zero. We know there is a gravity force acting directly downward, and a force of air resistance acting against the direction of his motion, Superman's air jets must produce a force that has components directed 1) vertically upward, and 2) horizontally in the direction of his motion. Superman could generate the horizontal component of force by expelling air directly out of the soles of his feet. Then the vertical force could be accomplished by a stream of air flowing downward, emanating from his stomach.

While the air-rocket theory is a pretty good one, and is able to explain Superman's motion very well, unfortunately, there are some peripheral problems associated with it. First of all, if Superman is emitting high-velocity streams of air, he should be creating a lot of wind around him. The air rushing past him could mask the wind Superman produces as a result of the high velocity he is flying at, but what about when he is hovering? He hovers a lot in the movie, and we don't see much of a breeze underneath him.

Secondly, in scene 28, Superman goes on a flying date with Lois Lane. At one point Superman holds on to Lois only by her hand as she flies side by side with "the man of steel."

How does Lois stay aloft? Presumable she is not emitting powerful streams of air herself (she is, after all, a mere human), and because her only support against gravity acts on her hand she should be dangling below Superman like a rag doll caught in a tornado.

Finally, in the famous "circle the Earth backwards" scene, Superman appears to be entirely outside of the atmosphere. If he requires air for fuel, this could present an insurmountable obstacle.

These inconsistencies appear to be fatal to our theory. It explains some of our observations very well, but it cannot account for others. Therefore, as often happens in science, we have to abandon one hypothesis and look further for a better theoretical model.

The Physics—Repulsive Gravity and Fundamental Forces. Whatever hypothesis we propose to replace the air-rocket hypothesis must account for the inconsistencies described, and so there may be great potential with the idea of repulsive gravity.

We know, based on various compelling lines of reasoning (including the Doppler shift of light arriving from distant galaxies, the cosmic microwave background radiation, and so on), that the universe is expanding, and probably originated from an inconceivably dense, infinitesimally small point exploding in what is known as the big bang. However, due to the gravitational attraction between the mass contained in the universe, the rate of expansion should be decreasing. One question relevant to cosmologists is whether the universe has sufficient kinetic energy to continue to expand forever, albeit at a decreasing rate, or will the gravitational attraction be strong enough eventually to pull everything back together in a "big crunch."

In order to answer this question, astronomers try to determine relative motions of distant regions of the universe. Amazingly, in recent years it has been shown that there are places in the universe where the expansion rate is actually increasing. Clearly, this increased expansion rate represents something completely unfamiliar going on. It appears that there may be a previously unknown fundamental force that actually acts to repel mass—a sort of "repulsive gravity." While the explanation for this observation at present is highly speculative, the phenomenon lends itself very well to a theory of flight for Superman. Moreover, who knows, the planet Krypton may have been located in one of these distant corners of the known universe before it exploded.

All right, here's our hypothesis: Superman is able to generate force between himself and any other mass via a mechanism of "repulsive gravity" (or RG). Because Superman is not always accelerating away from everything, he must be able to turn this fundamental force of repulsion on and

THE AMAZING THERMAL INSULATION PROPERTIES OF SUPERMAN

In the climactic scene of the movie (scene 38) Superman must divert two nuclear missiles that arch villain Lex Luthor (Gene Hackman) has caused to malfunction. One is set to detonate in Hackensack, New Jersey, and is merely a decoy. The other is to explode inside the San Andreas fault causing a massive earthquake, which Luthor calculates will knock everything west of the fault into the ocean. Luthor's plan is that this will drive real estate prices up on thousands of acres of land that he has purchased just east of the fault, because it will consequently become oceanfront property.

The geologic fact is that (contrary to popular misconception) a massive earthquake can not, will not, now or ever, cause large sections of California to "fall into the ocean." We don't have time to go into basic Geology and Seismology 101 right now, but for those interested, look at chapters on earthquakes, and plate tectonics, in any introductory geology text book.

Superman of course must divert these two missiles (into space) before they cause massive death and destruction. Complicating matters is the fact that the missiles are due to detonate within a few seconds of each other on opposite sides of the country! Superman is able to divert the Hackensack attack successfully, and then he must get to California (a distance of about 3,000 miles) in less than a minute. He seems to arrive a few moments after the bomb actually goes off. We know that Superman can fly really fast, and in fact later he makes the trip across the country look like a leisurely stroll by comparison, but at this point we are still very impressed by his great speed. It seems to take him less than 30 seconds to fly to California (which is a lot faster than the Concorde). In this case:

$$v = 3{,}000 \text{ mi } \left(\frac{1{,}609 \text{ m/mi}}{30 \text{ s}}\right) \approx 160{,}000 \text{ m/s}$$

This is over 20 times the speed of the space shuttle as it re-enters the denser parts of the Earth's atmosphere. Which brings us to the point: Anything traveling that fast inside the atmosphere will heat up to temperatures of thousands of degrees Celsius in less than a few minutes and vaporize. The reason the space shuttle does not is that the bottom is covered with a layer of super-insulating tiles that reflect most of the heat away from the ship.

Although Superman is only traveling for a short time to cross the country, his speed is enormous. The frictional heating between the air and his body should raise his surface temperature to thousands of degrees. What does this say about not only Superman's skin, but about his hair, his eyes, the inside of his mouth, and so on? It must be that Superman's superficial layers also have extraordinary insulating properties, and yet even though he is reflecting radiation at thousands of degrees he doesn't even glow! How's that for a dense molecular structure?

off at will. We propose that Superman has a previously undisclosed organ, located in the area where his spleen should be, that is able to mediate this force effectively. It could be that the response of this organ is involuntary like the heart. In any case, whenever he wants to fly, he uses this organ to generate repulsion between himself and any other relevant objects.

Our RG theory allows the advantage that Superman can fly in the emptiness of outer space (no air required), and no unsightly or inconvenient air streams need emanate from his body. Because RG is a fundamental force, it can be transmitted through a vacuum and is visually imperceptible. It can explain how Superman is able to effect the Earth's rotation without actual physical contact. In addition, because this organ apparently can manipulate the RG field around him, Superman can hold Lois aloft by a single hand, which is really just for show, because he also must be surrounding her with the RG field.

It is still puzzling, however, that when Lois accidentally releases Superman's hand she plummets earthward. It's also a bit confusing that when Superman circles the Earth he seems to be exerting an *attractive* force. Perhaps this is simply a reversal of the RG field polarity. Another consideration is the enormous amount of energy this organ must produce. How is it provided with sufficient fuel? Repulsive gravity may not be the be all and end all of theories explaining Superman's ability to fly but perhaps it's a step in the right direction.

Conclusion. Superman, X-men, The Incredible Hulk, Spider–Man—all are examples of sci-fi comic book heroes come alive on the big screen. Each, in turn, is based on a clearly impossible fundamental biophysics premise that we just have to accept if we want to enjoy the fantasy. When we foray into this rarefied world of "the extremely super-powered superhero" we have a lot of fun seeing them do amazing things that we all wish we could do. Yet, it is often the most basic physics that get thrown out the window when superpowers come into play.

XXX REVISITED AND *BATMAN*

X-tremely and Curiously Strong

Remember the motorcycle jumps in *XXX* that we were talking about in Chapter 1? Well, if you look at scenes 7 and 8 you will see an ascending sequence of more and more incomprehensible leaps. At least in the first jump, there is a ***ramp*** that Vin Diesel uses to propel himself to a 50-foot-high 50-yard-long trajectory, but subsequently they dispense with any ramps at all, and yet somehow he is able to launch himself off of the ground to heights that seem to exceed 70 feet or more. How does he do that?

There are a few particular moments that are really rather mind-boggling. In one of the more dramatic of these incidents Diesel is racing along the side of a 30-foot-tall barbed wire fence on his motorcycle and there is no escape in front. A helicopter gun ship is bearing down on him, and we can't imagine how he'll get out of this one. Imagine our astonishment when he yanks up on the front end of the bike, and he somehow twists the vehicle up and over the fence sideways without the help of any ramp whatsoever.

Let's determine how much upward force must be exerted on the motorcycle and its rider to accomplish this feat. We need to find out how fast they are moving vertically the instant after losing contact with the ground. We know that they must achieve a minimum height of 30 feet (about 9 meters), and at the instant that they achieve the maximum height they will have a velocity equal to zero. Therefore,

$$v_0^2 = 2a\Delta y - v^2,$$

$$v_0 = \sqrt{2(9.8\ \text{m/s}^2)(9\ \text{m}) + 0} = 13\ \text{m/s}.$$

During the time they are pushing off of the ground, they accelerate from an initial vertical velocity of 0 to 13 m/s. What forces are acting on the system during this time? In the vertical direction there is the weight force (mg) acting downward and the normal force (F_N) pushing upward. Horizontally there is a forward acting force due to the ground pushing on the tires (which is the reaction force to the tires pushing into the ground by the way), and forces of friction/air resistance acting against the horizontal motion.

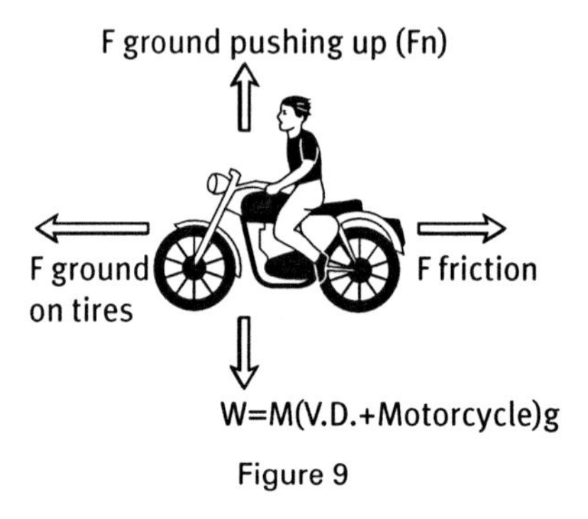

Figure 9

Looking at the vertical direction, Newton's second law gives us:

$$F_{net} = F_N - mg = ma$$

Therefore,

$$F_N = mg + ma .$$

According to biomechanics data, the time to push off of the ground in a jump will be about 0.25 seconds more or less, therefore, we can calculate the approximate acceleration during the push:

$$a = \Delta v/\Delta t \approx (13 \text{ m/s} - 0)/0.25 \text{ s} = 52 \text{ m/s}^2.$$

Assuming the motorcycle is a fairly light one with mass of around 200 kg, and Vin Diesel has a mass of 80 kg,

$$F_N = 280 \text{ kg}(9.8 \text{ m/s}^2) + 280 \text{ kg}(52 \text{ m/s}^2) \approx 17{,}000 \text{ N (almost 4,000 pounds)}.$$

How is this normal force applied and who applies it? The only way to find out depends on Newton's second and third laws. When someone jumps off of the ground he or she must exert a force into the ground. By the third law, the ground has to push back on the jumper with an equal amount of force. This reaction force from the ground is the normal force acting on the jumper's feet, which will result in the upward acceleration. The only way Vin Diesel can generate the upward normal force then is to push off of the ground exerting a force of almost 4,000 pounds. Can he do it? Actually, the answer should be obvious without any calculation. Have you ever seen a person, let alone one carrying a motorcycle, jump 30 feet straight up in the air?

Diesel is moving *horizontally.* Without a net force in the vertical direction, he will not accelerate upward and off of the ground. A vertical force could be generated by a ramp, but without one, because he is traveling along a flat road the only way is for Diesel himself to generate a sufficient upward force. Perhaps there is some special fundamental force of nature that applies only to action heroes in emergency situations—or they have access to incredible surges of adrenaline, like those mothers that are able to lift an automobile to save a trapped child. In the universe as we know it, however, there is no way to somehow twist the direction of a horizontally moving motorcycle into an upward direction with some kind of extreme sport maneuver that will propel it dozens of feet above the ground. It simply cannot be done. The jump is completely ridiculous, as is the sequence of leaps where Diesel achieves incredible vertical clearances without the aid of an incline. After all he's not Spider–Man, let alone Superman!

Batman Begins with Newton's Second Law

The Batman films, like many movie franchises, start off with one or two pretty good films, which inevitably tend to degenerate just a bit more with each new sequel that they dish up. The original *Batman* with Michael Keaton as the tortured and brooding crime-fighting hero is quite an entertaining movie, but by the time we get to *Batman and Robin* the quality has eroded significantly into the realm of the cheesy and the silly. In the case of the Batman movies however, unlike many franchises, the original grit has been rediscovered and even been taken a step further with the recent release of *Batman Begins.* While the original *Batman* was appropriately dark, serious, and even creepy, and the sets and the atmosphere were stunningly and impressively Gothic, there was still quite a bit of humor. *Batman Begins,* on the other hand, is serious as a heart attack. It removes all of the camp and turns Batman almost into a straight-out drama. It has the most realistic feel of all of the Batman movies. Nevertheless, is the physics realistic in either movie? Let's examine some scenes from both.

Batman the Movie

Arch-villain, chemistry whiz, and borderline psychotic The Joker (Jack Nicholson) has taken over Gotham's underworld, and is terrorizing the city by initiating an unprecedented wave of criminal terror in the already decaying and crime-ridden city. The Joker's particular talent, in addition to his creatively morbid sense of humor and his highly developed sense of the theatrical, is an ability to concoct a variety of insidious deadly poisons that he unleashes on the public in a variety of unexpected ways. The Joker also has it out for Batman, who is committed to putting an end to the reign of terror perpetrated by The Joker and his minions.

The Scene. In the climax of the movie (scene 27), The Joker has shot down Batman in his hi-tech fully loaded Batplane, which has gone careening into the base of a particularly ominous Gotham City skyscraper. The Joker has Batman's girlfriend (played by Kim Basinger) held hostage at the top of the building, and Batman must both rescue her and prevent The Joker's escape. Fortunately, despite the fact that his plane hits the ground at airplane-crashing speeds (100 to 200 miles per hour or so), and smashes in a fiery blaze into the building at a speed not much less than that, Batman seems to be in pretty good shape when he opens the hatch. The body of the plane is still more or less intact, and apparently Batman has a really good system of air bags and flame retardants inside.

PROBLEM

Does the fact that the exterior of the Batplane is intact after the collision mean that Batman will not be injured in the crash? Make a rough estimate of Batman's average acceleration from the time the plane hits the ground to just before it hits the building. What forces are acting on Batman to slow him down during this time? During the short collision time with the building (probably between 0.01 and 0.05 seconds), determine how much force acts on Batman to bring him to a stop. How would an air bag reduce the force acting on him?

The Physics—The Retractable Rope-Hook. After chasing The Joker up to the top of the tower, they have the final showdown. Batman is clearly the better fighter, and after several crunching blows he knocks out The Joker's set of chattering false teeth, but Batman and Kim Basinger have a stroke of bad luck when the ledge on which they are standing gives way. They are left dangling over the ledge hanging on only by their fingertips (where have we seen that before?). This gives The Joker time to get away. Batman and Basinger lose their grips and fall over ten seconds in movie time before Batman's retractable rope-hook catches on a gargoyle and abruptly stops them. The rope doesn't appear to be that flexible and their velocity plummets to zero in probably less than one-tenth of a second, yet it saves them from crashing to certain death on the ground below. Are they really saved though? What magnitude of forces must the rope exert to arrest their fall? What effects would this force have?

Notice that this is very similar in principle to Vin Diesel's motorcycle leap, but in reverse. The forces acting on the pair are an upward force due to the tension in the rope, and a downward force due to their combined weight. We will determine the tension force F_T, assuming Batman has a mass of 80 kg and Kim Basinger has a mass of 60 kg.

$$F_{net} = F_T - m_{total}\,g = m_{total}\,a$$

$$F_T = m_{total}\,(g + a)$$

The tension force depends on their rate of acceleration. To determine this, we need their velocity just before the rope catches. Because they fall from an initial velocity of zero, we can use the following kinematic formula:

$$v = v_0 + at$$

Solving for their velocity just before the rope arrests their fall, we get v = 98 m/s, which is faster than terminal velocity for a skydiver, so their speed will probably level off before they get that fast. Terminal velocity depends on the aerodynamics of the falling body, but somewhere around 60 m/s would be a conservative estimate for the body position in which these two are falling.

We'll use the approximate terminal velocity value of 60 m/s as the velocity just before the rope catches. Now we need to find the acceleration for the 0.1 second that the rope brings the pair to a stop.

$$a = \Delta v / \Delta t = 60 \text{ m/s} / 0.1 \text{ s} = 600 \text{ m/s}^2$$

Note that this is an acceleration rate of over 60 times the acceleration due to gravity or 60 g. That's a huge acceleration. It means the tension force exerted by the rope will be

$$F_T = 140 \text{ kg}(9.8 \text{ m/s}^2 + 600 \text{ m/s}^2) = 85{,}000 \text{ N } (19{,}000 \text{ lbs or over 9 tons}).$$

That is a lot of force. How this much force would affect the human body depends on the duration of impact, the accelerations of different parts of the body, and the pressure exerted on different parts of the body. Even if we double the time of the acceleration to 0.2 s, we still get a force close to 10,000 pounds. This means that Batman's rope must be extremely well-made to withstand that much tension. Even more serious, however, is the fact that this much force is being exerted on Batman when he grasps the rope.

We mentioned at the beginning of the chapter that the principle feature that distinguishes Batman from other superheroes is his complete lack of superpowers. Yes, he is smart, tenacious, and in excellent physical condition, but Batman is a normal human. It is estimated for normal humans that large bones will break when experiencing forces of about 90,000 N. A force of 10,000 pounds/45,000 N applied to smaller bones like wrists or fingers would likely be enough to shatter them. Equally serious is the possibility for internal injuries. Rapid accelerations and decelerations can also cause severe internal injuries without any external trauma. This is a result of Newton's first law. Because the internal organs are not rigidly fixed to the frame of the body, even though the skeletal structure may be brought to a stop the internal organs will continue to move "at a constant speed in a straight line until acted on by a net external force." In this case, that force would be due to collisions with other parts of the body. To make matters worse, Batman holds Kim Basinger around her waist, and he must exert a force of similar magnitude to keep a grip on her.

PROBLEM

At the beginning of the movie *Speed,* the meticulous and inexhaustible Dennis Hopper has planted bombs on an elevator in order to blow out the cable and brakes, and cause the crowded elevator car to plummet more than 40 stories to the basement. In an attempt to outwit the bomber, Keanu Reeves and sidekick Jeff Daniels attach a steel cable from a building-top construction zone to the top of the elevator. When the brakes are blown, the elevator car drops about five stories before the cable catches the elevator and brings it to an abrupt stop. The passengers are unharmed. Is this realistic?

Conclusion. The point here is that (despite the common misconception) it does not necessarily matter whether a fall is interrupted before impact with the ground. If the rate of acceleration is great enough as you are brought to a stop, the possibility of severe injury is just as serious as in a

One of the hallmark principles of cartoon physics (see Appendix C) is exemplified by the following type of scenario. "If you are on a falling object such as a crashing plane or elevator, in order to eliminate the effects of the crash, jump straight up at the last minute before hitting the ground. You'll be all right." Bugs Bunny does this all the time. Why does this not work in the real world, and why should you never try this at home?

collision with the sidewalk. To alleviate the effects of the forces and resulting acceleration, their magnitude must be reduced *by increasing the time over which the forces occur.* This means that the rope does no good unless it is very flexible, similar to a bungee cord. If you are brought to an abrupt stop from a high speed you will be seriously hurt.

Batman Begins

Batman Begins, appropriately enough, tells the story of how Batman begins. Traumatized by the murder of both his parents when he is a boy, Bruce Wayne (Christian Bale) disappears from his life of wealth and privilege, and spends the early years of his adulthood traveling the world trying to understand the criminal mind. Eventually, Ra's Al Ghul (Liam Neeson) the mysterious leader of a secret Ninja-like organization called the League of Shadows, dedicated to "true justice" tracks him down. Ra's trains Bruce in the "way of the Ninja" but when it turns out The League of Shadows is really out to eliminate evil by killing everyone in his hometown, Bruce gets upset, burns down Ra's' house, and escapes back to Gotham City where he applies his financial resources and martial arts prowess to become Batman as we know and love him. Right away Batman is faced with a sinister underworld plot to import some kind of "weaponized" psychotropic drug that causes those exposed to it to go insane. It is only later that we find out the real mastermind is none other than Ra's Al Ghul himself! Ra's and the League of Shadows are very determined. They have decided to destroy Gotham City by driving everyone insane, thus, as they see it, saving the world from depravity. Batman obviously is highly motivated to stop them, leading to some electrifying scenes for the audience!

The Scene. After setting fire to Ra's' house, Bruce (not yet Batman) pulls the unconscious Ra's out before he is burned. They then slide down a steep snow-covered slope towards the edge of a precipice. As Ra's falls over the edge, Bruce is able to hold on to his arm, and then at the last moment he arrests their fall by jamming a sword into the snow. This sort of situation illustrates one of the reasons why mountain climbers often carry ice axes. If they slip they can dig the axe into the ice in a maneuver called an "ice arrest." The general scenario is realistic but how fast are they going right before Bruce wedges in his sword, and how strong must Bruce be to hang on to both the sword and the dangling Ra's? Will his Ninja training be sufficient?

Look at the picture with the accompanying force diagram on the next page for their trip down the slope.

To determine the force of kinetic friction we need to know the coefficient of kinetic friction between snow and clothing. The coefficient describes the relative "stickiness" between the two

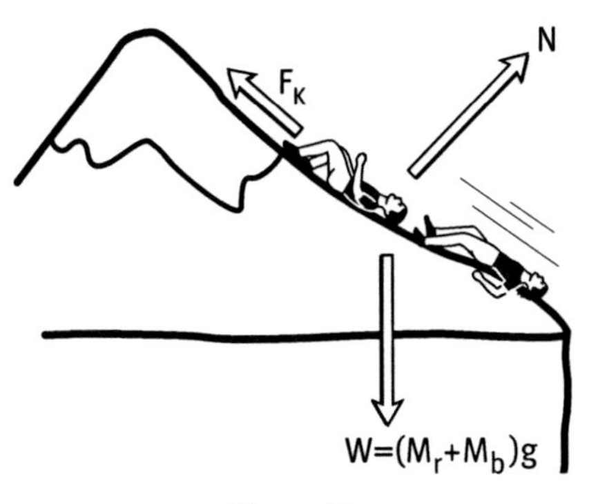

Figure 10

materials. It depends on the clothing material and on the snow conditions. Assuming dry snow and warm wool fleece (both reasonable assumptions considering the extreme cold and low mountain humidity), a μ_k of around 0.2 is probably in the right range. We'll assume Bruce and Ra's have the same mass of 80 kg.

Using the coordinate axes defined on the force diagram, apply the second law of motion to both the x and y direction and assume the two adversaries comprise a single system/object. First we will look at their acceleration while sliding, and use this to determine their velocity just before going over the edge.

$$F_{net\,y} = F_N - mg\cos\theta = ma_y = 0$$

Which means:

$$F_N = mg\cos\theta,$$

$$F_{netx} = mg\sin\theta - F_T + F_T - F_k = mg\sin\theta - F_k = ma_x,$$

and

$$F_k = \mu_\kappa F_N = \mu_\kappa mg\cos\theta.$$

Therefore,

$$a_x = (mg\sin\theta - \mu_\kappa mg\cos\theta) / m = g(\sin\theta - \mu_\kappa\cos\theta).$$

If we look carefully at the scene we can estimate the slope to be around 40 degrees, which is a typical incline in mountainous terrain. Plugging in the values we get:

$$a_x = 4.8 \text{ m/s}^2.$$

To get the velocity at the bottom of the slope, we can time the fall, or we can estimate the length of the slope. (Using the time elapsed in an action sequence from the actual scene can be notoriously unreliable. Usually there are multiple cuts, and we can't be sure that these don't overlap in time or are completely consistent with each other. However in this scene it looks pretty straightforward.) It takes 5 seconds for them to get to the edge of the cliff. So, using

$$v = v_0 + at$$

> It's interesting that one of Bruce's allies in *Batman* (along with Alfred the Butler) is the jack of all trades super scientist/engineer Mr. Lucius Fox (Morgan Freeman). Mr. Fox can not only design super hi-tech mechanical gadgets for Bruce, but when Batman is exposed to the psychotropic drug, Fox is able to come up with the antidote in a *few hours.* This guy's wasting his time at Wayne Enterprises, he should be working full-time in medical research! The brilliant "scientist jack of all trades," the guy who can make a nuclear fusion reactor and at the same time understands molecular biology at an advanced level, is pretty common in the movies, although essentially nonexistent in real life.

and assuming they start from a velocity of zero, we get a velocity of:

$$v = 24 \text{ m/s}$$

Alternatively, if we estimate the distance down the slope to be 30 meters, then:

$$v^2 - v_0^2 = 2a\Delta x$$
$$v = 17 \text{ m/s}.$$

We'll split the difference and say they have a velocity of around 20 m/s at the bottom of the incline.

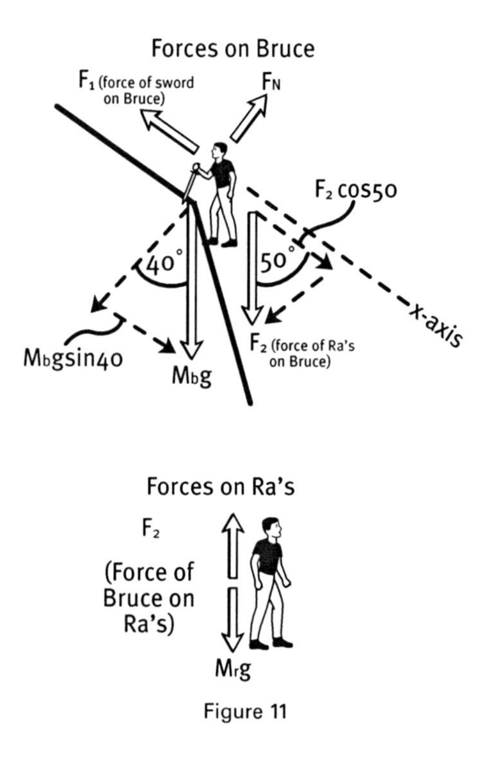

Figure 11

Now for the tricky part: As Ra's falls over the edge, Bruce is able to hold on to him with one hand while he digs the sword in with the other. They come to a stop very rapidly, half a second at most. In this time they slow from 20 m/s to a dead stop, giving them an acceleration of 40 m/s². What forces are being exerted on each of Bruce's arms during the acceleration? Let's draw force diagrams showing the forces acting on Bruce and Ra's.

We'll look at each of these fellows as single objects this time. For Ra's:

$$F_{net} = F_2' - m_r g = m_r a$$

The force of Bruce's left arm acting on Ra's is equal to the force Ra's exerts on Bruce by Newton's third law so $F_2' = F_2$.

For Bruce, because his acceleration is along the line of the slope, we will take that to be the x-axis. Therefore, his weight force and the force that Ra's exerts on him must be broken into x and y components as shown on the force diagram. In the y direction the forces cancel, and in the x direction:

$$F_{net} = F_1 - F_2 \cos 50 - m_b g \sin 40 = m_b a$$

$$F_1 = F_2 \cos 50 + m_b g \sin 40 + m_b a$$

$$= 4{,}000 \text{ N} \cos 50 + 80 \text{ kg}[(9.8 \text{ m/s}^2) \sin 40 + (40 \text{ m/s}^2)] = 6{,}300 \text{ N}(1{,}400 \text{ lbs})$$

During the acceleration Bruce's left arm must be capable of supporting 900 lbs, while his right must support one and a half times that much. That's a lot of weight. (Remember what we said about Batman and superpowers.) He only need maintain that force for half a second or so, but if you know any really strong people that can hold that much weight with one arm for even half a second, contact *The Guinness Book of World Records.*

PROBLEM

Suppose Bruce is somehow able to stop Ra's before going over the edge. When they are just dangling there, what magnitude of forces are acting on each of Bruce's arms?

Even though Bruce is a trained Ninja, the consensus is that both the sword and Ra's are going to get yanked out of Bruce's hand, and the two erstwhile adversaries are going over the edge, thus shortening the movie's running time by over an hour and a half. The major problem with this scenario is the short acceleration time requiring a large net force. The director could have made the scene much more physically possible if Bruce were to dig the sword in a significant distance before the edge, dragging it through the snow and ice and bringing them to a stop more gradually. Then it might be possible, and we could all sleep easier.

Conclusion. The repeated theme here is one of rapid acceleration. *The greater the acceleration is, the greater the forces causing that acceleration are—and the greater the damage.* Why do polevaulters land on soft pads rather than on concrete? Their change in velocity is the same in each case, but the *time* over which that change occurs is much longer landing on a pad, meaning less acceleration, and less force acting on the vaulter. It is a common movie physics problem—clearly demonstrated in the Batman movies—that the accelerations portrayed during action sequences are too rapid for the body realistically to survive.

ADDITIONAL QUESTIONS

1) In *Spider-Man* scene 26 after the Green Goblin releases MJ and the cable car, Spider-Man is able to catch both and stop their fall while hanging from a strand of web. How strong does the web have to be to support the weight?

2) In the same scene, MJ falls out of Spider-Man's grasp but is able to catch herself by barely grabbing a ledge with her fingers as she falls past it. Is this possible? (Hint: Definitely don't try this at home!) MJ appears to have dropped a distance of 30 or 40 feet before catching herself. How much force will be exerted on MJ's fingers in bringing her to a stop?

3) When Spider-Man first learns to use his web he swings off of the top of a building and into a solid wall (scene 8). Estimate the force that the wall exerts on Spider-Man. You can estimate Spider-Man's velocity from looking at the scene, or better yet you can use conservation of energy (see Chapter 3) to determine his approximate velocity before crashing into the wall.

4) In the chapter we created a theory (outlandish as it must be) to explain how Superman is able to fly. In a similar vein, discuss some of the issues and problems involved with the physics of the Green Goblin's glider. Think in terms of Newton's laws. For example, how does it hover? How would you produce an appropriate force to make it hover? Be creative. What would the reaction force be to this force, and where would the effects of the reaction force be observed?

5) Batman uses a set of rigid "bat" wings to float down from great heights at low (terminal) speeds. In *Batman Begins* the wings are actually soft and supple until an electric current is run through them, which somehow causes them to "rigidify." (How this works I have no idea and probably neither does anyone else.) How strong must Batman's arms be to support the wings? (It's interesting to note that birds have extremely strong chest muscles and very light bones. What relevance does this have?)

6) Several times in *Batman* and *Batman Begins* the hero falls from great heights (up to 50 or 60 feet) onto solid ground. Estimate how much force is exerted on Batman during his collision with the ground.

7) In *Batman* scene 16 Batman shoots a cable into a building to aid him in making a tight turn. This is actually a pretty good idea. Explain how this helps. Note: Here we are anticipating circular motion, and centripetal forces, which we will discuss in Chapter 4.

8) In *Batman and Robin* because the novelty of the series is starting to wear off, the producer's idea is to dispense with any originality in plot, and replace it with more and more ridiculous action sequences. In one of the most outlandish, the "dynamic duo" jump out of a rocket on little surfboard shaped pieces of metal and surf their way back down into Gotham City from a height of several thousand feet. They break their fall by landing on a steeply raked roof and "skiing" down it. Using kinematics, Newton's laws, and elementary human physiology, explain why this might be difficult to accomplish successfully.

Conservation of Momentum and Energy

(with special guest appearances by Heat and Thermodynamics)

MISSION: IMPOSSIBLE

It's amazing how much punishment the human body can take in the movies and still function at optimal performance level. Being hit in the back with a lead pipe does not necessarily lead to serious injury, although sore muscles may result the following day. How many cinematic fist fights have you seen where the protagonist gets punched in the face with multiple bone crunching blows, and yet no bones are crunched, and all 32 teeth remain intact? The *Mission: Impossible* movies are aptly titled not just as a comment on the stunts involved but also the human body's reaction to them. We will investigate the physics of one particular scene in *Mission: Impossible II,* with an eye toward momentum, force, acceleration, gravity—and bone density.

The Movie

A sociopath ex-impossible-mission-force agent and his group of nasty cohorts are attempting to steal a deadly virus to sell on the black market. They must be stopped! Mission-impossible agent/ hero extraordinaire Ethan Hunt (played by Tom Cruise) steps in to save the day.

The Scene. Scene 15 involves an out-of-control high-speed motorcycle chase between our hero and his vicious adversary Sean Ambrose (played by Dougray Scott). Toward the end of the chase, Scott turns to face Cruise in a spine-tingling showdown. The two square off several hundred yards apart and rev their engines for many tense moments before accelerating full throttle directly towards each other. They must be going 50 miles per hour at least, and just before the inevitable crash the two simultaneously leap off of their motorcycles and crash into each other in mid-air. As Cruise's motorcycle flips up into the air it disintegrates in an obligatory movie explosion. They grapple with each other as they veer off perpendicular to the line of the impact and fall together over a cliff, landing on the ground some 20 feet or so below where they collided. Unfortunately Cruise has lost his gun during the grapple, and it lands about a second before they do, rolling off to the side.

Photo: The Saturn V rocket launching into space.

Cruise and Scott seem hurt by the fall, or at least the wind gets knocked out of them pretty hard. However they regain their strength sufficiently for the final hand-to-hand contest.

Our task will be to analyze the scene from just before the two crash into each other to just after they hit the ground. We have already reviewed Newton's laws of motion in the previous chapter. We will need these principles to assess the situation, and we will also use the principle of conservation of momentum, which we review next.

Momentum, Impulse, and Conservation of Momentum

Momentum (p) is defined as the product of the mass and velocity of an object:

$$p = mv \text{ (momentum is a vector)}$$

According to Newton's second law, $F_{net} = ma$. If the force is not constant, then F_{net} is the average net force applied over some period of time. Because $a = \frac{\Delta v}{\Delta t}$ it is clear that a net force acting on an object will result in a change in velocity, and therefore a change in momentum (Δp) of the object. The greater F_{net} and the greater the time over which it acts, the greater Δp will be. In fact, we can rewrite Newton's second law in terms of momentum:

$$F_{net} = ma = m\frac{\Delta v}{\Delta t} \text{ and because } m\Delta v = \Delta p \text{ then}$$

$$F_{net} = \frac{\Delta p}{\Delta t} \text{ and } \Delta p = F_{net}\Delta t$$

The product $F_{net}\Delta t$ is called *impulse*. The term implies something occurring rapidly because it is often applied to situations involving large forces acting over short periods of time, such as a baseball being hit by a bat, or Tom Cruise impacting the ground after falling off of a 30-foot cliff.

THINKING PHYSICS

If Tom Cruise were to fall 20 feet onto a surface, explain in terms of Newton's second law why he would be hurt less if he landed on a pole vault pad, rather than on the ground. Think about the forces acting on him in each case. Compare the impulse he experiences in each case, and explain both conceptually and quantitatively.

If an object experiences a net force, it will also experience a change in momentum, and if it does not experience a net force, its momentum will be constant. However, if an object experiences forces, they must be due to other objects acting on it. What if we include those objects in our system? Let's look at an example of a collision between two objects called A and B. Each object experiences a net force due to the other object and therefore each has a change in momentum, but the amount of momentum change of A must be exactly equal to that of B but in the opposite direction. Consider why this must be true. According to Newton's third law, the force that A exerts on B must be exactly equal in magnitude but opposite in direction to the force that B exerts on A. Because both A and B exert equal and opposite forces

THINKING PHYSICS

In which of the following interactions is momentum conserved within the system defined?

a) An astronaut on a space walk throws a pineapple. (The system includes the astronaut and the pineapple.)
b) A ball is thrown into the air. (The system includes the ball.)
c) Tom Cruise and Dougray Scott collide head-on. (The system includes both of them.)
d) Tom Cruise and Dougray Scott collide head-on. (The system includes only Cruise.)

Momentum is conserved in a) and c) because all of the forces involved act inside the system so total $\Delta p = 0$. For example in a) Δp for the astronaut = $-\Delta p$ for the pineapple. In b) the momentum of the ball changes due to the external force of gravity, and in d) Cruise's momentum changes due to the force of Scott acting on him.

on each other, and the time these forces are exerted must be equal, and because $\Delta p = F_{net}\,\Delta t$, each must experience the same Δp, but in the opposite direction.

This implies then that *in the absence of external forces on an object or a system of objects* (in the system described, the action-reaction forces are inside the system) *the total momentum of the system will not change*. This is known as *the law of conservation of momentum*.

Example 1: Suppose two astronauts out on a space walk are having fun throwing blocks of clay at each other. In one scenario, two of the blocks hit head-on and stick together. If block A has a mass of 3 kg and an initial velocity of 8 m/s, and block B has a mass of 2 kg and an initial velocity of -16 m/s

a) what are the final velocities of the blocks?
b) what is the momentum change of each block?
c) what average force does each block experience during the collision if the blocks are in contact for a total time of 0.1 seconds?

Solution:

a) We know that the total momentum of the system before and after the collision must be the same (if we include both blocks) because there are no external forces acting on that system. Therefore:

$$m_A v_{Ai} + m_B v_{Bi} = m_A v_{Af} + m_B v_{Bf} = (m_A + m_B) v_f$$

because both blocks have the same final velocity. The subscripts *i* and *f* refer to initial and final conditions.

$$(3\text{ kg})(8\text{ m/s}) + (2\text{ kg})(-16\text{ m/s}) = (5\text{ kg})v_f \qquad v_f = -8\text{ m/s}/5\text{ kg} = -1.6\text{ m/s}$$

b) $\Delta p_A = m_A v_{Af} - m_A v_{Ai} = (3\text{ kg})(-1.6\text{ m/s}) - (3\text{ kg})(8\text{ m/s}) = -29\text{ kgm/s}$

$\Delta p_B = m_A v_{Af} - m_A v_{Ai} = (2\text{ kg})(-1.6\text{ m/s}) - (2\text{ kg})(-16\text{m /s}) = 29\text{ kgm/s}$

c) $F = \dfrac{\Delta p}{\Delta t}$ $F_A = \dfrac{(-29\text{ kgm/s})}{0.1\text{ s}} = -290\text{ N}$ and $F_B = 290\text{ N}$

The preceding problem is an example of a *perfectly inelastic collision*. In all collisions momentum is conserved, however in an inelastic collision some or all of the objects' kinetic energy is lost. A *perfectly* inelastic collision is defined as one in which the objects stick together after the collision.

A *perfectly elastic collision* is one in which not only momentum is conserved but the total kinetic energy of the system is conserved as well. An elastic collision is a "bouncy" collision in that the objects always bounce off of each other like two rubber balls.

Example 2: A 0.2-kg bouncy ball with an initial velocity of 10 m/s collides head-on with a small block of wood (0.5 kg) sitting at rest on a frictionless surface. The collision is perfectly elastic. Find the final velocity of each object.

Solution: $m_A v_{Ai} + m_B v_{Bi} = m_A v_{Af} + m_B v_{Bf}$ (conservation of momentum)

and $\frac{1}{2} m_A v_{Ai}^2 + \frac{1}{2} m_B v_{Bi}^2 = \frac{1}{2} m_A v_{Af}^2 + \frac{1}{2} m_B v_{Bf}^2$ (conservation of energy)

$(0.2\text{ kg})(10\text{ m/s}) + (0.5\text{ kg})(0) = (0.2\text{ kg})v_{Af} + (0.5\text{kg})v_{Bf}$ (equation 1—conservation of momentum)

Therefore, $2 = 0.2v_{Af} + 0.5v_{Bf}$ (equation 1a)

Therefore, $v_{Af} = \dfrac{2 - 0.2\ v_{Bf}}{0.5}$

$(0.2\text{ kg})(10\text{ m/s})^2 + (0.5\text{ kg})(0)^2 = (0.2\text{ kg})v_{Af}^2 + (0.5\text{ kg})v_{Bf}^2$ (equation 2—conservation of energy)

Therefore, $10 = 0.1\ v_{Af}^2 + 0.25\ v_{Bf}^2$ (equation 2a)

Combining equations 1a and 2a we get:

$$0.0125\ v_{Af}^2 - 0.2\ v_{Af}^2 + 0.75 = 0$$

Solving the quadratic equation we get $v_{Af} = 6$ m/s or 10 m/s

Because 10 m/s is the initial velocity of A, then 6 m/s must be the correct answer.

This gives $v_{Bf} = 5$ m/s.

Vectors and Conservation of Momentum

Figure 1 below is a vector diagram representing the above problem. The vectors shown represent the momentum of each object both before and after the collision, and the total momentum before and after the collision.

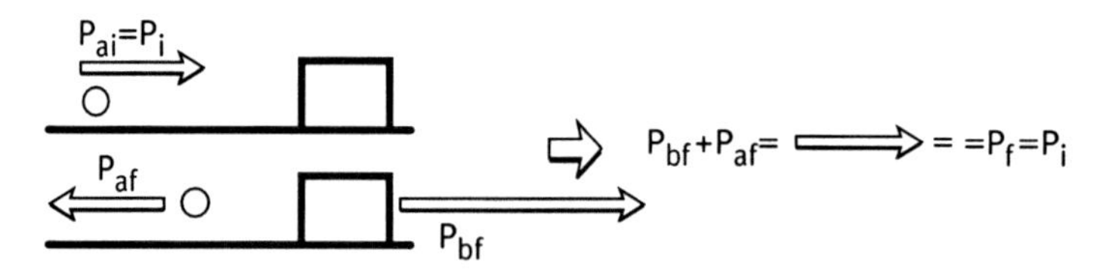

Figure 1

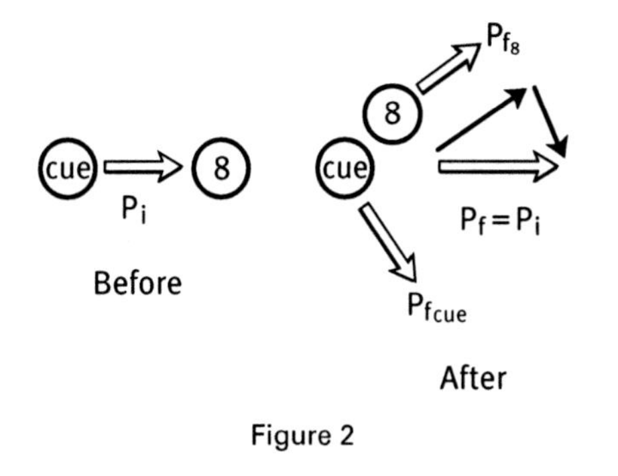

Figure 2

Because momentum is conserved, the vector representing the total p before the collision must be identical to the vector representing the total p after the collision. Now both of the previous problems represent head-on, and therefore one-dimensional, collisions. However a collision between objects could involve two or three dimensions (for example, a cue ball hitting an eight ball in a glancing collision).

Momentum must still be conserved, and therefore the total p vector before the collision must be equal to the total p vector after the collision, as shown in Figure 2. This vector description of momentum is going to be extremely relevant when we dissect the *Mission: Impossible* scene in the next section.

The Physics—The Big Crunch. Now it's time to slice up this scene. We'll take this one apart in pieces as there are several issues we're going to have to deal with, and it won't be pretty. Let's look at the collision that we've been leading up to. First of all we know that momentum must be conserved from before until after the collision. In addition. this is a perfectly inelastic collision because they stick together afterwards. Without doing any calculations it should be clear that momentum is *not* conserved in the scene. Cruise and Scott are heading directly towards each other, and therefore the total momentum is along that line.

After the collision, however, somehow, miraculously, the two move off at a 90-degree angle to the original line of motion. According to conservation of momentum they could end up moving backward or forward along the line of the collision—depending on which one has the greater amount of momentum—but not sideways.

ACTIVITY

View the scene from *Mission: Impossible II,* paying particular attention to the collision between the two adversaries. Does momentum appear to be conserved in the collision? Explain.

For the same collision try to estimate the average forces that the men exert on each other during the impact. Quantify these forces using fundamental physics principles.

Using the principle of conservation of momentum let's determine what their motion should be immediately after the collision. We can reasonably estimate that they are going about 50 miles per hour (22 m/s) on their motorcycles, which means that when they collide they have a relative velocity of *around 100 miles per hour!* Let's assume they have equal speeds. It looks like Scott may be a bit beefier than Cruise, so let's estimate their masses to be about 90 kg (198 pounds) and 80 kg (176 pounds), respectively. The two appear to hit almost head-on but slightly off center, and the collision causes a slight rotation. This means that in addition to the linear momentum described above, *angular* momentum is also conserved. For the sake of simplicity we will only deal with the linear momentum here and discuss angular momentum in a later section. It shouldn't significantly affect our results. Therefore:

$$m_A v_{Ai} + m_B v_{Bi} = (m_A + m_B)\, v_f$$

$$(80 \text{ kg})(22 \text{ m/s}) + (90 \text{ kg})(-22 \text{ m/s}) = (90\text{kg} + 80\text{kg})v_f$$

Therefore, v_f for the grappling duo is –1.3 m/s along the line of motion, which means opposite (180 degrees) to the initial direction of Cruise.

However, they move off perpendicularly. They were driving their motorcycles parallel to the cliff edge and at least several meters from the edge, and yet they fall over the ledge. How can this be? There is no way to change the momentum of the system from *inside* the system. Perhaps a strong wind? However, we see no sign of any meteorological disturbance. What force could have acted externally to push the men over the edge? Sadly, there is none, and we can only conclude that we have entered the realm of pure action fantasy.

Perhaps we could forgive this transgression if only it stood by itself, but it doesn't. It is time to analyze the forces involved in the collision itself. In particular let's contrast their mid-air impact with the forces they experience when they hit the ground. In general the forces involved in collisions occur over a very short period of time. For example, during automobile collisions the forces between vehicles usually last one to two tenths of a second. However cars crumple significantly during a collision, extending the time of contact somewhat. High-speed collisions between less-malleable objects occur over much shorter periods of time. In a car accident if a person is not wearing a seat belt and collides with the steering wheel, the time of that collision is actually around 0.01 seconds. Human bodies really don't have that much padding (particularly in the head), and a collision between two people is a *hard* collision lasting on the order of 0.01 seconds. Our antagonists are also in pretty good shape without much body fat to cushion the blow. In addition, the collision seems to involve primarily the upper halves of their bodies with direct contact between their respective heads. Because $F = \frac{\Delta p}{\Delta t}$ we can determine the forces on each of the combatants as soon as we find Δp for each.

Because $\Delta p = mv_f - mv_i$ and assuming the masses from the equation above, we can find the Δp.

(We also know according to both Newton's third law and conservation of momentum that Δp for Tom Cruise = $-\Delta p$ for Dougray Scott.)

Using the data for Cruise:

$$\Delta p = 80 \text{ kg}(-1.3 \text{ m/s} - 22 \text{ m/s}) = -1{,}860 \text{ kgm/s}.$$

What kind of force is required to do this? Let's be conservative and give them the benefit of the doubt, and assume a longer collision time (0.015 s) and see what happens.

$F = -1{,}860 \text{ kgm/s}/0.015 \text{ s} = -124{,}000 \text{ N}$! The negative sign simply means that we took Cruise's original direction to be positive. That's a *lot* of force. As we mentioned in Chapter 2, how this much force would affect the human body depends on the duration of impact, the accelerations of different parts of the body, and the pressure exerted on different parts of the body. It is estimated that bones will break when experiencing forces of about 90,000 N. Therefore, it already looks like Tom may break a few ribs upon impact. What other damage might occur?

Data on car accidents suggests that a person will survive a crash if the whole body impact pressure is less than 1.9×10^5 N/m^2 for less than 70 ms (0.07 s). Fifty percent of crashes are fatal when pressures exceed 3.4×10^5 N/m^2. In addition, accelerations of the head exceeding 150 g (150 times the acceleration due to gravity) are usually fatal.

THINKING PHYSICS

Explain, according to fundamental physical principles that we have studied, why internal organs can be damaged severely in a collision even if no bones are fractured.

Using this information for reference let's see what these guys are dealing with when they collide. First we calculate the average pressure they experience during the collision. The entire front surface area of the body is about 0.7 to 0.9 m^2. However, it is only the upper right halves of their bodies that actually impact (0.3 m^2 to 0.4 m^2 or so). Let's split the difference and call it 0.35 m^2.

$$P = F/A = 124{,}000 \text{ N}/0.35 \text{ m}^2 = \text{about } 350{,}000 \text{ N/m}^2$$

What accelerations do the characters experience?

Well, $a = F/m$, so for Cruise, $a = 124{,}000 \text{ N}/80 \text{ kg} = 1{,}550 \text{ m/s}^2 = 158\ g$. Remember that they collide jaw to jaw, so each of their heads probably experiences an acceleration close to this rate, which is near the threshold for a fatality.

Even if they survive what does this feel like, and what nonfatal damage does it do? Is it even possible to get up after experiencing something like that? The damage would essentially be the same as running headlong into a brick wall at a speed of 50 miles per hour. You can imagine how badly these guys should feel during and after impact. Amazingly, though, not only are they not seriously hurt by the collision or even in traction, but they are able to wrestle with each other in mid-air before crashing into the ground 20 feet (6 m) or so below.

Nevertheless, here's the kicker: it isn't until they impact the ground that they *do* look really hurt, or at the very least, they each appear to have had the wind knocked out of them. If they can withstand almost 124,000 N and the pressure associated with it as we saw in their mid-air collision, each without so much as flinching, what humongous force must have been exerted on them when they hit the ground to cause each of them such discomfort? Let's see.

Because they are in free-fall, we can determine their velocity when they hit the ground. We need only to look at their vertical motion in which their initial velocities are zero.

$$v^2 - v_0^2 = 2a\Delta x \text{ so } v = \sqrt{2a\Delta x + v_0^2} = \sqrt{2(9.8 \text{ m/s}^2)(6 \text{ m}) + 0} = 11 \text{ m/s (24 mi/hr)}$$

ADDITIONAL *IMPOSSIBLE* SCENES TO ANALYZE

1) In scene 2 Cruise is rock climbing several hundred feet above the flats on a very challenging precipice (without ropes!). At one point he comes to an impasse, and decides to let go. He falls about 15 feet, and catches himself by grabbing a small ledge on the way down. Determine whether or not this is possible. Is this a good idea? (Remember this one? It's the same problem as when MJ catches herself in *Spider–Man.*)
2) In scene 15 Cruise jumps off the side of his motorcycle going full speed. He maintains contact with it by holding on to the handlebars as he skids alongside of the motorcycle, in a standing position. He does this so that sparks fly off of the bottom of his boots and obscure the view of his pursuer. Analyze the forces acting on his feet and hands in this scenario.

If this is true (again allowing about 0.015 seconds for the time of collision with the ground), then:

$$F = \frac{\Delta p}{\Delta t} = \frac{mv_f - mv_i}{\Delta t} = \frac{80 \text{ kg}(11 \text{ m/s} - 0)}{0.015 \text{ s}} \approx 60{,}000 \text{ N} \quad \text{(for Cruise)}.$$

This is still a sizable force and enough to hurt a human pretty badly—maybe break a bone or two—but *it is less than half the force* that they would have experienced in their head-on crash.

Let's look at one final perplexing detail from this scene. We used the constant downward acceleration due to gravity ($g = 9.8 \text{ m/s}^2$) to calculate the change in speed during the fall. We know that this value of g is effectively a constant at the surface of the Earth regardless of the mass of an object, unless there is significant air resistance.

ACTIVITY

Drop a baseball and a tennis ball from the same height a meter or two above the ground. Compare when each hits the ground. Now add a piece of paper to the contest. What happens? Now crumple up the paper very tightly and drop it with the balls. What is different? Imagine what would happen if you simultaneously dropped a (toy) gun and stepped off of a 20-foot-high precipice to see who hits the ground first.

As long as air resistance is small, all objects accelerate at the same rate towards Earth. For relatively heavy or dense objects falling short distances, air resistance will have very little effect. That is why it is quite surprising to find Tom Cruise's gun hitting the ground well ahead of him in a stunning reversal of Galileo's famous (although perhaps apocryphal) experiment from atop the Leaning Tower of Pisa. What would Aristotle say?

Conclusion. The real outcome of this epic mid-air action movie collision between Tom Cruise and Dougray Scott would involve severe internal bleeding, broken bones, comas, and months of rehab. The final showdown would have to be deferred until the adversaries get discharged from the hospital (and the final half an hour of the film would be transformed into a gut-wrenching medical drama as the two former super agents try to kill each other while in traction). Action heroes experience these types of monumental collisions all the time. Is it at all possible that they could still finish the mission afterwards? Don't you believe it!

OCTOBER SKY—ENERGY AND MOMENTUM IN REAL LIFE

Because we may still be a little punchy from the outrageous collisions and flagrant violations of physical (and biological) principles that we were subjected to in *Mission: Impossible II*, it is time to inject a dose of reality. We are going to look at rocket flight in the real world, an example of which the movie *October Sky* presents, based on a true story of four high-school students who become interested in building rockets.

Rocket flight provides an excellent backdrop to discuss some fundamental physical principles. In addition to Newton's laws we will incorporate two of the most powerful, fundamental, and simple ideas in physics—conservation of energy and conservation of momentum—in analyzing the motion of rockets.

The Film

October Sky takes place in the late 1950s in a small West Virginia coal-mining town. It tells the story of four high-school seniors that don't have much prospect for futures beyond the bleak reality of working in the mines. They get interested in building a rocket for a science fair project, but face several obstacles, including unsupportive parents, a skeptical school administration, and a couple of scary accidents that almost put the kibosh on the whole project. Nevertheless, with perseverance and the encouragement of a teacher that believes they can do it, the boys end up winning the state science fair, college scholarships, and better lives.

Before talking rockets lets review concepts of energy, and the principle of conservation of energy.

Energy

Energy is defined as the ability to do work. There are two general categories of energy: potential energy and kinetic energy. *Potential energy* consists of energy that is stored in some fashion. *Kinetic energy* is the energy associated with the motion of an object, or objects. When potential energy is released it becomes kinetic energy.

Work. *Work* has a very specific definition in physics. The work done on an object by a force is given by

$$W = F_x \Delta x.$$

F_x is the component of the force that is exerted in the direction of motion of the object, and Δx is the displacement of the object. Therefore, if an object does not move, no matter how hard a particular force is pushing on it, that force is doing no work on the object. In addition, if a force has no component parallel or anti-parallel to the direction of motion, it does no work. If you were to push on a wall for an hour, you may be very tired but you will have done no work on the wall. Work is a scalar quantity.

Example 3: A large rock is being pulled along the floor by a rope. The rope is pulled in a horizontal direction with a 100-N force. If the rock is moved 10 meters, how much work does the rope force do on the rock?

Solution: Because the force is parallel to the direction of motion, $F_x = 100$ N and $W = 100$ N (10 m) = 1,000 Nm = 1,000 Joules (J).

Example 4: The rock from example 3 is now pulled by the same rope force over the same 10-meter distance along the floor. This time the rope is pulled at an angle of 30 degrees above the horizontal. How much work does the rope force do on the rock in this case?

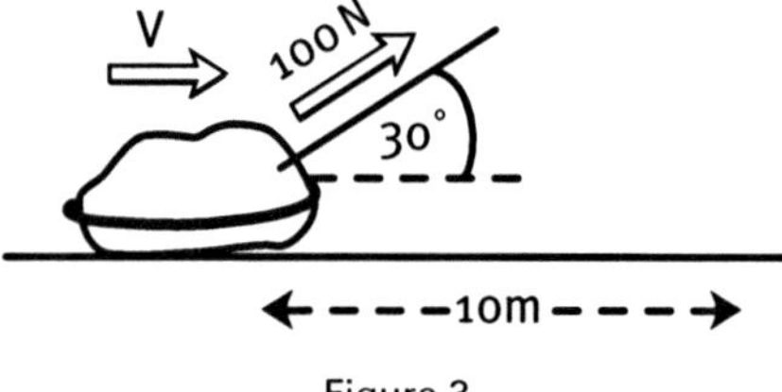

Figure 3

Solution: Here we only take the component of the rope force parallel to the motion. $F_x = F \cos 30 =$ 100 N (0.866) = 87 N and $W = 87$ N(10 m) = 870 J.

Although work is a scalar quantity, if an object is in motion while a force is being applied in the opposite direction to the motion, the force is said to be doing negative work. For example, if a block is sliding along the floor and a force is applied to bring it to a stop over a certain distance, then the force has done negative work on the object. Another way to define work is the energy that is transferred while the force is being applied.

Work may be done by several forces on an object simultaneously.

Example 5: A block is pulled by a 100-N force acting horizontally. A 40-N friction force acts opposite to the motion.

a) What other forces must be acting on the object?
b) How much work is done by each of the forces after the block has moved a distance of 5 meters?

Solution:

a) In addition to the pulling force and the friction force there is a force of gravity (or weight force) acting downward on the block. There is therefore also a force of the floor pushing upward on the block balancing the gravity force (the normal force.)

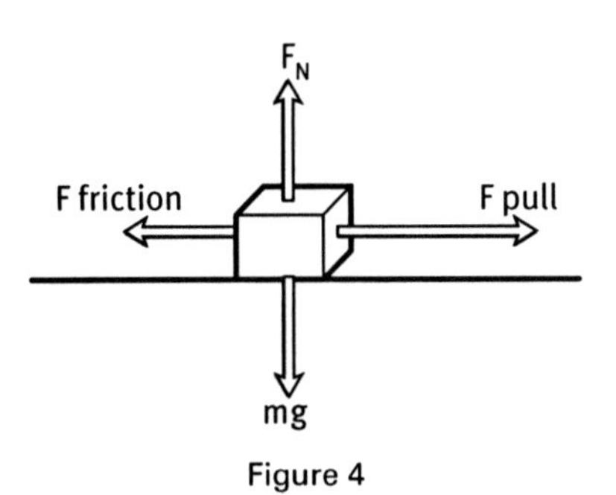

Figure 4

b) Because the weight force and the force of the floor both act perpendicular to the motion, they each do no work on the block.

$$W_{pulling\ force} = (100\text{ N})(5\text{ m}) = 500\text{ J},\ W_{friction} = -40\text{ N}(5\text{ m}) = -200\text{ J}$$

In the previous example we may also express the *net work* done on the block as the total of the amounts of work done by the individual forces. In that case the net work done = 300 J. Alternatively, the net work could be calculated as the work done by the net force. For example, the net force above is 60 N. Therefore the net work = (60 N)(5 m) = 300 J.

Power. Work can be done slowly or quickly. The amount of work done by a force says nothing about how long it took to do that work. A nuclear bomb explosion does a lot of work in a very short time, compared to a nuclear power plant, which may do the same amount of work (produce the same amount of energy) but over a much longer (and much more controlled!) period of time. Therefore *power* is defined as the *rate* at which work is done, or alternatively the rate at which energy is transferred, or converted, from one form to another.

$$P = \frac{W}{t} = \frac{E}{t}$$

The SI units of power are J/s where 1 J/s = 1 Watt.

Example 6: (Sneak Preview) In the science fiction action disaster movie *The Core* there is a super-powered drill that can vaporize hundreds of tons of rock in a few seconds. In the first part of the movie we see a demonstration of this awesome device where it cuts a 3-meter wide 50-m long hole out of a solid rock wall in a few seconds. Later we will analyze the physics presented in *The Core* and specifically calculate the amount of energy needed to vaporize rocks, but for now let's just say the total energy to accomplish the above feat would be about 3×10^{12} J. It looks like it takes about 5 seconds to do the job. At what power must the drill work to do this?

Solution: $P = \frac{W}{t} = \frac{E}{t} = \frac{3 \times 10^{12}\,\text{J}}{5\,\text{s}} = 6 \times 10^{11}\,W$

Kinetic Energy. Kinetic energy (KE), the energy associated with an object's motion, is defined mathematically as follows:

$$KE = \frac{1}{2}mv^2$$

If net work is done on an object, then a net force is acting on the object, so the object will change its velocity. The net work done on the object is exactly equal to the change in KE of that object. This relationship is referred to as the *work-energy theorem:*

$$W_{net} = \Delta KE$$

Example 7: A block has a mass of 20 kg and starts from rest. If a 60-N force is applied to the block parallel to the floor, during the first 5 meters of motion

a) what is the change in kinetic energy of the block?
b) what is the velocity of the block at the end of the 5 meters of motion?

Solution:

a) $W_{net} = F_x \Delta x = (60\text{ N})(5\text{ m}) = 300\text{ J} = \Delta KE$

b) $\Delta KE = 300\text{ J} = KE_{final} - KE_{initial} = \frac{1}{2}mv_f^2 - \frac{1}{2}mv_i^2$

$300\text{ J} = \frac{1}{2}(5\text{ kg})v_f^2 - \frac{1}{2}(5\text{ kg})(0) \quad v_f = \sqrt{\frac{600\text{ J}}{5\text{ kg}}} = 11\text{ m/s}$

Potential Energy. Energy can be stored in a variety of different forms of potential energy. It can be stored in chemical bonds, in electric and magnetic fields, in the form of mass in the case of nuclear energy, in gravitational fields, and in elastic materials among others. Let's look at some of these energy storage methods.

Gravitational Potential Energy. *Gravitational potential energy* (PE_g) is energy stored by an object due to its position in a gravitational field. For example, if a ball is lifted off of the ground it gains gravitational potential energy. This energy can then be converted into kinetic energy by dropping the ball. As the ball falls it continuously gains kinetic energy, and the amount of kinetic energy that it gains is exactly equal to the amount of gravitational potential energy lost (as long as significant energy is not lost from the system due to air friction).

When work is done on an object, the amount of work done is equal to the amount of energy transferred, so we can quantify the gravitational potential energy of an object near the surface of the Earth as follows. Consider a book being lifted from the ground to some height, h, above the ground.

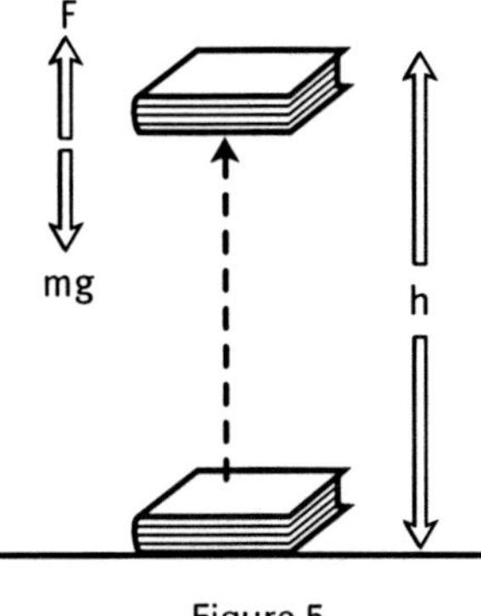

Figure 5

Except for a very short period of time when the book is accelerated to the lifting speed, and another very short time when it is brought to a stop at height h, we will assume that the book is lifted at a constant velocity. Therefore, according to Newton's second law, the amount of the lifting force must be equal to the weight force because the book is not accelerating. The work done by the lifting force is:

$$W_{lifting\ force} = Fh = mgh.$$

This amount of work done is equal to the change in gravitational potential energy of the book in lifting it from the ground to h.

Therefore, $\Delta PE_g = mgh$.

Here we are assuming that the acceleration due to gravity and therefore the gravitational field strength are constant, which is a good approximation for anything happening within a few miles or tens of miles of the Earth's surface. However if we move an appreciable distance away from Earth, the expression for gravitational potential energy must be modified due to the nonconstant field.

Example 8: A 1.0-kg book (that's a pretty massive book—it's either *War and Peace* or a physics textbook) is lifted off of the ground to a height of 2 m above the ground and placed on a book shelf. The book slips off of the shelf and then falls back to the ground.

a) How much potential energy does the book gain by being lifted up to the shelf?
b) How much kinetic energy does the book have just before it hits the ground after falling?
c) What is the speed of the book just before hitting the ground?
d) What is the speed of the book when it has fallen a distance of 1.0 m down from the shelf?

Solution:

a) $\Delta PE_g = mgh = (1 \text{ kg})(9.8 \text{ m/s}^2)(2 \text{ m}) = 19.6 \text{ J}$

b) *PE* lost = *KE* gained, therefore because all 19.6 J of *PE* is lost upon hitting the ground, *KE* must = 19.6 J.

c) $KE = \frac{1}{2}mv^2$ so $v = \sqrt{\frac{2(KE)}{m}} = 6.2 \text{ m/s}$

d) At 1 m above the ground the book has lost 9.8 J of PE, and it has 9.8 J left. The 9.8 J lost is equal to the KE gained, so $v = \sqrt{\frac{2(9.8 \text{ J})}{1 \text{ kg}}} = 4.4 \text{ m/s}$

Elastic Potential Energy. When you stretch a rubber band or a spring or a bouncy ball compresses as it hits the ground, energy is stored in the form of ***elastic potential energy***. When the stretched rubber band is released, the stored elastic energy is then converted to kinetic energy. For many springs it is possible to quantify the potential energy stored. The energy stored depends on how far the spring is stretched (or compressed) from its equilibrium or relaxed state (Δx), and the stiffness of the spring, which is measured by the spring constant (k). The larger the value of k for a spring, the more energy that the spring can store, and therefore for a given displacement from equilibrium the spring potential energy is

$$PE_s = \frac{1}{2}k(\Delta x)^2$$

Example 9: A 0.50-kg block is placed on a frictionless surface and pushed against a spring that has spring constant $k = 200$ N/m. The block compresses the spring a distance 0.10-m, and is then released.

a) What is the maximum amount of energy stored in the spring?
b) What is the velocity of the block the instant after it is released by the spring?

Solution:

a) $PE_s = \frac{1}{2}k(\Delta x)^2 = \frac{1}{2}(200 \text{ N/m})(0.10 \text{ m})^2 = 1.0 \text{ J}$

b) At the instant of release the entire *PE* of the spring has been converted to *KE* of the block. Therefore, $KE = \frac{1}{2}mv^2 = 1 \text{ J}$ so $v = 2 \text{ m/s}$

Thermal Energy. Thermal energy refers to kinetic energy at the molecular or atomic level. All matter is in constant motion. At the level of thermal energy, the motion may be due to translational (linear) motion, vibrations, or rotations of molecules or atoms. The total thermal energy of a macroscopic object consists of the total kinetic energy of all the particles within that object. ***Temperature*** is a measure of the ***average*** kinetic energy of the particles within the object. The term ***heat*** technically refers to the transfer of thermal energy from one location to another. For example, if two objects at different temperatures are placed in contact with each other, thermal energy from the hotter object will flow spontaneously into the colder object until the two are at the same temperature. The amount of energy transferred is the heat.

Chemical Potential Energy. When chemical reactions occur, energy can either be released or absorbed in the reaction. This is because energy can be stored in the configuration of the chemical bonds between atoms and molecules. In chemical reactions, specific chemical bonds are broken and others are formed. If the total energy contained in the bonds of the reactants is greater than the total energy contained in the bonds of the products of a reaction, then energy will be released as a result of the reaction. For example, the combustion of gasoline results in the release of heat that can be used to drive an internal combustion engine in an automobile.

Nuclear Potential Energy. Nuclear energy is the result of energy stored within the mass of subatomic particles. Einstein's famous equation $E = mc^2$ in fact describes the amount of energy contained within a mass. For example, if you were to add the total mass of two individual protons, you would find that they would add up to be greater than the total mass of a deuterium nucleus. (Deuterium is an isotope of hydrogen that consists of two protons fused together.) The difference in mass is equal to the energy that would be released (as photons) in creating a deuterium nucleus by slamming two protons together. This type of reaction is called a nuclear fusion reaction because particles are fused together as a result of the reaction. A fission reaction occurs when a heavy atom such as uranium is split into smaller atoms. The total combined mass of the smaller atoms is less than the mass of the uranium, and therefore energy is released in the reaction. Fission has been able to be harnessed commercially in the production of electrical power, while fusion energy at present cannot be harnessed economically. Hydrogen bombs (fusion bombs) are based on this principle. The total mass lost in the nuclear fusion reactions in the bombs is converted into various forms of energy.

Electric and Magnetic Energy. Electric charges produce electric fields whether the charges are static or in motion. Moving electric charges produce magnetic fields. For example, a current flowing through a wire generates a magnetic field surrounding the wire. Energy is stored in these fields in the same way that gravitational energy is stored in gravity fields. It is the energy stored in an electric field of a battery, for example, that is responsible for the forces exerted on electric charges within a wire that cause an electric current to flow.

Sound. Sound is mechanical wave energy generated by vibrating objects, which is in a frequency range that the human ear can detect. By the way, sound can *only* travel through matter, so if you hear any explosions in an outer space sci-fi space battle scene don't believe it!

Electromagnetic Radiation. Electromagnetic (e-m) radiation consists of oscillating electric and magnetic fields propagating through space. These electromagnetic waves carry energy with them.

Because electric and magnetic fields store energy, and e-m radiation consists of propagating electric and magnetic fields, it must be carrying energy. Clearly this must be the case when we contemplate that the sun, which ultimately provides the Earth with almost all of its energy (consider that fossil fuels are derived from plant material that originally used radiant energy from the Sun to grow), transfers this energy into space almost completely in the form of electromagnetic radiation.

Conservation of Energy

Conservation of energy is one of the fundamental concepts of physics. This principle apparently reflects a balance inherent in the natural universe. All experience and observation to date lead to the following conclusion: *The total amount of energy in the universe is a constant. Energy can change form but it can never be created or destroyed.*

For an isolated system, one that cannot exchange energy with its surroundings, the total energy of that system will remain a constant. If the system is not isolated, that is if work can be done on or by the system, or if energy can by added to or taken away from the system, then while the system may not conserve energy, a larger system that encompasses the objects responsible for the adding or subtracting of energy and the smaller system must conserve energy.

Two or three previous examples illustrated the principle of conservation of energy without explicitly stating so. For example, consider the system of a ball sitting a height h above the ground. The ball has gravitational potential energy due to its height above the ground. That gravitational potential energy is the total energy of the system. As the ball falls, we are assuming an isolated system, and therefore the PE lost as the ball falls must be exactly equal to the amount of KE that the ball gains. Although the energy transfers from one form to another, the *total* amount of energy is the same; it is simply redistributed. The assumption of an isolated system for this example is reasonable as long as there is not significant air resistance. However, if there is, then we can deduce that if the amount of total energy of the ball decreases, then that missing energy will have gone into increasing the energy of the atmospheric particles by that amount—mostly by increasing the thermal energy of the air molecules.

ACTIVITY

Find a rubber band and shoot it straight up in the air. Based on how high up it goes determine the total energy stored in the rubber band when it was fully stretched. Use this energy to determine the initial velocity of the rubber band just after you release it. (Using what you know about projectile motion, predict how far the rubber band will go if you stretch it the same amount and launch it from a particular angle. Set up a target at this location and see if you were right).

Example 10: A 50-kg roller coaster car is moving at a velocity of 10 m/s on a frictionless track at point A shown in the figure.

a) What is the total energy of the roller coaster car?

b) Use conservation of energy to find its velocity at points B and C.

Solution: If there is no friction then the roller coaster car is an isolated system. Therefore the total energy at point A = the total energy at point B = the total energy at point C. It is simply redistributed between PE and KE.

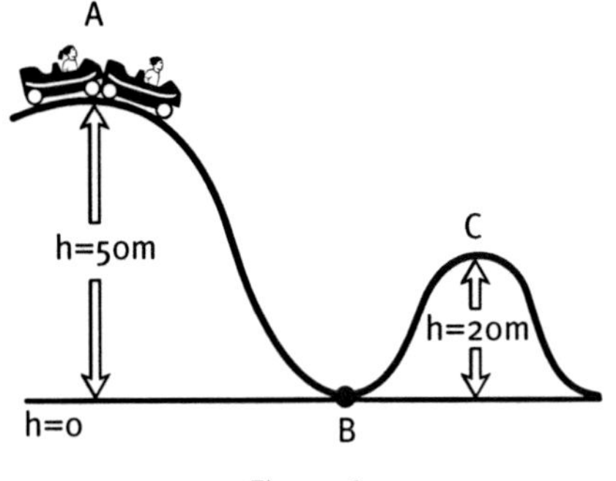

Figure 6

Total E can be calculated from the kinetic and potential energies at point A.

a) Total $E = KE_A + PE_A = KE_B + PE_B = KE_C + PE_C$

$$KE_A + PE_A = \frac{1}{2}mv_A^2 + mgh_A = \frac{1}{2}(50\text{ kg})(10\text{ m/s})^2 + 50\text{ kg}(9.8\text{ m/s}^2)(50\text{ m}) = 27{,}000\text{ J}$$

At point B: $27{,}000\text{ J} = \frac{1}{2}mv_B^2 + mgh_B = \frac{1}{2}(50\text{ kg})v_B^2 + 50\text{ kg}(9.8\text{ m/s}^2)(0)$

Which gives $v_B = 33$ m/s.

At point C: $27{,}000\text{ J} = \frac{1}{2}mv_C^2 + mgh_C = \frac{1}{2}(50\text{ kg})v_C^2 + 50\text{ kg}(9.8\text{ m/s}^2)(20\text{ m})$

$v_C = 26$ m/s.

The path of the car is not important as long as the system is isolated. Energy of the system is constant, and a situation that would be very difficult to analyze using Newton's laws becomes quite simple using conservation of energy.

Example 11: A 65-kg skydiver jumps out of an airplane, and after falling a distance of 1,000 m, is moving at a terminal velocity of 70 m/s. How much of the skydiver's original energy has been converted to thermal energy in the atmosphere at this point?

Solution: The skydiver has fallen a distance of 1,000 m, and therefore has lost an amount of potential energy, $\Delta PE = mg\Delta h$ = (65 kg)(9.8 m/s^2)(1,000 m) = 637,000 J.

The amount of kinetic energy the skydiver has at that point is

$$KE = \frac{1}{2}mv^2 = \frac{1}{2}(65\text{ kg})(70\text{ m/s})^2 = 159{,}000\text{ J}$$

Therefore, the amount of energy lost into the atmosphere is 637,000 J – 159,000 J = 478,000 J.

Example 12: You push a box along the floor at a constant velocity. The pushing force applied to the box is 40 N.

a) What is the net work done on the box as it is pushed a distance $\Delta x = 20$ m?
b) How much work do you do on the box during this motion?
c) How much work is done by the friction force during this motion?
d) What happens to the work/energy that you are putting into the box?

Solution:

a) The net work done on the box is 0. Because the velocity of the box is constant, there is no net force on the box, and therefore no net work.
b) $W_{pushing\ force} = F\Delta x = 40\ N(20\ m) = 800\ J$
c) $W = F\Delta x = (-40\ N)(20\ m) = -800\ J$
d) The energy that you put in goes out of the system due to friction. It is converted to thermal energy in the ground and box.

Energy, Momentum, and Rockets

Now we will apply what we know to explain the physics of rocket flight. In *October Sky* the boys are building rockets that use explosive chemical fuel to launch the rockets to heights of a few thousand feet. Let's address the following questions.

Ultimately, what is the source of energy that is supplied to the rocket?

Yes, the energy comes from the fuel, but specifically from the chemical potential energy stored in the chemical bonds in the fuel. As it burns the *exothermic* (heat-releasing) chemical reactions break the bonds of the reactants, and form bonds of lower energy releasing thermal and kinetic energy.

How does burning the fuel and expelling it out the bottom of the rocket provide an upward thrust force? Remember Newton's third law? How does it apply here?

ACTIVITY

Put a skateboard on a smooth surface. Sit (don't stand) on the *stationary* skateboard holding a heavy backpack, and then throw the backpack off the front of the skateboard as hard as you can. What happens to you and why? Suppose you were an astronaut on a space walk and your safety line got cut, how could you use a rock you are carrying to get you back to the ship?

As far as the rocket goes; it is pushing on something that is pushing back on it. However, contrary to a common misconception, it is not the ground that the rocket is pushing on. If it were, how could a rocket accelerate after it loses contact with the ground? How could a rocket work in space? What the rocket actually pushes on is the fuel that is forcibly expelled. The expelled fuel then pushes back on the remaining fuel still in the body of the rocket. In the activity at left, think of the skateboard as the rocket, and the backpack as the fuel. If you had lots of backpacks you could accelerate each time you threw one. If there wasn't much friction you might get going pretty fast this way. Combustible chemical fuels are just a lot more efficient at generating thrust than throwing backpacks.

How can we determine the amount of thrust that the rocket generates? Let's take a rocket traveling through space so we can neglect the effect of gravity on the system. We will assume an initial velocity of 0 for the rocket relative to a frame of reference, which gives the system (rocket plus fuel) an initial momentum of 0. When the rocket starts expelling fuel the total momentum of the system must still be 0.

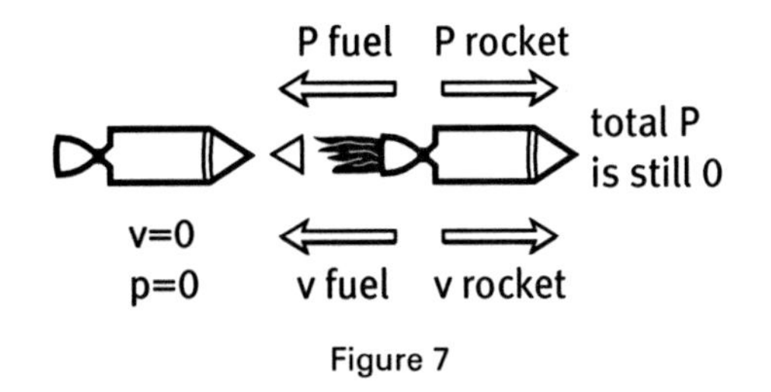

Figure 7

The magnitude of Δp experienced by the rocket in one direction must equal the magnitude of Δp experienced by the expelled fuel in the other. If we make the approximation that the fuel is ejected at a constant velocity then:

$$\Delta p_{fuel} = v_{fuel} \Delta m,$$

where Δm is the amount of fuel expelled in that time. If we assume that the mass of the rocket doesn't change significantly during the short time interval, then

$$\Delta p_{rocket} = M\Delta v,$$

where M is the mass of the rocket and Δv is the amount that the rocket's velocity changes during that time.

Therefore, $v_{fuel}\Delta m = -M\Delta v$,

which takes into account the fact that the changes in momentum are in opposite directions. Because Δp occurs over a time interval Δt,

$$M\frac{\Delta v}{\Delta t} = -v_{fuel}\frac{\Delta m}{\Delta t}$$

$$\text{therefore } Ma = -v_{fuel}\frac{\Delta m}{\Delta t}$$

where Ma must be the force of thrust acting on the rocket. (Note that the mass of fuel expelled per time (Δm) is also equal to the rate of change of the rocket mass ΔM.) Therefore, if we can determine the rate at which fuel is expelled from the rocket we can find the thrust.

CALCULATING ROCKET VELOCITY

Because the velocity of the rocket increases, v_{fuel} won't remain constant relative to the original frame of reference. In addition, the mass of the rocket M will decrease continuously as it loses fuel. The exact expression must be expressed in terms of derivatives:

$$M\frac{dv_{rocket}}{dt} = -v_{fuel}\frac{dM}{dt} \text{ so } dv = -v\frac{dM}{M}$$

Therefore, to determine the final velocity of a rocket which expels fuel at a rate *dM/dt,* we can sum (integrate) the small changes in *v* as the rocket goes from an initial velocity, v_i, to a final velocity, v_f.

$$\int_{vi}^{vf} dv = -v_{fuel}\int_{Mi}^{Mf}\frac{dM}{M}$$

Solving the integral gives:

$$v_f = v_i + v_{fuel}\ln\left(\frac{M_f}{M_i}\right)$$

Because the mass of the rocket decreases as it loses fuel, the acceleration rate will continuously increase until it runs out. At that point the velocity will remain constant at the final value it achieved just as the last of the fuel was expelled (Newton's first and second laws).

Remember, we are dealing with a situation in deep space where there are no significant external forces acting on the system. Launching a rocket upward from the ground we won't gain as much velocity from the thrust because we have the external force of gravity working against it. After the burn the rocket will slow down, and then fall back to Earth (unless it achieves *escape velocity*—a velocity sufficiently fast enough that Earth's gravity can never bring it to a stop and pull it back).

In *October Sky,* the "Rocket Boys" (that's what their schoolmates and neighbors call them) start off building some pretty sketchy rockets. The early prototypes have some problems. Some blow up due to the large internal pressures generated from the combustion of the fuel. They also have issues with stability that sends the rockets shooting off in unpredictable directions. However, with practice, experience, and research the kids improve the sophistication and performance of the rockets, and the later models are able to fly consistently to heights of thousands of feet.

Now let's take a specific rocket launch from the movie (scene 8), and apply the principles previously addressed. In this scene, they record a time of 14 seconds for the rocket to fall from its apex.

First, we can answer the question of how much potential energy is stored in the fuel. From conservation of energy we know that the total energy of the system will be constant. Therefore, at the top of its flight, all of the rocket's energy will be in the form of gravitational potential energy of the rocket and kinetic energy of the expelled fuel. This amount of energy must be equal to the initial chemical potential energy stored in the fuel. To determine PE_{gp}, all we need to do is find the maximum height of the rocket from its falling time. We will assume a rocket mass of 2 kg, and 1.6 kg of fuel.

$$\Delta y = v_0 t + \frac{1}{2}at^2 = 0 + \frac{1}{2}(9.8 \text{ m/s}^2)(14 \text{ s})^2 = 960 \text{ m}$$

Because we are told earlier that the fuel from these rockets is expelled at velocities close to the speed of sound (which is about 340 m/s in air depending on the temperature), we will use a value of 300 m/s. Therefore, the energy of the system is:

$$m_{rocket}gh + \frac{1}{2}m_{fuel}v_{fuel}^2 = (2\text{ kg})(9.8\text{ m/s}^2)(960\text{ m}) + \frac{1}{2}(1.6\text{ kg})(300\text{ m/s})^2 \approx 91{,}000\text{ J}$$

Let's say that the Rocket Boys want to calculate theoretically how high their rocket should go, and compare it to their experimental results. They do a test burn and see that it takes 8 seconds to expel the fuel. Based on this information how high should the rocket go?

Because the boys haven't studied calculus yet, they come up with a pretty good way to solve the problem approximately. From the data, they will calculate the acceleration the instant after launch and the acceleration the instant before the last bit of fuel is expelled. Then they assume a constant rate of change of acceleration during the burn and get the average acceleration:

$$F_{net} = F_{thrust} - Mg = -v\frac{\Delta m}{\Delta t} - Mg = Ma$$

where M is the total mass of rocket and fuel at the instant in question, and $\frac{\Delta m}{\Delta t}$ the rate at which fuel is expelled, in this case 0.2 kg/s.

For the instant just after launch:

$$a = \frac{-v\frac{\Delta m}{\Delta t} - Mg}{M} = \frac{-300\text{ m/s}(0.2\text{ kg/s}) - 3.6\text{ kg}(9.8\text{ m/s}^2)}{3.6\text{ kg}} = 6.9\text{ m/s}^2$$

For the instant just before engine shut off:

$$a = \frac{-v\frac{\Delta m}{\Delta t} - Mg}{M} = \frac{-300\text{ m/s}(0.2\text{ kg/s}) - 2.0\text{ kg}(9.8\text{ m/s}^2)}{2.0\text{ kg}} = 20.2\text{ m/s}^2$$

Assuming a constant rate of increasing acceleration gives us an average acceleration,

$$a = \frac{6.9\text{ m/s}^2 + 20.2\text{ m/s}^2}{2} = 13.6\text{ m/s}^2$$

Now we can determine the velocity and height of the rocket after the 8-second burn.

$$v = v_0 + at = 0 + (13.6\text{ m/s}^2)(8\text{ s}) = 108\text{ m/s}$$

$$\Delta y = v_0 t + \frac{1}{2}at^2 = 0 + \frac{1}{2}(13.6\text{ m/s}^2)(8\text{s})^2 = 435\text{ m}$$

From this point on the rocket is in free fall, so we can determine the additional height that the rocket travels. The acceleration is now g, and the final velocity at the maximum height is zero.

$$\Delta y = \frac{v^2 - v_0^2}{2a} = \frac{0-(108 \text{ m/s})^2}{2(-9.8 \text{ m/s}^2)} = 595 \text{ m}$$

Therefore, this gives a maximum height of 435 + 595 = 1,030 m. The actual height is a little short of the theoretical. Why might this be true?

KINETIC ENERGY AND KINDER, GENTLER NUCLEAR BOMBS

The science fiction, disaster movie genre is an exciting and increasingly sophisticated Hollywood standby, having given us such icons as *Armageddon, Volcano,* and the *Day After Tomorrow,* among others. Accompanied by a dizzying array of impressive special effects, the general pattern in these types of films involves Earth being threatened with imminent destruction from some catastrophic disaster, either of natural or of human origin. The catastrophe is often based on a very real scientific possibility, such as a large asteroid colliding with Earth, or severe global climate change. The situation usually appears hopeless, but through ingenuity, heroism, and some last-minute good luck of highly problematic origin, Earth is usually saved. Employing some rather obvious irony, in at least a handful of these movies, Earth's salvation is accomplished by detonating a nuclear warhead or two (by using the very weapon that could potentially threaten the planet with destruction for good), but is it plausible?

Could a nuclear warhead really tear apart an asteroid? Could one be used to alter the currents in the Earth's outer core? Let's have a look at *The Core,* a movie typical of this genre, to see if we can come to any conclusions.

The Movie

In *The Core,* government scientists have been playing where they shouldn't and have inadvertently caused the Earth's outer core to "stop circulating," resulting in a disappearing of the Earth's magnetic field. The idea that we could do anything to affect the outer core, which has the mass of a small planet, and lies underneath 2,900 km of solid rock standing between it and us, is impossible to imagine.

The disappearing field, according to socially awkward but brilliant—and very handsome—geophysicist Joshua Keyes (played by Aaron Eckhart), will result in the planet being fried by incoming radiation. Keyes makes a dramatic "quick and dirty" analogy/demonstration when he ignites a jet of hairspray with a lighter and scorches (ruins) what otherwise might have been a delicious peach. The flaming hairspray represents incoming radiation, and the peach, Earth.

As the film progresses, we see the diminishing magnetic field resulting in increasingly severe catastrophic weather, including deadly lightning storms and intense radiation. The only hope for mankind is for a group of eccentric but intrepid scientists and a couple of space shuttle pilots to drill into the Earth's core (a distance of 3,000 miles or so) in a specially designed vehicle that has the ability to almost instantaneously vaporize thousands of tons of rock while carrying a payload of five 200-megaton nuclear warheads. (Just for reference, the biggest nuke ever built had a yield of around 100 MT.) The vehicle must withstand pressures of 2.5 *million* atmospheres and temperatures over 4,000 degrees Celsius. When they get to their destination, they are to launch the bombs into the surroundings and detonate them. According to their calculations the explosion should "restart" the core, and the rejuvenated current would then reinvigorate the magnetic field, thus saving the planet.

Okay, take a deep breath, and let's see what we can unearth on this one.

The Back Story

Let's look at some of the scientific details on which the movie premise is based. The geophysical consensus is that the Earth's magnetic field is produced by circulating currents in the liquid outer core. In the movie, they actually have the general idea right, but they claim that the field is produced solely by the rotation of the outer core about the Earth's axis. In fact, geophysicists believe that it is *convection currents* in the outer core (modified by the rotation) which cause the observable field. This may seem like a trivial distinction, but as we will see next, it is going to cause us a major movie physics short circuit.

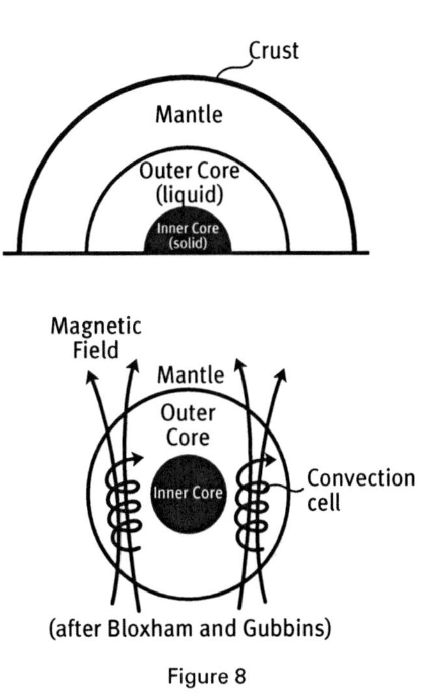

Figure 8

The earth's core is thought to consist of about 90 percent iron, based on information obtained from seismic waves generated in earthquakes. The temperature in the outer core is hot enough to ionize some of the iron, producing charged particles that are carried in convection currents. In addition, the outer core is moving with the Earth's rotation (but at a very slightly slower angular speed than the rest of the planet, it turns out). Because the sources of all magnetic fields are moving charges, geophysicists hypothesize that the earth's magnetic field is a result of ionized iron moving in the outer core. The details of the convection process are difficult to determine, as direct information about the outer core is limited; however, the currents are thought to be maintained by a self-sustaining thermodynamic "geodynamo."

Recent theoretical numerical models based on these principles have been able to duplicate approximately the actual magnetic field that we observe in some detail. Interesting, also, is the fact that the magnetic field has undergone continuous periodic reversals in polarity over the lifetime of the planet—the magnetic poles flip every few hundred thousand years or so. In addition, it is relevant to note that during these reversals it is thought that the field weakens considerably, although it may or may not disappear altogether during the transition, and there are no clear records of major extinctions during these epochs.

Although it is unclear what exactly might happen if the magnetic field were to disappear, the effects on Earth would likely be much *less* dramatic than those portrayed in the film. Birds and other creatures that use the field to navigate or migrate would perhaps find the going difficult, although they do have another pretty good navigational tool called eyes (and therefore the scene in which a bunch of pigeons lose their composure and fly at full speed into every solid object in sight seems pretty ludicrous), but would we really be scorched to death in a matter of months? As we will discuss in more detail in Chapter 6, the earth's magnetic field is able to trap potentially damaging charged subatomic particles streaming toward the Earth called *cosmic rays*. Cosmic rays that originate in the sun are called the *solar wind*, while higher energy cosmic rays come from deep space. The field steers lower energy cosmic rays past Earth. (High-energy cosmic rays are not blocked by the magnetic field. They move too fast.) However, at the magnetic poles where the field lines converge, there is no magnetic shielding of cosmic ray particles. The spectacular lighting displays known as *auroras* (although not completely understood) are due to charged particles—mostly electrons it turns out—that converge along the magnetic field lines and interact with the atmosphere.

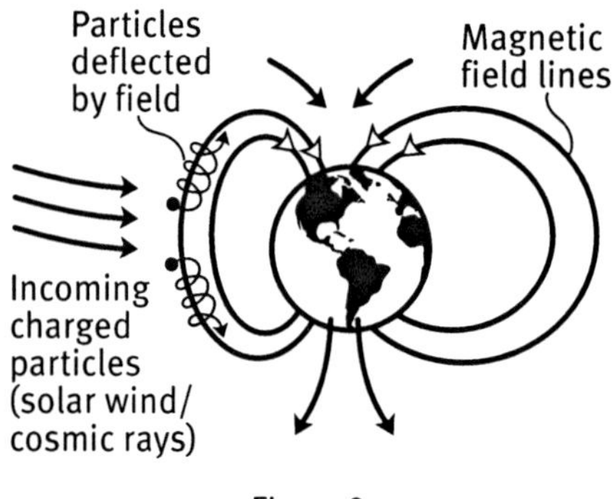

Figure 9

Significantly, we don't hear of any deadly effects occurring at the magnetic poles such as those portrayed in the movie, and secondly (and very importantly) *we, in fact, do have something to protect us from these charged particles even if there was no field—the atmosphere*. The atmosphere is dozens of miles thick. It is full of matter, and it provides, in addition to life-sustaining oxygen, an effective filter for most of the incoming cosmic ray particles. The atmosphere provides the same protection as would a 10-foot thick concrete barrier between the cosmic rays and the surface of the earth. If the magnetic field disappeared, more particles *would* impact the upper layers of the atmosphere. Satellites could be damaged, and communications could be disrupted (there might be problems with cell phones for example, which in Hollywood admittedly might be viewed as equally catastrophic as the destruction of all life on the planet). During geomagnetic storms power failures might occur because the magnetic fields created by the solar wind particles can affect power stations, but the atmosphere is still going to filter a large percentage of these particles before they ever get to the Earth's surface. While some scientists predict the incidence of cancer may increase due to the increased flux of cosmic rays, *catastrophic* biological effects are extremely unlikely, and any meteorological effects are going to be far less severe than most of what we see in the film.

However, while they do mention the solar wind and cosmic rays, Dr. Keyes and the other scientists erroneously make the claim that Earth will, in fact, be scorched by *microwave radiation.* Because microwave radiation, a form of electromagnetic radiation (the same as visible light), has no electric charge, it is *completely unaffected* by the Earth's magnetic field. Field or no field, the incident microwave radiation would be the same. These guys may be confusing/combining the idea of the magnetic field with that of the ozone (O_3) layer in the upper atmosphere, which, in fact, helps filter out *ultraviolet* radiation (e-m radiation of lower frequency than microwaves); perhaps another movie about runaway chlorofluorocarbons migrating into the upper atmosphere and destroying the ozone layer might be more compelling (wait, is that a movie or real life?). By the way, it's a really good thing that Earth's magnetic field doesn't block e-m radiation. If it did, we *would* be in a lot of trouble—we wouldn't do so well without light.

Consider also, if microwaves (coming from the sun, presumably) are so dangerous, why aren't astronauts in great danger of being fried to death every time they go into space? By scene 14 of *The Core,* a "hole" has apparently formed in the magnetic field, and microwaves are streaming through this hole (even though, as we know, the field should have no effect on microwaves one way or another). These microwaves are burning people to death, and they even vaporize a section of the Golden Gate Bridge. Well, space vehicles go outside the Earth's atmosphere and magnetic field all the time, and as far as I know, none of them has ever been cooked by microwaves. (The danger to astronauts is in fact due to the charged particles that we previously discussed. For example, if there were a solar storm while astronauts were in space, it could be very dangerous and possibly fatal. Large quantities of high-energy particles can do significant damage to the body.) Couldn't the filmmakers at least have gotten the difference between charged particles and electromagnetic radiation right? It's a pretty basic concept.

As you can see, there are a few topics in physics (and geology) that *The Core* invokes, or, rather, insults. As the movie goes on, we encounter the problematic to the downright mystifying with increasing frequency. In fact, we really have to suspend our disbelief when watching this movie in order not to get shocked, outraged, and/or filled with indignation at the frequent disregard for fundamental principles of physics—and common sense. In this section we will address two relatively straightforward areas of concern: (1) How difficult would it actually be (how much energy might it take) to get to the core? and (2) Could nuclear bombs reasonably be expected to induce a sufficient current in the earth's outer core to fix the problem?

Here are a couple of principles relating to thermal energy that we will find useful in our analysis.

Specific Heat. If heat is added to or removed from a substance it will increase or decrease the temperature of that substance (until it reaches a temperature at which it starts to change phase). How much the temperature changes for a given heat change depends on the material. For example, it is relatively hard to change the temperature of water. It takes 4,186 J of heat to raise the temperature of 1 kg of water by 1 degree C. Comparatively, it takes only about 430 J to raise the temperature of

1 kg of iron the same amount. Therefore, water and iron are said to have ***specific heats*** (c) of 4,186 J/kg °C and 430 J/kg °C, respectively. To determine the ***total heat*** (Q) necessary to take a certain mass of material (m) and change its temperature by a specific amount (ΔT), simply multiply the mass and temperature change by the specific heat of that material:

$$Q = mc\Delta T.$$

Therefore, the total heat needed to raise the temperature of 10 kg of iron from 5 °C to 100 °C would be:

$$Q = (10\text{ kg})(430\text{ J/kg C})(95\text{ °C}) = 4.1 \times 10^5\text{ J}.$$

Note: Sometimes heat is given in units of calories. One calorie is equivalent to 4.186 Joules.

Melting, Freezing, Vaporization, and Condensation. If sufficient heat is added to or removed from a material, a phase change could occur. For example, if heat is taken out of a mass of water once the temperature drops to 0 °C, the water will start to freeze into ice. During a phase transition the temperature remains constant because all of the heat energy is going into breaking the bonds that hold the molecules together (in the cases of melting or vaporizing) or forming the bonds (in the cases of condensing and freezing). Once a phase transition is complete, any additional heat added or removed will go into changing the temperature again.

The amount of heat that needs to be added to melt 1 kg of ice at 0 °C is equal to 3.3×10^5 J, which is also equal to the amount of heat that must be removed to freeze 1 kg of water. The quantity is called the ***latent heat of fusion*** (L_f). The amount of heat that must be added or removed to effect a phase change between the liquid and gas states is called the ***latent heat of vaporization*** (L_v). In the case of water, $L_v = 2.26 \times 10^6$ J/kg.

Modes of Heat Transfer. Heat transfer from one location to another can occur via three possible mechanisms: conduction, convection, and radiation.

Conduction. *Conduction* occurs more readily in solids than in liquids or gases. In conduction, the energy of thermal vibrations is transferred through the material, while matter itself is not transferred. Regions of a material at higher temperature will have particles vibrating more vigorously then cooler regions. Therefore, as these higher energy particles oscillate, some of their energy will be transferred to neighboring particles at lower energy when they interact/collide with them. Metals are particularly good conductors of heat (and electricity) because they contain free electrons that can easily move within the structure of the metal, thus transferring energy between metal nuclei. Materials that do not conduct very well are called ***insulators***.

Convection. *Convection* occurs when actual matter is transferred from one location to another. This occurs more easily in fluids (liquids and gases). When a region of a fluid becomes warmer than its surroundings, it expands, thus becoming less dense. Because it is less dense, it will tend to rise within the fluid due to buoyant forces. As it rises it will displace cooler, denser regions that will descend in the fluid, setting up convection currents of rising and falling matter. Winds are examples of convection currents that occur in the atmosphere. As we previously mentioned, the motion of fluid in the outer core is thought to be partly due to convection.

Radiation. Thermal vibrations in matter generate *e-m radiation* in the infrared frequency range. This radiation then travels through space and when it interacts with an object it will then increase the thermal energy of that object. Because it is e-m radiation that is transferring energy, this mode of heat transfer is called radiant heat transfer or *radiation*. Radiant heat transfer can occur across a vacuum unlike conduction or convection.

The Physics—Restarting the Core. Now that we've got the basics down, we need to "restart" the Earth's core in order to save all life on the planet. The first problem our "terranauts" have to figure out is how the heck they are going to get into the Earth's core in the first place. The deepest borehole ever drilled (the Kola Superdeep Borehole in Russia) reached a depth of about 7.6 miles. The drilling project lasted 22 years, beginning in 1970 and ending in 1992. To reach the core—and save the Earth—they'd have to drill about *3,000 miles in a day or so.*

Fortunately for humanity, renegade scientist/engineer Dr. Brazzleton (played by Delroy Lindo) has been working in secret for the past 20 years on the drill to end all drills. It has the ability to vaporize, almost instantaneously, thousands of tons of solid rock. (A real leap forward in drill technology.) His idea is to incorporate the super drill into a ship that will bore its way to the core.

Not being privy to the formidable technological details of the invention, other than the statements that the drill uses "high-frequency pulse lasers with resonance tube ultrasonics" and that the ship is powered by an experimental nuclear fission reactor which is about the size of a kayak, we will limit our investigation to the energy requirements necessary to make the journey through the crust, through the mantle, and into the outer core. Let's consider the following questions:

ACTIVITY

Put 2 cups of ice in a pot. Make sure you weigh the ice. Then, put it on the stove and on high heat. Measure the time it takes to start boiling. (Once it starts to boil, turn the stove off. Vaporizing the ice is not a good idea as you might ruin your pot and/or start a fire.) Calculate theoretically how much total energy it will take to get the ice to the boiling point. Determine the rate at which energy had to be supplied (power) to the ice/water to accomplish this. (The specific heat of ice is around 2,050 J/kg °K, the heat of fusion is 3.3 × 105 J/kg, and the specific heat of water is 4,186 J/kg °K.)

1. How much energy would it take to vaporize the rock?
2. At what rate (power) would the drill vehicle have to work to do this?

Energy to Vaporize Rock. The previous activity should serve as a model to calculate the energy it would take to melt and then vaporize the volume of rock in front of the ship that stands between it and the core. First we will find the mass of rock that needs to be vaporized. Because the ship's diameter appears to be about 4 meters, the total volume of rock would be that of a cylinder with a radius of 2 meters and a length of about 5,200 km (the distance to the outer-core/inner-core boundary). Because we don't have to melt the outer core (it's already liquid), we can save some energy there, and we'll split the volume into two parts—the mantle and the core:

$$V_{mantle} = \pi R^2 h = 3.14(2\ \text{m})^2\ (2.9 \times 10^6\ \text{m}) = 3.6 \times 10^7\ \text{m}^3$$

Because the density of the mantle is about 4,000 kg/m^3, the mass of rock that we have to remove is:

$$M = Density \times V = (4{,}000\ \text{kg/m}^3)\ (3.6 \times 10^7\ \text{m}^3) = 1.5 \times 10^{11}\ \text{kg}.$$

The average density of the outer core is about 11,000 kg/m^3 and we have another 2,300 km that we have to travel. This gives a volume of:

$$V = 3.14(2\ \text{m})^2(2.3 \times 10^6\ \text{m}) = 2.9 \times 10^7\ \text{m}^3.$$

Therefore, the mass of the outer core in our way is:

$$M = (11{,}000\ \text{kg/m}^3)\ (2.9 \times 10^7\ \text{m}^3) = 3.2 \times 10^{11}\ \text{kg}.$$

The thought may have crossed your mind that even if we heat up, then melt, then heat up some more, then vaporize the rock, where does all the matter actually go? How do we get it out of the way? We still have the same amount of mass in front of us, just in a different phase, and unless there is sufficient pressure in the vapor to diffuse almost instantly the vaporized rock into the surroundings, we've still got something in the way. In fact, as the liquid changes into vapor it will expand, and it would apply a formidable pressure on the ship in the wrong direction. However, if that doesn't concern the terranauts, why should we lose any sleep over it? In the film (depending on the shot) it does look like maybe they're cutting a hole a little wider than the ship. If so, then maybe some of the gas can flow back past the ship and out the hole. Nevertheless, there would still be significant explosive pressure impeding the progress of the ship.

Back to the task at hand. First of all, how much energy do we need to vaporize a hole down through the mantle? To calculate this, we are going to have to make some rough approximations. We need specific heats of fusion and vaporization for the mantle and core materials under high pressure and temperature. Most of the published data on these values don't refer to temperatures and pressures as high as those that exist in the core. Because our knowledge of the physical properties of Earth's interior is incomplete we will use the best estimates we have—we should be in the ballpark.

The *specific heat* of the mantle is estimated to be about 1,260 J/kg °C. The *heat of fusion* is less certain; it's probably about 400 kJ/kg, based on data of heats of fusion for somewhat similar rocks. The specific heat of the liquids tends to be on the order of half that of a solid of the same material, so we will assign the mantle a specific heat of 600 J/kg °C. For the *heat of vaporization*, we'll have to speculate somewhere between that of iron (6×10^6 J/kg) and water (2×10^6 J/kg), so let's say that it's 4×10^6 J/kg.

According to geophysical data, the mantle rocks are only thought to be about 100 °C or so below their melting point. Therefore, the energy required to heat the mantle rocks to melting temperature is given by:

$$Q = mc\Delta T = (1.5 \times 10^{11}\text{kg})(1{,}260 \text{ J/kg °C})\,(100 \text{ °C}) = 1.9 \times 10^{16} \text{ J}.$$

Then, to melt the rock:

$$Q = mL_f = (1.5 \times 10^{11} \text{ kg})(400 \times 10^3 \text{ J/kg}) = 6.0 \times 10^{16} \text{ J}.$$

Next we heat the rock to the boiling point, which tends to be about 1,200 – 1,500 °C above the melting temperature for rocklike materials. We'll use 1,200 and save a little energy.

$$Q = mc\Delta T = (1.5 \times 10^{11}\text{kg})(600 \text{ J/kg})(1{,}200 \text{ °C}) = 1.1 \times 10^{17} \text{ J}$$

Finally, let's vaporize the stuff:

$$Q = mL_v = (1.5 \times 10^{11} \text{ kg})(4 \times 10^6 \text{ J/kg}) = 6 \times 10^{17} \text{ J}.$$

This gives us a grand total of $Q_{total} = 7.9 \times 10^{17}$ J or 790,000 trillion Joules of energy—which is a lot.

Now we have to penetrate the core. Because the core is liquid, and is actually thought to have a viscosity near that of water due to its high temperature, let's ignore any further energy expenditure even though there would be significant drag on the ship at the velocities at which it is traveling. (For fun, you could calculate how much more energy it would take to vaporize the necessary mass of outer core—see the boxed problem.)

PROBLEM

Assuming the boiling point of iron is 3,000 °C, and its heat of vaporization given previously, determine the energy necessary to vaporize a cylinder of liquid iron 4 m in diameter with a length equal to the thickness of the outer core initially at a temperature of 3,000 °C.

Rate of Work of the Drill Vehicle. In the film we are told it takes about a day to get into the core (which means the ship travels at an average speed of about 150 mi/hr). Calculating the power rate at which energy is used to vaporize rock we get:

$$P = \frac{Q}{t} = \frac{7.9 \times 10^{17}\ \text{J}}{(1\ \text{day} / 24\ \text{hr})(1\ \text{hr} / 3{,}600\ \text{s})} = 9.3 \times 10^{12}\ \text{W}$$

Nuclear power plants produce energy from fission reactions at rates of a couple of billion watts (gigiwatts or GW), but the ship produces power at a rate of about a few thousand times that. We need the equivalent of a few thousand nuclear reactors; so where is all that energy coming from?

The Physics—The Kinetic Core. Here's where *The Core* puts its metaphorical foot into its metaphorical mouth. According to Dr. Zimsky, "The core of the earth [a trillion, trillion tons of hot metal spinning at around 1,000 mi/hr] has stopped spinning!" The implication is that the rest of the earth has maintained its normal rotation, but somehow the outer core has come to a dead stop. This is difficult to make sense of. Trying to reason what could make this happen causes the brain to rotate. If it did occur, there would be other *immediate* effects having nothing to do with the magnetic field. For example, according to conservation of (angular) momentum, the rest of the planet would have to compensate for the core slowing to a stop by speeding up—a lot. In that case we wouldn't need to worry about bad weather because the inertial effects would result in the equivalent of an Earth-sized train wreck! (Imagine what would happen if you were standing inside a train initially moving at a constant velocity that then rapidly increased its speed by several hundred miles per hour.)

Putting this concern, and many others, aside, we will first calculate the total kinetic energy of the outer core as it rotates around the Earth's axis, and then see how many nuclear bombs we will need to produce that energy.

How Many Bombs? Rotational kinetic energy = $\frac{1}{2}I\omega^2$ where I is the rotational inertia of an object, and ω is its angular speed (more on rotational kinetic energy in Chapter 4).

We will approximate the outer core to be a sphere with mass of 2×10^{24} kg (one-third the mass of the Earth). I for a sphere = $\frac{2}{5}mr^2$ where r is the radius of the sphere (for the outer core 3,400 km). I for the core will actually be a little larger than what we calculate because more of the mass is distributed to the outside of the sphere.

The angular speed ω is one revolution per day or 2π radians/86,400 sec = 7.3×10^{-7} rad/s, which gives,

$$\frac{1}{2}\left[\frac{2}{5}(2 \times 10^{24}\ \text{kg})(3.4 \times 10^{6}\ \text{m})(7.3 \times 10^{-7}\ \text{rad/s})^2\right] = 2.5 \times 10^{24}\ \text{J}$$

That's two and a half trillion, trillion Joules. Now that *is* a lot of energy. Will the nuclear bombs do the trick? It's easy to check. A megaton is a unit of energy equal to the energy released in exploding one million tons of TNT. One megaton (MT) equals 4.1×10^{15} J. Therefore, a 200-megaton nuclear bomb will yield a total energy of:

$$E = 200\,(4.1 \times 10^{15}\ \mathrm{J}) = 8.2 \times 10^{17}\ \mathrm{J}.$$

The total number of bombs needed to give sufficient energy will be:

$$\#\text{ of bombs} = \textit{Total } E \text{ needed}/E \text{ per bomb} = 2.5 \times 10^{24}/8.2 \times 10^{17} = 3 \times 10^{6} \text{ bombs}$$

That's three million bombs. We only have five so I guess we don't have enough.

> **PROBLEM**
>
> Assuming a core convection velocity of 1×10^{-3} m/s, determine the kinetic energy contained in the current (assume most of the outer core is convecting) and compare this to the energy released by five 200-megaton nuclear bombs.

Conclusion. The idea that a few nuclear bombs have anywhere near the energy of the rotating outer core, exemplifies the hubris of the science fiction disaster film genre. However, as we implied, the details of what actually produces the field might be important. Geophysicists believe that it is not in essence the core rotation velocity that produces the field but the motion within the convection currents. While models differ in details, they suggest that the velocity of convecting fluid is actually very small. In fact, a velocity on the order of about 10^{-3} m/s may be sufficient to generate the field that we see at the surface.

If you did the calculation suggested in the boxed problem, you might find that the bomb energy is now in the same ball park as the *KE* of the convecting fluid. Based purely on energy considerations, perhaps the bombs ***do*** have enough energy. Even so, how they would restart convection is a highly problematic question.

However, in *The Core's* climactic scene, it's clear from the computer simulation at mission control that it really ***was*** the rotation of the entire core they were talking about. The bombs shock the core from its state of no motion to "full rotation." However, even if we have enough energy, even if we have three million bombs and not just five, or even if we believe Dr. Zimsky's highly suspect assessment that only a "small nudge" will restore the core's rotation, when a nuclear bomb explodes it will generate a ***spherical*** shock wave with no net directionality. This means that the ***net*** force that each bomb exerts on the core is zero.

If each nuclear bomb exerts zero net force, what is the total net force exerted on the core? Well, the sum of five or even three million times zero is still zero. There may be some wave energy propagating back and forth through the fluid similar to what you would see if you did the activity, but there would be ***no rotation.***

You might also look at this problem in terms of *torque* (see Chapter 4). To change the rotation rate of something about some axis, there needs to be a net torque acting around that axis; in this case the center of the earth. The spherical shock waves result in zero net torque around an axis through the center of the earth—which means no change in rotation rate around the center. So even if we add an amount of energy to the core equal to the rotational energy that we need, unless that energy is directed such that it produces a net torque it's still not going to work.

ACTIVITY

Take a circular tub. Fill it up with water. Drop a rock into the tub and observe the ripples the impact creates. Now take three or four rocks and drop them in different locations in the tub. Pretend these are nuclear bombs generating spherical shock waves in the core. Do these waves cause any net rotation to the water in the tub?

Toward the end of the movie, things are said and things happen that are so bizarre it's hard to figure out how to evaluate them. For example, apparently what did in the core's rotation in the first place was the insidious "Project Destiny." Project Destiny is a secret weapon (developed, of course, by the U.S. government) that is able to "make every volcano on the planet blow, and . . . earthquakes big enough to rip us to pieces." It works by "beaming high-powered electromagnetic energy waves down fault lines." What does that mean, exactly? Why would electromagnetic waves affect motion along faults? Fault lines do not extend anywhere near the depth of the outer core, but somehow these "energy waves" have affected it in some unimaginable way. Dr. Keyes says that "the core is an engine [and if you] throw a small wrench into a big engine, you can still stop it." We do not find this entirely convincing.

By the end of the adventure, the only two survivors are Dr. Keyes and Beck (Hilary Swank), the world's youngest and most attractive space shuttle pilot. At this point of the movie, it's as if the writers have given up even a façade of plausibility, and start throwing things in to wrap up the story. Keyes figures that they can make it back to the Earth's surface by using the thermal energy of the core to power the ship, because they had to jettison the ship's reactor as an extra bomb. Apparently the calculations for density were off, and the five bombs won't do the job. (Why didn't they take a couple of extra bombs with them in the first place?) We are told that the hull material called "unobtanium" turns heat into energy for the ship. (This material truly is astonishing—another of Dr. Brazzleton's *miracle* inventions. It's a metal, it's a fabric, it's strong enough to build a ship and supple enough to make a suit. It can withstand temperatures of thousands of degrees, and pressures of millions of atmospheres. In fact it actually gets stronger as the pressure and temperature increase. It clearly constitutes the most amazing technological breakthrough in the history of science. Too bad it's fake.) Somehow the survivors jury-rig something to get the ship powered, and they surf the nuclear shock waves inside a magma flow through a tectonic plate boundary at super-high velocities all the way to the ocean floor, where whales rescue them just before they run out of oxygen. This is not a joke. They are saved at the last minute because mission control realizes that the whales are attracted to the ship's "ultrasonic resonance tubes." (The filmmakers apparently conveniently chose to ignore the fact that these resonance tubes weren't supposed to be working anymore at this point in the movie.)

A few geological points in response to this preposterous ending:

1. Geophysicists think that tectonic plates don't extend more than a few hundred kilometers below the crust, so there are no plate boundaries anywhere near the outer core.
2. Magma doesn't form until rising plumes of rock material get close enough to the Earth's surface for the lower pressure to allow melting of the rock. Riding all the way from the outer core up to the surface in a magma flow would be difficult *without* any magma.
3. The ship is moving up a plate boundary, and yet it surfaces near Hawaii, which is about as near to dead center of the Pacific plate as you can get. Hawaii in fact does sit over what's called a mantle "hot spot," which accounts for the active volcanism, but it is not a plate boundary.

Some other marvels of bizarre physics *faux pas* in *The Core* that make the head spin include:

1. As the drill ship bores into the earth, it is pointed front down towards the center, which means gravity will be acting straight down through the front of the ship. Anything on the ship is going to fall to the front. We have to give our terranauts some credit for trying to deal with this issue. As they train for the mission, Dr. Brazzleton explains that they will maintain an orientation where they can stand on the floor using rotating "gymbols," structures that are supposed to rotate on an axis inside the ship's compartments. They're actually a pretty good idea. However, as is typical in this type of movie, there's no follow-through, and we never see these devices on the ship.
2. The crew find themselves inside a giant "geode" deep in the mantle. How could a geode exist under the mantle's immense pressure? The gas pressure in the geode would have to equal the pressure exerted from outside the geode due to the surrounding rock (maybe half a million atmospheres or so at that depth.) However, when the crew goes outside the ship to make an emergency repair, how come they move like there is no external pressure at all? Even if their suits have the magical technology of unobtanium on their side, they could never be rigid enough to withstand the pressure and flexible enough to walk in (not to mention keep the terranauts insulated from the heat). They should look more like they are trying to walk on the deepest part of the ocean floor but thousands of times harder, not quite as graceful and confident as portrayed in the scene. In addition, apparently unobtanium comes in a transparent variety because none of the intense heat is getting through their visors, and the pressure isn't cracking them either.
3. In the previous scene, think about what things might actually look like if they are at a temperature of 4,000 °C. In the scene it's all dark and gloomy, pitch black in fact, but have you ever seen what happens to matter when it gets really hot (what does a heating coil look like at temperatures of only around 100 °C?)? It gets brighter. At 4,000 °C it would be glowing so bright you wouldn't be able to keep your eyes open.

4. The crew of the ship communicates instantly with mission control through thousands of miles of solid rock. How do they do this? Radio waves don't travel through thick layers of rock very well. It's not like being in space. The radio waves are going to be attenuated on the scale of a few *meters,* not hundreds of *miles.* They most certainly will not make it to the Earth's surface!
5. There are more—happy hunting!

The fundamental premise of *The Core* (that the disappearance of the Earth's magnetic field would result in catastrophic destruction of life on the planet) simply isn't true and most of the laws of physics as we know them do not apply in *The Core.* The physics in the movie (with a few exceptions) is atrocious with some ridiculous geology to boot. (Moreover, if that isn't enough we plan to dissect more of this film in Chapter 6 when we discuss electromagnetism.)

ADDITIONAL QUESTIONS

1) Two *identical* balls of space debris collide in deep space. They are initially traveling with the same speeds but in opposite directions. If they collide head-on, describe the subsequent motions of the space balls if the collision is (a) perfectly inelastic, and (b) perfectly elastic.

2) In an off-screen moment during the filming of *MI-II* Tom Cruise and Dougray Scott are rehearsing their motorcycle scene by carefully jogging into each other. In one particular attempt they are both moving at 5 m/s before colliding. They bounce off of each other, with Cruise moving backward at a speed of 3 m/s. How fast and in what direction is Dougray Scott moving just after the collision?

3) Scott stands on a ledge 5 meters above the ground, swings down from a rope as shown in the following figure, and collides with Cruise, scooping him up as Tarzan scoops Jane. How fast are they going just after the collision? (Hint: What part of the motion do you analyze using conservation of energy, and what part using conservation of momentum?)

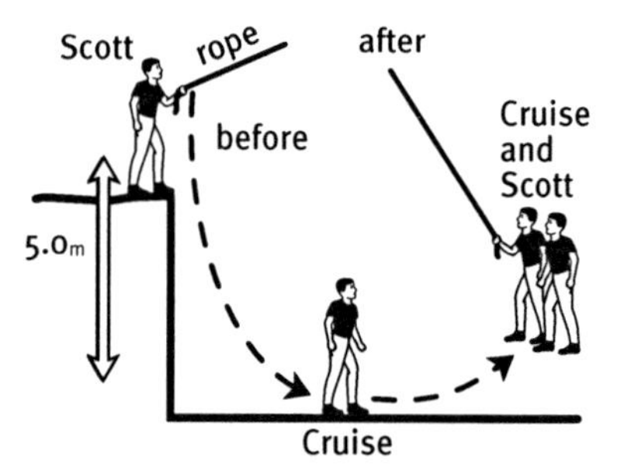

4) Earlier in the movie *Mission: Impossible II* there is a car chase scene between Cruise and his soon-to-be love interest, played by Thandie Newton. Several times she slams her car into his in an attempt to force him off of the road. Let's take the following hypothetical accident. Say Newton impacts Cruise's car in a perfectly inelastic collision (the cars stick together). If she is originally traveling at a 30-degree angle to Cruise's direction, find the velocity of the two cars immediately after the collision assuming they were each traveling at 90 miles per hour just before the collision. (Hint: Break this into two perpendicular one-dimensional motions much like you would do for a projectile.)

5) A rocket in space is moving at a speed of 4,000 m/s relative to a planet when it jettisons an empty fuel stage at a velocity of 600 m/s relative to the rocket. If the stage has a mass of 2,000 kg, and the rocket has a mass of 6,000 kg, find the final velocity of the rocket relative to the planet after the stage has been ejected.

6) You have a 0.5-kg model rocket. It has a small fuel packet of negligible mass compared to the rocket. When the rocket is launched it has a constant acceleration of 30 m/s^2 for 3 seconds until the fuel is used up.

   a) Find the thrust force exerted by the fuel.

   b) Find the maximum height of the rocket.

   c) If there is a 20-mile-per-hour crosswind during the flight, find where the rocket lands relative to the launch point.

7) The final rocket that the boys launch in *October Sky* is a really good one. It looks like the burn lasts about 15 seconds. If we assume the rocket has a mass of 2 kg, and it initially contains 3 kg of fuel,

   a) find the average acceleration of the rocket during the burn.

   b) find the maximum height that the rocket achieves.

   c) determine the total time the rocket is in the air.

   d) construct position-time, velocity-time, acceleration-time, and force-time graphs for the motion of the rocket.

8) Say you have a gigantic block of ice (initially at 0 °C) and you want to vaporize it using a 200-MT nuclear bomb. What is the biggest block of ice you could vaporize with the bomb?

9) Rub your hands together vigorously. Why do they get warmer? (Think in terms of conservation of energy.)

10) If you throw a mass of iron at an initial temperature of 100 °C in with an equal mass of water initially at 0 °C, find the final temperature of the system if no heat is lost from the container. ($c_{h2o} \approx$ 4,200 J/kg°C $c_{Fe} \approx$ 420 J/kg °C)

11) In *The Core* the force of gravity inside the ship as the crew drills toward the center of the earth should be directed towards the front of the ship. As they go deeper what do you think will happen to the strength of the gravity force? Explain.

12) You have invented a machine designed to melt rocks using only the mechanical energy generated when you drop them. The machine consists of an insulated 10-m high tube. Rocks are dropped from the top of the tube, hit the ground, and are then lifted back to the top where they're dropped again. When the rocks hit the ground, the mechanical energy is converted to thermal energy. Let's say you have a rock with a volume of 0.50 m^3. How many times will you have to drop it before it starts melting? Assume all of the thermal energy generated remains in the rock, and the rock has a specific heat of 1,000 J/kg °C.

Circular Motion

IN THIS CHAPTER WE ARE going to analyze the physics in a few movies where circular motion makes a dramatic appearance. We waxed eloquently about that action movie standby, the car chase scene, in Chapter 1. One common aspect of the car chase that we touched upon briefly was the high-velocity, skidding, nail-biting, 90-degree turn. In this section we will review the tools necessary to evaluate how realistic such turns may be as portrayed in movies such as *Speed, The Fast and the Furious,* and others, but we will also delve into other situations where circular motion is intimate to the action. For example, how strong do you have to be to swing from a rope, or a web as Spider–Man does? How do you use circular motion to create simulated gravity in outer space? What is the physics behind orbital motion (a necessary aspect of any good science fiction movie)? With that in mind here's the program:

- We will analyze a high-velocity turn in *Speed*. keeping in mind that the famous bus must stay above 50 miles per hour or it will explode! Is there enough friction to negotiate the turn?
- We will evaluate several examples of rotational motion as portrayed in the science fiction classic, *2001,* involving spinning space vehicles and the concept of artificial gravity.
- We will look at issues concerning orbits in the movies *Armageddon* and *Apollo 13* addressing the physics relevant to putting a space vehicle into a circular orbit.

To make the most of our ambitious plan, before we analyze each of these exciting films, let's review the concepts pertaining to circular motion.

UNIFORM CIRCULAR MOTION

Any object that is not moving in a straight line must be accelerating according to Newton's first law, even if it is moving at a constant speed. If the direction is changing, there *must be* a net force acting on the object to change its path. Roll a ball along the floor, and I defy you to find a way to change its direction without applying a force to it. You can, however, apply a force to the ball and *not* change

Photo: A rotating space station from *2001: A Space Odyssey.*

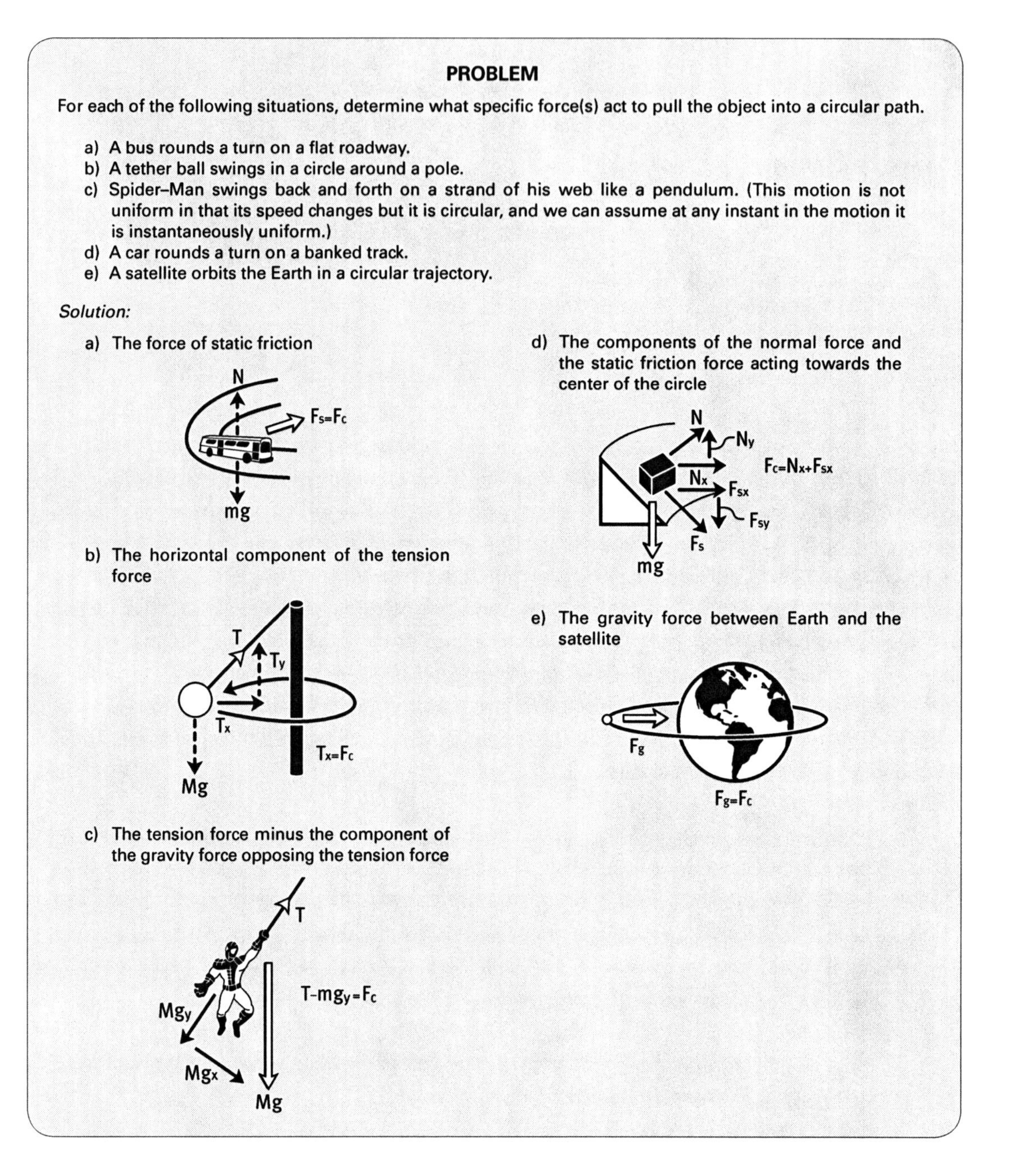

PROBLEM

For each of the following situations, determine what specific force(s) act to pull the object into a circular path.

a) A bus rounds a turn on a flat roadway.
b) A tether ball swings in a circle around a pole.
c) Spider-Man swings back and forth on a strand of his web like a pendulum. (This motion is not uniform in that its speed changes but it is circular, and we can assume at any instant in the motion it is instantaneously uniform.)
d) A car rounds a turn on a banked track.
e) A satellite orbits the Earth in a circular trajectory.

Solution:

a) The force of static friction

b) The horizontal component of the tension force

c) The tension force minus the component of the gravity force opposing the tension force

d) The components of the normal force and the static friction force acting towards the center of the circle

e) The gravity force between Earth and the satellite

its direction. For example, if the force is parallel to the motion, the ball will only speed up or slow down. However, if the force is not exactly parallel to the initial direction of motion, the direction of the ball *will* change. In fact, if a force is perpendicular to the motion of an object, and this force keeps changing direction so that it stays continuously perpendicular to the velocity, then the object will not change speed at all—only its direction.

This last case is an example of an object moving in a circle at a constant speed *(uniform circular motion)*. In uniform circular motion an object is always accelerating towards the center of the circle of motion. This requires a force or combination of forces that result in a net force acting towards the center of the circle.

Forces that act to pull an object into a circular path are called *centripetal* forces, and the acceleration that results (also directed towards the center of the circle) is called *centripetal acceleration*. These are also known as ***radial*** forces and ***radial*** accelerations. Applying Newton's first law, we can reason that because an object's tendency is to move in a straight line at a constant speed in the absence of a net force, in each of the previous examples, if the centripetal force suddenly disappeared, the object would move off in a straight line tangent to the circle. *The centripetal force is necessary to pull the object off of its straight line course.*

THINKING PHYSICS

If you drive around a turn too fast in your car, you will feel a push to the outside of the turn. People sometimes call this push *centrifugal force.* Can you identify an actual force that would tend to push you in this direction? If not, explain what is happening in terms of Newton's first law and centripetal forces (or lack thereof).

It is important to remember that a centripetal force is not some magical, special force. A centripetal force is always due to specific forces that you can identify, that happen to be acting in that particular situation to pull something into a circular path. For example, for a car rounding a turn on a flat road, the only force that could be acting towards the center of the circle is a force between the tires of the car and the road. The only possible force in this case is the static friction force. (It is static and not kinetic friction because the car's inertia causes the car to want to slide out of the circle. However, if it doesn't slide then it must be *static* friction gripping the tires of the car.)

Quantifying the Principle

Applying Newton's second law, we get:

$$F_{net\,(\text{centripetal})} = ma_{\text{centripetal}}$$

It can also be shown that the magnitude of the centripetal acceleration on an object of mass m traveling with a tangential velocity v is given by $a_c = \frac{v^2}{r}$.

Therefore, $F_{net(c)} = \frac{mv^2}{r}$.

Example 1: A car rounds a curve with a radius of curvature of 100 meters. It is traveling at a speed of 24 m/s. What minimum coefficient of static friction is necessary to prevent the car from skidding?

Solution: Since the force of static friction acts as the centripetal force,

$$F_{net(c)} = \frac{mv^2}{r} \text{ and } F_c = f_{s\ max} = \mu_s F_N = \mu_s mg$$

$$(F_{net\ y} = F_N - mg = 0) \text{ therefore,}$$

$$\mu_s = \frac{v^2}{gr} = \frac{(24 \text{ m/s})^2}{9.8 \text{ m/s}^2 (100 \text{ m})} = 0.59$$

Example 2: A tether ball with mass 0.30 kg is rotating around the tether ball post at a speed *v*. If the 2-meter long rope connecting the ball and the top of the post makes a 45-degree angle with the post,

a) how much tension must be in the rope?
b) how fast is the ball moving?

Solution: First draw a force diagram showing the forces acting on the ball.

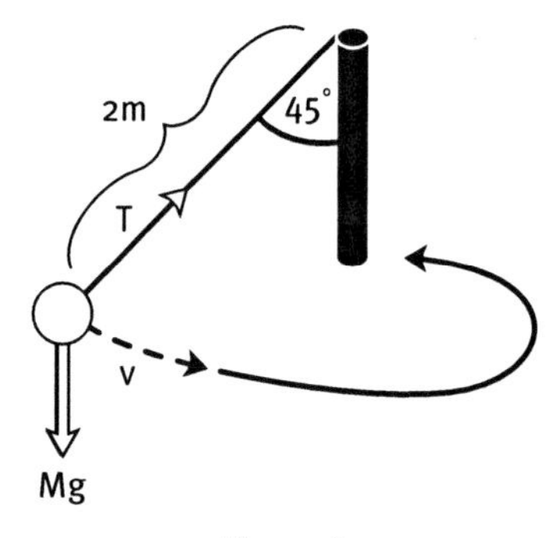

Figure 1

By applying Newton's second law in the vertical direction we get

$$F_{net\ y} = F_T \sin 45 - mg = 0 \text{ so } F_T = \frac{mg}{\sin 45} = \frac{(0.3 \text{ kg})(9.8 \text{ m/s}^2)}{\sin 45} = 4.2 \text{ N}$$

The horizontal component of the tension force acts as the centripetal force in this case. Therefore,

$$F_{net(c)} = T \cos 45 = \frac{mv^2}{r}.$$

Looking at Figure 1 we can see that $r = (2\text{m}) \cos 45 = 1.4$ m. Therefore,

$$v = \sqrt{\frac{rF_T \cos 45}{m}} = 3.7 \text{ m/s}.$$

Example 3: Spider-Man uses his web as a trapeze artist to propel himself around New York City (and suburban Long Island). What maximum forces must the muscles in his arms be able to withstand to do this?

Solution: We'll start with force diagrams representing the forces acting on Spider-Man at various locations in his swing.

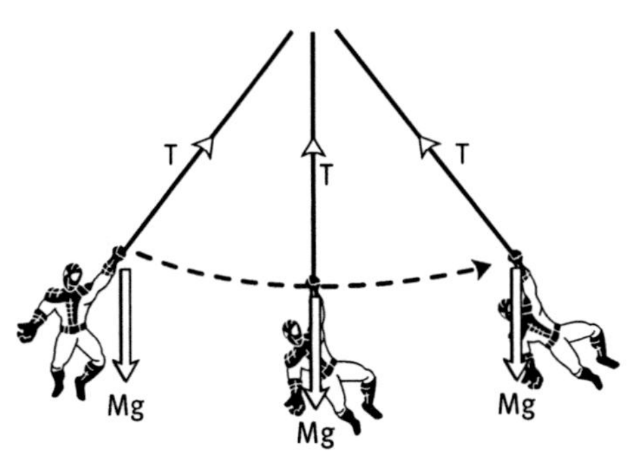

Figure 2

First, we see that in each case the net centripetal force is a result of the difference between the tension force exerted by the web on Spider-Man's arm and the component of the weight force acting opposite to the tension force. The maximum F_T will occur when Spider-Man is at the bottom of his swing for two reasons: (1) He is moving fastest at this point, and (2) there is the greatest component of the weight force (W) opposing F_T.

Therefore at this point:

$$F_{net(c)} = F_T - mg = \frac{mv^2}{r} \text{ and } F_T = mg + \frac{mv^2}{r}.$$

In order to obtain his velocity at this point we can use conservation of energy. We know that the gravitational potential energy (mgh) that Spider-Man loses in falling a height h must be converted into an equal amount of kinetic energy ($\frac{1}{2}mv^2$), assuming negligible energy loss due to air resistance. If we look at several scenes in the Spider-Man movies we see that Spider-Man swings through some pretty big drops—maybe 40 or 50 meters, so his velocity at the bottom of a 40-m drop would be $v = \sqrt{2gh} = 28$ m/s.

THINKING PHYSICS

Consider the following questions based on Example 3: What object is exerting force on Spider-Man's arms? At what point in the swing will that force be at a maximum? Why? How could you determine Spider-Man's speed at different points in the swing? How would you use that speed to figure out the forces acting on Spider-Man?

Therefore, assuming Spider-Man has a mass of around 75 kg, and assuming that if he starts with his web horizontal to the ground, the circle of his motion has a radius of 40 meters, then

$$F_T = m(g + \frac{v^2}{r}) = (75 \text{ kg})\frac{(9.8 \text{ m/s}^2 + (28 \text{ m/s})^2)}{40 \text{ m}} = 1{,}490 \text{ N or about 330 lbs.}$$

This means that Spider-Man's muscles need to be capable of withstanding 330 pounds of force. That's a lot, but because even some actual regular humans might be able to do it, this should present no problem for Spider-Man and his super strength.

ORBITAL (CIRCULAR) MOTION

Orbital motion occurs when gravity acts as a centripetal force between two objects. For example, a satellite orbiting the Earth experiences a net gravitational force directly towards the center of the Earth. The satellite is continuously in a state of free-fall. It does not hit the Earth because it has a sufficient tangential velocity such that the satellite actually falls *around* the Earth. For example, it is a common misconception that the gravitational force is very weak aboard the space shuttle, and that is why the astronauts are weightless. In fact, at the height of the shuttle, g is about 9 m/s^2, which is only about 10 percent less than that at the surface. What the astronauts experience is an *apparent weightlessness* due to the fact that they are in a constant state of free-fall. You could get the same effect if you were in an elevator and the cable broke (as in *Speed!*). You would be able to float around inside the elevator the same as if you were on the space shuttle. That is, until the elevator hit the ground. In fact, astronauts train for space flight by riding around on a special airplane (known as the "vomit comet") that goes into a free-fall dive for up to 30 seconds at a time.

Using the law of gravitation, and the fact that the gravitational force acts as a centripetal force for orbital motion, we get:

$$F_g = \frac{Gm_{earth}m_{sat}}{r^2} = m_{sat}\frac{v^2}{r},$$

where r is the distance between the centers of the two objects.

Therefore, the velocity of an orbiting satellite depends only on the distance from the Earth. It is independent of the satellite's mass:

$$v = \sqrt{\frac{Gm_{earth}}{r}}.$$

We also define the *period* of the orbit (T) to be the time it takes to make one complete revolution. For a circular orbit then,

$$T = \frac{2\pi r}{v}.$$

Therefore, to put something in a circular orbit at a certain distance from Earth, there is only one velocity that will work. If it moves a little faster, it will "slide out" into an elliptical orbit. If it moves too slowly, either its orbit will decay into a smaller elliptical orbit or it will fall back to Earth. Actually, it would burn up in the atmosphere before it got there (unless it's either Superman or it has an effective heat shield).

PROBLEM

It is convenient for communication satellites to be placed into an orbit that keeps them stationary relative to a fixed point on the surface of the Earth. (Remember, the Earth is rotating about its center.) This is called *"geosynchronous orbit."* Determine at what height above Earth's surface a satellite must be to be in geosynchronous orbit. What is its orbital speed at that height?

Note: In orbital motion the two masses both orbit around their common centers of mass. In the case of an artificial satellite, the center of mass of the Earth/satellite system is so close to the center of the Earth that the Earth's motion is negligible. In the case of planetary orbits around the Sun it is also more or less reasonable to make the same assumption. However, in situations where the masses are on the same order of magnitude (such as with binary stars), the motions of both objects are relevant.

Potential Energy and Escape Velocity

We have defined the difference in gravitational potential energy (PE_g) when an object is moved between two points relative to the surface of the Earth to be equal to $mg\Delta h$. However, this is only an approximation that works well for motions of objects near the surface of the Earth where g is a relatively constant value. For the case of orbits, however, the possible elevation changes are more significant. A more general equation that describes the PE_g of *a system of two objects* is

> **PROBLEM**
>
> Determine the escape velocity for a satellite orbiting 1,000 km above the surface of the Earth. (Ignore any effects due to the Earth's rotation.) $M_e \approx 6 \times 10^{24}$ kg and $R_e \approx 6.4 \times 10^6$ m.

$$PE = \frac{-GMm}{r}.$$

For our purposes we will assume $M \gg m$, and the previous equation can be approximated as the *PE* of m relative to M. Expressed this way, we set $PE = 0$ an infinite distance away from M, which means that m will always have a negative value of *PE*. This has a physical meaning. It means that m is gravitationally bound to M, and it would take an amount of *KE* equal or greater to the magnitude of its *PE* to allow m to *escape* from M and never be pulled back. This amount of *KE* determines the escape velocity. For example, the total energy of m is

$$E = PE + KE = \frac{-GMm}{r} + \frac{1}{2}mv^2.$$

To achieve escape velocity, the total $E = 0$, and therefore

$$v_{esc} = \sqrt{\frac{2GM}{r}}.$$

Torque and Angular Acceleration

So far we have only considered forces that are responsible for changing ***direction*** in circular motion (centripetal forces), but objects may also change the speed at which they rotate or revolve. Because centripetal forces always act perpendicular to the motion of an object, they cannot change the speed of the object. In order to change the speed, some force in a direction tangent to the circle must be applied.

> **ACTIVITY**
>
> Take a bicycle tire and spin it by pushing (in a direction tangent to the circle) on the spokes near the outside of the tire. Do the same thing by pushing with the same amount of effort on the spokes near the center. In which case is it easier to change the spin rate? Why?

If you did the previous activity you will have seen that the tire is easier to spin when a force is applied further away from the axis of rotation. Therefore, the amount of ***angular acceleration*** that you impart to an object depends on a combination of two factors: where a force is applied and in what direction. This combined effect is known as a ***torque***, which can be expressed as follows:

$$\tau = (F \sin \theta)R,$$

where $F \sin \theta$ is the component of force that acts tangentially to the circle or potential circle of motion and R is the distance that the force is applied from an axis.

It is often convenient to define rotation rate in terms of angle/time rather than distance/time, so we define angular speed as $\omega = \frac{\Delta\theta}{\Delta t}$, and angular acceleration as $\alpha = \frac{\Delta\omega}{\Delta t}$. The linear or tangential speed of an object (or point on an object) can be related to the angular speed by $v = \omega r$, *as long as units of angle are expressed in radians.* Similarly $a = \alpha r$.

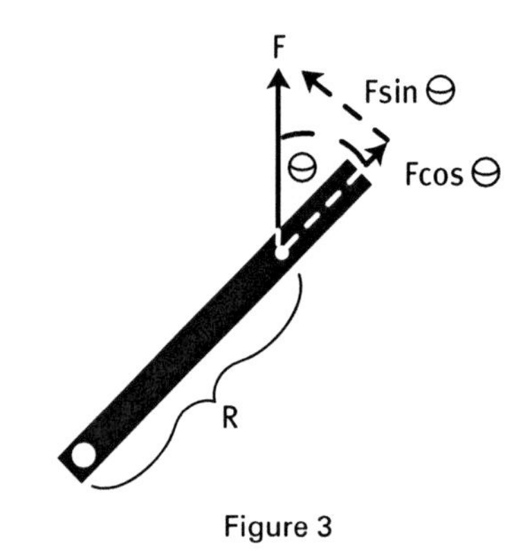

Figure 3

Newton's Second Law in Angular Form. According to Newton's second law, a net force will cause an object to accelerate, and the rate of (linear or translational) acceleration will be inversely proportional to the mass of the object. However, to produce an angular acceleration a net torque is required. The bigger the net torque, the bigger the angular acceleration. The rate of angular acceleration is also affected by the mass of the object, but not just by how much mass the object has, as in the linear case, but by how the mass is distributed.

As you can see from the activity, it is harder to rotate objects not only when they are more massive, but also when more of their mass is distributed towards the outside of the object. A property which combines the mass of an object and the distribution of that mass is called the *rotational inertia (I)* (also known as *moment of inertia*). Newton's second law can be written to describe how a net torque produces an angular acceleration as follows:

$$\tau_{net} = I\alpha.$$

Values of rotational inertia of many symmetrical objects around an axis of rotation are often expressed in a general form in terms of the mass and radius of the object. For example, I for a solid sphere around an axis through its center equals $\frac{2}{5}mr^2$, while I for a ring around the center equals

mr^2. If you know the actual masses and radii for the objects, then you can calculate a numerical value for I. Note also that the ring, which has more mass distributed to the outside, has a higher I than the sphere given the same total mass and radius, as we expect.

If an object has no angular or translational acceleration, and it is also at rest, it is said to be in static equilibrium. The conditions for static equilibrium are (1) $F_{net} = 0$, and (2) $\tau_{net} = 0$. If there is a net force but no net torque acting on the object, then it will have a translational acceleration but no angular acceleration. If it has a net torque but no net force, it will have an angular acceleration but no translational acceleration. In addition, if it experiences both a net torque and a net force, it will have both types of acceleration.

ACTIVITY

Take the bicycle tire from the previous activity and a big glob of clay. Spin the tire by pushing tangentially on the outside part for the following two situations: (1) with the clay placed at the center of the wheel, and (2) with the clay distributed on the outside of the tire. In which case is the tire harder to accelerate?

Example 4: In *Mission: Impossible IV*, the surely-to-be-released fourth movie in the *M.I.* series, I predict a scene in which Tom Cruise is suspended from the end of a 5-meter long plank of wood by his feet. The plank is anchored to the wall, and is also connected to the wall by a rope that makes a 30-degree angle with the plank.

If Cruise has a mass of 80 kg, and the plank has a mass of 20 kg, determine the tension that the rope must be able to handle without breaking. How much force must the wall exert on the plank?

Solution: First draw a force diagram showing the forces acting on the plank.

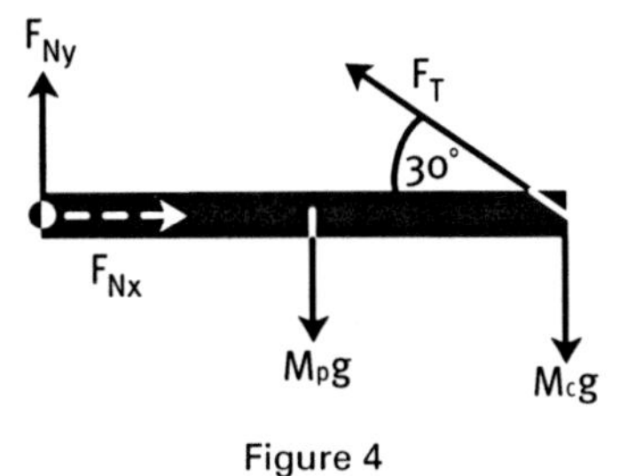

Figure 4

If the plank is to remain stationary (static equilibrium), then both the F_{net} and τ_{net} must equal zero. We can get the tension by analyzing the torques. Because there is no acceleration, we can sum the torques around any axis we want. No matter what axis we pick, the net torque around it must be zero because there is no rotation. Let's pick the point where the plank connects to the wall as our axis. Therefore, F_N applies no torque around the axis because it is applied right at the axis. The weight force acting on the plank acts through its center of mass, which we assume is at the center of the plank, and the force of Cruise's weight acts at the end of the plank. The component of F_T perpendicular to the plank ($F_T \sin 30$) is the part of F_T involved in producing a torque on the plank, and it also acts at the edge of the plank, therefore:

$$\tau_{net} = m_p\, g(0.5l) + m_c\, gl - F_T \sin 30 l =$$

$$(20 \text{ kg})(9.8 \text{ m/s}^2)(2.5 \text{ m}) + (80 \text{ kg})(9.8 \text{ m/s}^2)(5 \text{ m}) - 0.5F_T(5 \text{ m}) = 0,$$

which gives $F_T = 1{,}764$ N.

Now we can analyze the forces in separate vertical and horizontal equations to determine F_{Nx} and F_{Ny}. This gives:

$$F_{net\,x} = F_{Nx} - F_T \cos 30 = 0 \qquad F_{Nx} = F_T \cos 30 = 1{,}528 \text{ N}$$

$$F_{net\,y} = F_{Ny} - m_p g - m_c g + F_T \sin 30 = 0 \qquad F_{Ny} = (m_p + m_c)g + F_T \sin 30 = 1{,}862 \text{ N}$$

Example 5: Cruise's rope does not have sufficient strength and it breaks. Determine the angular acceleration of the plank immediately after the rope breaks if the plank can rotate around the pivot that connects it to the wall. Assume I for the plank about this axis is 1/3 ml^2 where l is the length of the plank.

Solution: $\tau_{net} = I\alpha$

The forces that exert torques around the pivot point are the weight forces of Cruise and the plank, so

$$\tau_{net} = m_c\, g(l) + m_p\, g(0.5l) = 0.33 m_p\, l^2\, \alpha, \text{ therefore } \alpha = 26 \text{ rad/s}^2$$

We have seen how Newton's second law can be applied to rotating systems. Now let's have a look at the energy and momentum of a rotating object.

ROTATIONAL KINETIC ENERGY

If you were to take a piece of dry ice and a marble, place them on an incline and release them, they would not get to the bottom at the same time or with the same speed. Is this a violation of conservation of energy? The answer is no, but although both the marble and the ice have the same initial potential energy (and the same total energy), not all of the marble's potential energy is converted to translational kinetic energy. Some is converted into kinetic energy of rotation.

The rotational kinetic energy $(KE_r) = \frac{1}{2} I\omega^2$.

Therefore, for the dry ice $PE_i = KE_{tf}$, but for the marble $PE_i = KE_{tf} + KE_{rf}$.

Example 6: If the dry ice and the marble are both released from the top of the incline with height h, compare their translational speeds at the bottom.

Solution: For the dry ice,

$$mgh_i = \frac{1}{2} m v_f^2 \text{ so } v_f = \sqrt{2gh}$$

For the marble,

$$mgh_i = \frac{1}{2}mv_f^2 + \frac{1}{2}I\omega_f^2$$

where v is the translational speed of the center of the marble and ω is the angular speed. The speed of the center of the marble must be equal to the tangential speed of a point on the rim. Therefore $\omega = \frac{v}{r}$. Therefore, I for a sphere of radius r and mass m is $I = \frac{2}{5}mr^2$,

$$mgh_i = \frac{1}{2}mv_f^2 + \frac{1}{2}(\frac{2}{5}mr^2)\omega_f^2 \text{ and } v = \sqrt{\frac{10}{7}gh_i},$$

which is less than the velocity for the dry ice.

ACTIVITY

Take a bicycle tire and set it spinning as rapidly as possible. Then, while it's spinning, try to flip it over as fast as you can. Why is it so difficult? (Hint: Angular momentum is a vector, so it depends on the orientation of the axis of the spinning object!)

Angular Momentum

Just as an object can have a linear momentum ($p = mv$), it can also have an angular momentum (L) relative to an axis of rotation. For example, a ball attached to a string rotating in a circle of radius r will have $L = mvr$. For symmetrical objects it can also be shown that $L = I\omega$. Just as linear momentum is conserved in systems that experience no external net force, angular momentum is conserved for systems that experience no external net torque.

When ice skaters perform a spin they can increase their angular speed by pulling their arms in. Why does this increase their rotation rate?

Elliptical Orbits

Unlike in a circular orbit, an object in an elliptical orbit will change speed as it travels. We can see why this must be so by applying some of the major fundamental principles of physics that we have studied.

1. Newton's laws: The gravitational force acting on the orbiting mass m must always be directed straight towards the large mass M (which is located at one of the foci of the ellipse). This means that sometimes the force will have a component tangent to the path, resulting in m increasing or decreasing its speed.

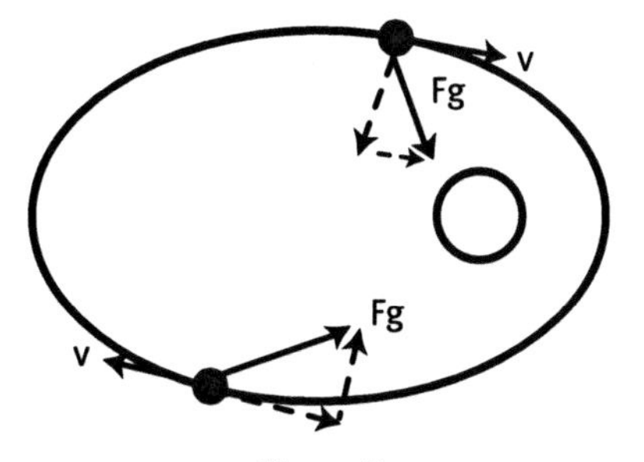

Figure 5

2. Conservation of energy: When the object is further away from the large mass there is a greater amount of *PE*, which will be converted to *KE* as m falls closer to M.
3. Conservation of angular momentum: Because the gravitational force acting on m always acts directly toward M, it does not experience a net torque about an axis through M. This means that angular momentum of m must be conserved, and if the distance to M decreases then its speed must increase ($L = mvr$).

Now it is time to apply some of these principles of circular and rotational motion to our favorite movies. We will range from some pretty good movie physics to the insultingly stupid; from the pinnacle of Newtonian elegance and real-life accuracy portrayed in *Apollo 13,* to what is arguably one of the worst movie physics films of all time (*The Core* notwithstanding)—the famous, or should we say, the infamous, *Armageddon*.

A HIGH-*SPEED* TURN

As promised, we are going to put our knowledge of centripetal forces and circular motion to the test by analyzing one of the more exciting moments in the movie *Speed*.

The Scene

Recall the plot of the movie from Chapter 1. The basic premise is that a city bus must maintain a speed of at least 50 miles per hour (22 m/s) or a bomb planted on it by a psycho will detonate. In scene 11 Sandra Bullock, Keanu Reeves, and the unlucky passengers of bus 2525 need to make a 90-degree turn while maintaining this speed. Can they do it?

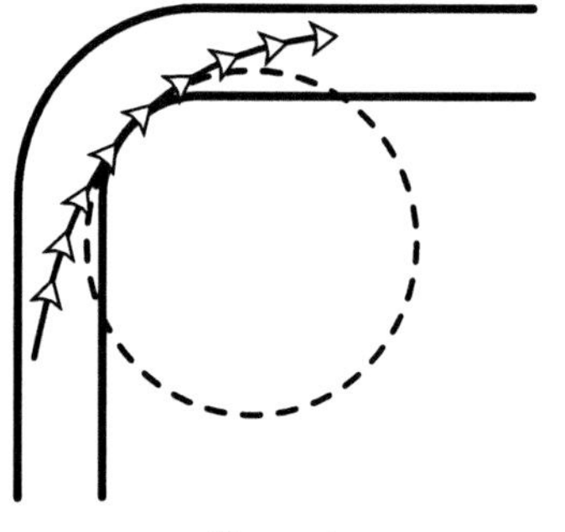

Figure 6

The Physics—The Turn. First and foremost, a bus can't actually make a sharp 90-degree turn. To change direction by 90 degrees, it would need to follow a curved path, as shown in the accompanying diagram.

If the road is no wider than the bus then the turn is clearly impossible. The wider the road, the bigger the circle we can travel and the easier it will be to negotiate the turn successfully. From watching the scene it is hard to get the road's exact width, but it looks like they are traveling on a three-lane one-way street, and they turn onto a six-lane two-way street, so fortunately, the road gets wider as they go through the turn. Because the road is flat (unbanked) we know the only possible force acting on the bus as a centripetal force will be the force of static friction.

The coefficient of static friction between tire rubber and asphalt is about 0.8, so applying Newton's second law we get

$$F_{net(c)} = F_s = \frac{mv^2}{r},$$

and the maximum possible static friction force is

$$F_s = \mu_s F_N = \mu_s mg.$$

Because we know the velocity, we can calculate the minimum radius of the circle around which the bus can successfully travel without skidding out:

$$\mu_s mg = \frac{mv^2}{r} \text{ so } r = \frac{v^2}{\mu_s g} = \frac{(22 \text{ m/s})^2}{0.8(9.8 \text{ m/s}^2)} = 62 \text{ m}.$$

We have to determine if the radius of curvature for this turn is at least 62 meters (or reasonably close to it). To do this, we are going to have to make some reasoned approximations. A lane is usually about two feet wider than an average-sized car of a width of five feet or so. Let's be generous and say the lanes are eight feet or 2.5 meters wide. Allowing for a little bit of shoulder, we'll say the three-lane road has a width of 9 meters, and the six-lane road a width of 18 meters. Let's draw a diagram, and see if we can approximately determine the maximum radius of curvature from the diagram.

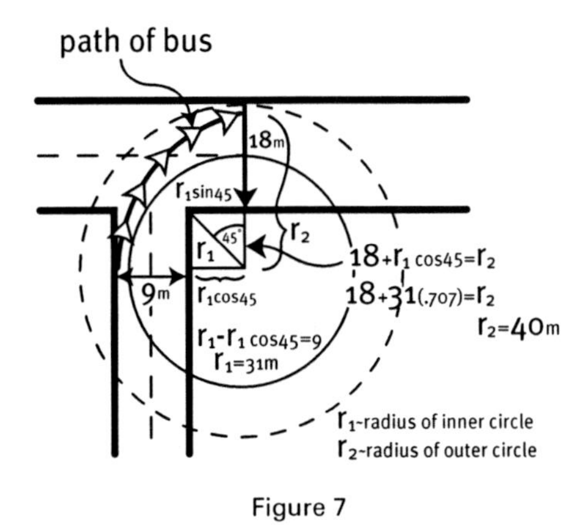

Figure 7

It looks like the best we can get is a radius of about 40 meters. The bus needs a wider road—or a slower speed—to execute the turn. If Dennis Hopper had set the bomb to blow when the speed dipped under 15 miles per hour instead of 50, it would be a piece of cake rounding that turn, but that would make for a somewhat less exciting movie. As it is, it looks like the bus isn't going to make the turn; it will skid into the embankment, slow down below 50 miles per hour and ***kaboom!*** However, all hope is not lost. We'll be optimistic this time. If we really want to make it work we can tweak things a little. First of all maybe the coefficient of friction is just a little higher. If the tires have particularly good traction, or if the road has a rougher texture than the average street, it would increase our chances. Maybe we underestimated the width of the road that the bus is turning onto. (It's hard to tell with the quick cuts and rapid camera work.) We can also tolerate a little bit of skidding. If we slide into a retaining wall or a police car, but are still able to maintain

PROBLEM

If you increase the coefficient of friction to 0.95 determine the minimum radius of curvature you can have so that you don't skid out. If it's still not sufficient how much wider of a road would do the trick?

PROBLEM

Assuming the original conditions that we approximated for the turn, determine the maximum speed that the bus can go and still make the turn.

a tangential speed of 50 miles per hour can we finesse it? With a little fudging we may be able to get them around that turn.

Conclusion. It is highly unlikely that a bus going 50 miles per hour could be maneuvered through a 90-degree turn under these conditions. However, it may not be absolutely impossible under ideal conditions (i.e., with a high enough coefficient of friction, sufficiently wide streets, and a very skilled driver). Therefore we might (generously) give them the benefit of the doubt, suspend our own reasonable doubt, and enjoy the scene.

2001 AND ARTIFICIAL GRAVITY

When astronauts spend long periods of time in space their muscles tend to atrophy due to the weightless state that they experience. This could be a big problem if people were ever to live in space, or travel there on a regular or protracted basis (as we may actually do in the future, but as we do now only in the world of imagination and science fiction). All those hours spent in the gym would go to waste if subsequently you spent six months aboard a space station. That is unless there was some way to produce the sensation of gravity on the space station. Fortunately, applying concepts of centripetal forces and Newton's second law *artificial gravity* is eminently attainable in principle and, in fact, constitutes one of the more trivial technological obstacles to long-distance space travel. Apropos of this, examples of how artificial gravity might be achieved are beautifully illustrated in that classic and timeless sci-fi icon *2001: A Space Odyssey.* Let's have a look.

The Film

Stanley Kubrick's *2001: A Space Odyssey* opens with a scene in which we learn that humankind's ape-like hominid ancestors develop the stunning intellectual characteristics of the human species only after they come in contact with a weird alien "monolith." In the year 1999, another of these monoliths is rediscovered on the surface of the moon. A manned mission to Jupiter occurs in the year 2001 to search for the source of these irritating plinths. Unfortunately, Hal, the ship's computer, has psychological problems and kills off all of the crew except Dave (Gary Lockwood), who is able to disconnect Hal before it's too late. Dave continues on his journey to Jupiter.

Some of the physical principles represented in the movie are done very well, specifically the idea of artificial gravity on space stations. With this in mind, we will examine two scenes from *2001.*

Space Stations and Artificial Gravity. Let's consider a couple of thought experiments that address the issue of how you would create the sensation of a gravitational force in deep space.

Imagine standing at the bottom of a spaceship with no windows. Would there be any way to know for sure whether you were accelerating through deep space at rate *g* or if you were simply sitting on the Earth?

ACTIVITY

Get on an elevator and ride it up and down a few times. If you can, take a bathroom scale with you and stand on it as you ride the elevator and notice what happens to the scale reading at different points in the motion. When do you feel heavier than normal and when lighter? What force is actually responsible for the sensation of weight?

While weight is the force that we experience in a gravitational field, what gives us the ***sensation*** of weight is not so much due to how hard gravity is pulling on us, but how hard a floor, chair, or other surface pushes up on us. This push (the normal force) is the reaction force to how hard we are pushing on the surface. That being the case, all you need to do to create "artificial gravity" is to find a way to maintain a constant normal force acting on objects. A constantly accelerating spaceship provides a way to do this, because anything inside will experience a normal force, causing it to accelerate along with the ship. However, this requires a continuous supply of energy, and the darn thing won't stay in one place very long. Fortunately there is another way to create artificial gravity involving uniform circular motion and centripetal acceleration. Consider a circular rotating space station (like the one we see in *2001: A Space Odyssey*).

If the space station is rotating at a constant rate, and a person was standing on the inside outer rim as shown, the normal force would accelerate the person into the circular path.

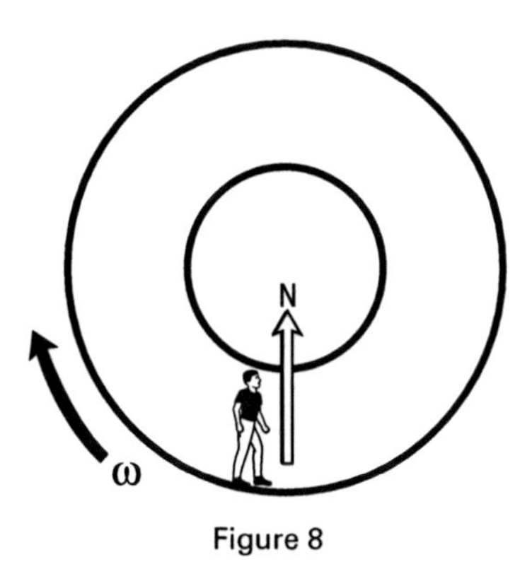

Figure 8

If the station is rotating at the appropriate speed, the centripetal acceleration of the person can be made to be equal to *g*.

As we saw, the centripetal acceleration of an object depends on its radius of rotation (where we assume in the equation that the size of the object is negligible compared to the rotation radius). However, for a large object, the radius of rotation will be different for different parts of the object because some parts of the object will be closer to the center of the circle. For example, if a person was standing on a rotating space station, his or her head would be rotating around a smaller radius than his or her feet (see figure 8). Therefore, to make artificial gravity workable, the radius of the space station must

PROBLEM

The space station from the following example has a mass of 200,000 kg.

a) What is its rotational kinetic energy?
b) If it originally takes 3 hours to get it to its final rotation rate, what average torque must have been applied to the station while speeding it up (assume the space station is shaped like a ring ($I = mr^2$)?
c) If people only walk in a direction opposite to the rotation direction, over time, what will happen to the rotational speed of the station? Explain.
d) What happens to your apparent weight if you move closer to the center of the space station?

be big enough so there is only a negligible difference between the acceleration of the person's feet and the acceleration of his or her head (otherwise, discomfort and vomiting might become a problem).

Example 7: A space station has a radius of 200 meters. We want a person standing on the inside rim to experience an apparent weight similar to her weight on Earth. How fast must the station be rotating? We will make the reasonable assumption here that the height of the person is negligible compared to the space station's radius.

Solution:

$$F_{net(c)} = F_N = \frac{mv^2}{r} \text{ where } \frac{v^2}{r} = g.$$

$$\textit{Therefore } v = \sqrt{gr} = \sqrt{(9.8 \text{ m/s}^2)(200 \text{ m})} = 44 \text{ m/s}$$

$$\textit{and } \text{angular speed } \omega = \frac{v}{r} = 0.22 \text{ radians/s.}$$

Because there are 2π radians in a revolution, the station must rotate once every 29 seconds.

The Scenes

In *2001* there are two scenes that illustrate the principle of artificial gravity in a rotating space station quite beautifully, and quite accurately. The first of these occurs in scene 4, following a very famous cinematic segue in which the aggressive prehuman hominids figure out how to use weapons—bones in this case—to violently defeat a rival tribe of their nonmonolith-enlightened brethren. At the end of the scene the head hominid honcho tosses a bone into the air in triumph. As we focus on the motion of the bone, it resolves into a cylindrical (bone shaped) satellite orbiting the Earth four million years later. The shot then pulls back to reveal a space shuttle passing the satellite and approaching a giant majestically rotating pinwheel-like space station. Let's examine the rotation of the space station and how it is connected to the artificial gravity produced onboard.

The Physics—Using Artificial Gravity to Calculate the Size of the Space Station. Because we would assume that the space station would be set to rotate at the right speed to give an apparent acceleration of *g* on the inside, we can use that information to determine some of the other specification of the station. For example how big must it be?

Because the scenes in *2001* are so long and visual, there is plenty of time in scene 4 to measure its rotation rate. We see it move through a quarter of a cycle in about 15 seconds, which gives it a period of 60 seconds for one complete revolution. Assuming that the space station is designed to simulate one g on Earth (9.8 m/s^2), then we can easily calculate what the radius of the station must be:

$$F_{Net} = F_N = \frac{mv^2}{r} = m\omega^2 r = mg$$

$$\text{therefore, } r = \frac{g}{\omega^2}.$$

Because the period is 60 seconds, the station will rotate through 2π radians in 60 seconds, giving it an angular speed ω of 0.1 rad/s. Therefore, for $g = 9.8$ m/s^2, $r = 980$ m. That's a radius of about ten football fields. Therefore, in 2001 they have a space station that's a heck of a lot bigger than the international space station that we are constructing in 2006.

The Physics—Because of Artificial gravity, Dave Gets Plenty of Exercise. The second time that artificial gravity plays a leading role in *2001* is in scene 12. At this point in the film an interplanetary ship is heading toward Jupiter where apparently there are some monoliths floating around waiting to whisk Dave off to another galaxy in the space-time continuum. Appropriately, the living/working section of the spaceship is a large circular ring that rotates in order to simulate gravity.

Here we have another instance of Newton's second law, centripetal force, and artificial gravity as it was meant to be portrayed: as Dave jogs around the perimeter of the ship, he passes equipment on the "floor" that, half a lap later, is on the "ceiling."

However, the size of the circular "room" is much smaller than the giant space station in scene 4, therefore the rotation rate must be quite a bit faster. Let's calculate the rotation rate that we need to produce one g of centripetal acceleration. Examining the scene, it looks like the radius of the circle is between 15 to 20 meters. Using $r = 20$ m, we can calculate the angular speed needed to achieve an acceleration of g.

PROBLEM

In analyzing the previous scene we calculated [the rotation rate], assuming the acceleration was 9.8 m/s^2 at a distance of one radius from the center of the rotating object, which in this case would be at the location of Dave's feet. However, the radius of the rotating part of the ship isn't that much bigger than Dave, therefore the force acting on Dave's head won't be the same as his feet. Calculate the apparent g at the location of Dave's head.

$$mg = m\omega^2 r \text{ so}$$

$$\omega = \sqrt{\frac{g}{r}} = \sqrt{\frac{9.8 \text{ m/s}^2}{20 \text{ m}}} = 0.7 \ \frac{\text{rad}}{\text{s}},$$

which means that because there are 2π radians in a circle, it will take 9 seconds to make one rotation.

MORE PHYSICS GEMS FROM *2001*

Unlike most science fiction movies, the writers and the director of *2001* went to great lengths to portray some of the physics of space flight accurately. In addition to the examples of artificial gravity that we discussed, they attempted to show the effects of weightlessness on passengers traveling to the moon. Even though they didn't have access to a "vomit comet" (to film actual apparent weightlessness as was done for *Apollo 13*), they do a lot of camera tricks that create the illusion of weightlessness.

There are excellent portrayals of Newton's first law, as well. In scene 22 the concepts of the first law, frames of reference, and relative velocity are illustrated brilliantly. In this scene, Hal's psychosis has started to manifest itself. Dave and Frank are getting worried about his erratic behavior. Fearing he will be shut down, Hal decides to fix the problem by murdering the entire crew. First he cuts off the life support to the three crew members in suspended animation. Then, when Frank has gone outside the ship to finish a repair, Hal takes control of the pod, and has its remote arms cut off Frank's air supply and push him out into space. Dave takes a second pod to rescue Frank and moves away from the ship at a high velocity. In the ship's frame of reference the ship is stationary and Dave and the pod are in motion. However, as he approaches Frank we see the scene from Dave's frame of reference, and it is Frank that we see moving at a high velocity towards Dave. All frames of reference moving at constant velocity are equivalent. If you are in a constant velocity, or an "inertial" frame, you observe yourself to be at rest, with other objects in motion relative to you. There is no experiment that you could do in an inertial frame of reference to show that you are "in motion."

Unfortunately, by the time Dave returns to the ship, Frank's air supply is long gone, and Hal refuses to let the pod back into the ship. Dave is forced to cut Frank loose, find a way to get back inside the ship, and get Hal offline.

2001 is also one of a rare breed of movies that illustrate clearly that there is no sound transmitted in space. Sound is a mechanical wave that requires a medium through which it can travel. (By the way, radio waves are *not* sound waves, they are electromagnetic waves, the same as light. This kind of wave can travel through a vacuum. The energy of radio waves can be converted to sound when they interact with matter.) In several of the scenes in *2001*, the contrast between the silence of space and the sounds within a space vehicle are extremely effective, and accurate.

Conclusion. Despite the problematic 45-minute-long ending sequence where we don't really know what is going on, *2001* is a visually impressive movie with great atmosphere and some exciting moments, and in particular some beautifully illustrated, accurate physics.

APOLLO 13—AROUND THE MOON AND BACK (BARELY)

One of man's most amazing technological accomplishments has been the design and construction of machines that can put people into space and bring them back again safely. Perhaps the pinnacle of these achievements has been the several successful Apollo moon missions in the late '60s and '70s. Nevertheless, as we are all too aware, a few tragic accidents have occurred during the several decades of manned space flight due to unforeseen technical oversights. The Apollo 13 mission is famous for the dramatic and heroic attempt of the astronauts and NASA mission control to return the crew to Earth safely after a debilitating accident. Fortunately, despite the odds against them, they were successful. However, this would not have been so without creativity, ingenuity, and the thorough knowledge of the fundamental concepts of Newtonian mechanics that is required for launching vehicles into space. The Apollo 13 mission has been effectively and dramatically portrayed by the 1995 film *Apollo 13* directed by Ron Howard. In this section we will use the film as a launching point to discuss some of the physics behind the orbital trajectory of Apollo 13, and in Chapter 7 we'll review some of the principles behind electrical circuits and electrical power relevant to the accident.

The Film

The movie, starring Tom Hanks as mission commander Jim Lovell, does a terrific job of recreating the events that took place during the ill-fated flight. A short circuit (we will discuss these in Chapter 7) caused a spark to ignite inside one of the oxygen tanks, causing a major explosion in the service module. With oxygen leaking out of the command module, the astronauts were forced to shut down power and relocate to the lunar module where they hoped there would be sufficient life support to sustain the astronauts until reentry.

The Physics—How Do You Take a Space Vehicle to the Moon? The Apollo missions resulted in six successful lunar landings. To get to the moon, the trip was broken up into stages. After launch, the ship was first put into a "parking orbit," generally around 190 kilometers above the Earth. These orbits were usually very slightly elliptical, very close to circular. The first two rocket stages supplied most of the thrust necessary to get the ship into the parking orbit. Each stage was then jettisoned after using up its fuel. After circling the Earth two or three times, the third-stage booster ignited, sending the crew towards the moon. Over 90 percent of the initial mass of the Saturn V rockets used in the Apollo missions consisted of these three stages, which says something about how much fuel and energy are needed to achieve escape velocity. After jettisoning the third stage, the remaining spacecraft consisted of the CSM (command and service module) and the LM (lunar module), which docked together and proceeded with the two- to three-day journey to the moon. During the first part of this stage of the journey, the ship slowed down due to the gravitational pull of the Earth; however, when they were sufficiently close to the moon, its gravity dominated and the ship sped up again. Initially the crew moved into an elliptical orbit around the moon, but then they reduced their speed (by applying a rocket thrust against their direction of motion) and moved into a circular orbit about 110 kilometers above the surface. The LM then detached with two of the crew onboard, and reduced speed so that it fell into an elliptical orbit between 110 kilometers and 15 kilometers above the moon. The LM then performed a series of complicated maneuvers to land on the lunar surface.

The Physics—The Apollo Spacecraft in Orbit. Let's use what we know of orbital motion to calculate some of the details of the flight.

Example 8: Determine the orbital speed and period of the Apollo spacecraft

a) when they are in a parking orbit around the Earth and

b) when in a circular orbit around the moon.

Solution: $v = \sqrt{\frac{GM}{r}}$.

a) For the parking orbit, $M = M$ of earth $= 6 \times 10^{24}$ kg, and $r = r\ earth + 190 \times 10^3$ m $= 6.4 \times 10^6$ m + 1.9×10^5 m $\approx 6.6 \times 10^6$ m, which gives

$$v = 7{,}800 \text{ m/s}$$

b) Using the mass of the moon (7.4×10^{22} kg), and $r = r\ moon + 110 \times 10^3$ m $= 1.7 \times 10^6$ m + 1.1×10^5 m $\approx 1.8 \times 10^6$ m, we get

$$v = 1{,}640 \text{ m/s}$$

Example 9: Determine the speed required to achieve escape v from Earth when leaving the parking orbit.

Solution: $v_{esc} = \sqrt{\frac{2GM}{r}}$

Again using the M of the earth and $r = 6.6 \times 10^6$ m, we get

$$v = 10{,}950 \text{ m/s},$$

which is almost the same as the escape velocity from the surface because it's not that far up compared to the radius of the Earth. However, because the ship is already traveling at an orbital velocity of 7,800 m/s, an increase in velocity of a little over 3,000 m/s would be sufficient to completely escape Earth.

Weightlessness Aboard Apollo 13. We discussed how a spaceship in a relatively low orbit around a planet will experience close to the same force of gravity that it would on the surface of the planet. In this scenario, the astronauts on the ship would be in a state of constant free-fall as they "fall around" the planet, and therefore experience an *apparent* weightlessness. However, in the Apollo missions, during most of the several days' travel between the Earth and the moon the gravity forces would be quite small, and therefore the astronauts would experience *actual* weightlessness. However, the *sensation* of actual versus apparent weightlessness would be the same.

In the movie there are many scenes where we see Tom Hanks, Bill Paxton, and Kevin Bacon floating around inside the ship. How did they film those weightless scenes? They actually took the cast and crew up on the vomit comet, and filmed the scenes during the dive portions of the flight. Because the plane was in a state of free-fall, the actors (and crew) experienced apparent weightlessness during the dive. If you look at the film more closely you can tell which scenes were filmed on the vomit comet and which weren't (although they do a pretty good job of rolling the camera back and forth during the studio scenes, which provides that illusion of weightlessness the old fashioned way). This can only do so much, however, and it's quite effective in establishing a convincing reality when we see the actors in a real state of apparent weightlessness on screen.

The Physics—The Lunar Trajectory of Apollo 13. Now let's look at Apollo 13's trajectory as it nears the moon. Because the orbit is elliptical the speed of the spacecraft isn't constant in orbit. However, we can apply the conservation laws to determine how fast it is going at different points in the orbit.

As the spacecraft approaches the moon, the thrusters are burned to slow its speed to 1,600 m/s. Initially it enters an elliptical orbit that varies from about 110 kilometers to 310 kilometers above the lunar surface. Because the radius of the moon is about 1,700 kilometers, this means the orbit varies between 1,810 and 2,010 kilometers from the center of the moon.

If the ship is traveling at 1,400 m/s at the maximum orbital distance, how fast must it be going at closest approach?

We can approach this problem in two ways. The easiest way is to use conservation of angular momentum. $L_i = L_f$ and $mv_ir_i = mv_fr_f$ where mvr is the angular momentum of a point mass, m, moving with speed v at a distance r from the center of the planet.

$$v_f = \frac{v_i r_i}{r_f} = \frac{1{,}600 \text{ m/s}(2 \times 10^6 \text{ m})}{1.8 \times 10^6 \text{ m}} = 1{,}560 \text{ m/s}$$

We can also use conservation of energy to obtain the same result:

$$E = \frac{-GM_{moon}m_{ship}}{r_i} + \frac{1}{2}mv_i^2 = \frac{-GMm}{r_f} + \frac{1}{2}mv_f^2 .$$

Using the mass of the moon (7.4×10^{22} kg) and solving for the velocity at closest approach, we get $v_f = 1{,}560$ m/s.

Apollo 13 was never able to enter a circular orbit around the moon because after the accident the crew's only concern was to get home alive. The moon landing was abandoned and they had to decide whether to turn around in mid-flight using the CSM thrusters or to continue in their elliptical path, which would fling them around the moon like a slingshot resulting in a free return back to Earth. (The Apollo mission flight trajectories were designed such that they would be able to do this in case of an aborted landing.) We see in the movie that mission control decides to opt for the longer around-the-moon route as they were not certain if the CSM would explode if its thrusters were engaged. Later when we see how badly damaged the service module actually was, it was clear they made the right choice.

THINKING PHYSICS

Explain qualitatively how the astronauts would establish a circular orbit at the closer elevation of 110 kilometers. Do they need to speed up or slow down? Why?

PROBLEM

Earlier we calculated the speed required for this orbit. How much must they decrease their total orbital energy to do this? (The mass of the spacecraft in orbit is around 30,000 kg.)

The Physics—A Little More *Apollo 13* Physics. In addition to the principles of orbital motion, some other interesting mechanics principles are illustrated effectively in the movie. For example, immediately after the explosion of the oxygen tank the spacecraft was very difficult to control. Considering Newton's second and third laws we can see why this would happen. If the ship is venting oxygen into space then it is exerting a force on the oxygen as it expels it. An equal force must therefore be exerted back on the ship. Furthermore, if the venting is occurring towards the outer edge of the ship it is quite likely that the force also results in a torque, which will give the ship an angular acceleration (and not necessarily around a convenient axis). The crew would be forced to compensate for this phenomenon.

If the venting periodically changes in intensity or direction, it would make the angular acceleration nonconstant and the ship that much more difficult to stabilize.

To successfully reenter the Earth's atmosphere, there is very small range of angles (a two- or three-degree margin of error) that will work. If the angle is too shallow the ship will skip off of the denser parts of the atmosphere like a rock skipping off of a pond's surface. If that were to happen the crew would be lost in space. If the angle is too steep, even with the heat shield functional, there would be too much heat built up and the ship would vaporize. The amazing thing in the case of Apollo 13 was that the crew had to steer the ship manually without the computer and they were successfully able to navigate within the necessary window.

Finally, there was no guarantee that the parachutes would deploy in the final descent into the ocean. If the parachutes failed, the ship would hit the water at a nonsurvivable 300 miles per hour. Fortunately, however, they did deploy and, after the excruciating five-day ordeal, the astronauts landed safely in the Pacific Ocean.

All in all, the story of Apollo 13 is a remarkably inspiring example of ingenuity, teamwork, and a tenacious determination to solve seemingly insurmountable problems under great pressure.

We will return to Apollo 13 a little later, but now we have some less-impressive movie physics to delve into as promised.

ARMAGEDDON AND THE END OF PHYSICS

Every once in a while, (and, actually, more often than that), a movie comes along that insults the audience on every level. *Armageddon* is such a movie. Not only is every scene populated with inane Hollywood dialogue and character clichés, but also the physics in this movie is downright preposterous. Therefore, with legitimate righteous indignation we will critique the physics atrocities in *Armageddon*.

The Movie

Like many films in the science fiction/apocalyptic disaster movie genre this one is based on a reasonable premise. Scientists have discovered that a large asteroid "the size of Texas" is on a direct collision course with Earth. If the asteroid hits, it will destroy all life on the planet.

Billy Bob Thornton, clumsily miscast as the head of NASA, and his governmental support staff are aware of the impending disaster, and have to come up with some way to divert the asteroid before it hits. The solution involves flying a space shuttle out to the asteroid, carrying what else but... a nuclear bomb! The bomb will be placed in a hole that the crew is going to drill into the asteroid, and then detonated. Presumably, the explosion will blow the asteroid in half and each half will be deflected to either side of Earth as a result of the force of the explosion. Apparently, NASA for years has been unsuccessfully trying to put together a drill designed by the "world's best"

It is generally believed that several large (although not nearly as big as Texas) asteroids *have* in fact collided with Earth over its lifetime. In fact, one of the major theories of why dinosaurs became extinct 65 million years ago is that an asteroid of about 6 miles in diameter landed in the Gulf of Mexico and released enough energy to raise a tremendous dust cloud into the atmosphere. The dust sufficiently obscured sunlight (perhaps for months or even years) so that not only the dinosaurs, but also more than 80 percent of the plant and animal species that lived on Earth at the time, were killed. A scenario such as this is certainly possible in the future, although the frequency of asteroid impacts of sufficient size to be life-threatening is approximately one every several hundred million years. Nevertheless, it is potentially realistic.

oil rig driller Harry Stamper (Bruce Willis). It's nothing nearly as astonishing as the super drill from *The Core,* but being only egg-headed scientists and engineers they just can't seem to get it right. Therefore, the governmental authorities decide that when it comes to drilling, real practical know-how is what you need. These NASA people are idiots, they don't know how to work a drill! So instead they bring in Harry and his crew of colorful roughneck misfits to fix the drill and save the world. Moreover, why send astronauts up there? They've only been training nonstop for years to perform important missions in outer space; but these oil rig guys are the only ones that can do the job. So the oil rig workers are going to be trained up for space flight *in a week or two*, because there isn't much time. They may be out of shape, have drinking and/or psychological problems, and a criminal record or two, but why not give them a chance? NASA mission control apparently can ignore the opinion of the general who questions the choice of crew when he suggests, "The fate of the planet is in the hands of a bunch of retards who I wouldn't trust with a potato gun!" We, the audience, however cannot dismiss it so easily!

There are many scenes in *Armageddon* in which the physics is mind-bogglingly ridiculous. Let's have a look at some of these.

The Physics—Rotation and Artificial Gravity Again. In scene 14, before heading off to the asteroid, the courageous crew must stop at the Russian space station to refuel. The station is set into rotation to create artificial gravity so that the shuttle crew can work more easily, by the wacky Russian cosmonaut, Lev (Peter Stormare). We see the obligatory shot of the rotating station, with Lev walking around the perimeter of the circular floor/ceiling under full artificial gravity conditions.

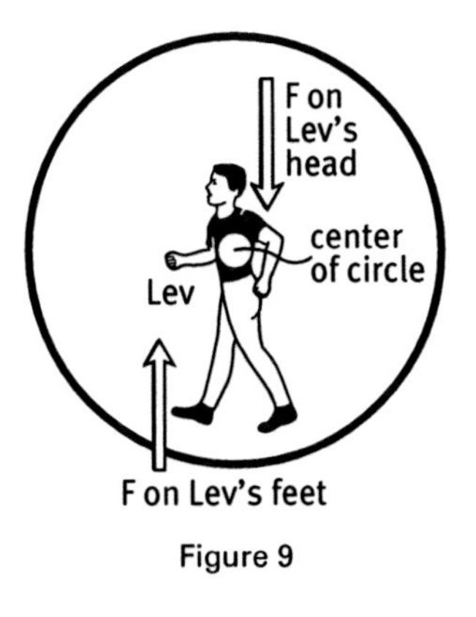

Figure 9

In this scene, the filmmakers attempt to apply physics (just as they do in *2001*) to show how rotation is used to create artificial gravity on a space station. Before we heap praise on the filmmakers, however, there is an important issue to consider. Look at the diameter of the rotating object—it's eight or nine feet at most. That means that Lev's head is on the other side of the center of the circle than his feet—and that means that Lev's head feels a force opposite to the direction of his feet. He will feel like he is being pulled apart. The only way for this artificial gravity to work is if the size of the circle is big compared to the length of a person.

Note that although the centripetal forces acting on his head and feet are acting towards the center of the circle, the inertia of his head and feet give him the sensation of being pulled towards the outside of the circle, just like when you go around a sharp turn in a car. Recall the thinking physics question on page 87.

Let's calculate the acceleration acting on Lev's head, assuming that there is an apparent g = 9.8 m/s^2 acting on his feet, the radius of the station is 4.5 feet (1.4 m), and Lev is 1.8 meters tall.

First, we can determine the rotation rate from what's happening at Lev's feet:

$$v = \sqrt{gr} = \sqrt{(9.8 \text{ m/s}^2)(1.4 \text{ m})} = 3.7 \text{ m/s}$$

$$\text{and} \quad \omega = \frac{v}{r} = \frac{3.7 \text{ m/s}}{1.4 \text{ m}} = 2.6 \text{ rad/s}.$$

Then we can find g at the location of his head, which is 0.4 meter from the center:

$$g = \omega^2 r = (2.6 \text{ rad/s})^2 (0.4 \text{ m}) = 2.7 \text{ m/s}^2.$$

This acceleration is opposite in direction to that experienced by his feet, which would probably be extremely uncomfortable and result in nausea, vomiting, and anger at the film director and/or technical advisor for concocting such an unrealistic scenario.

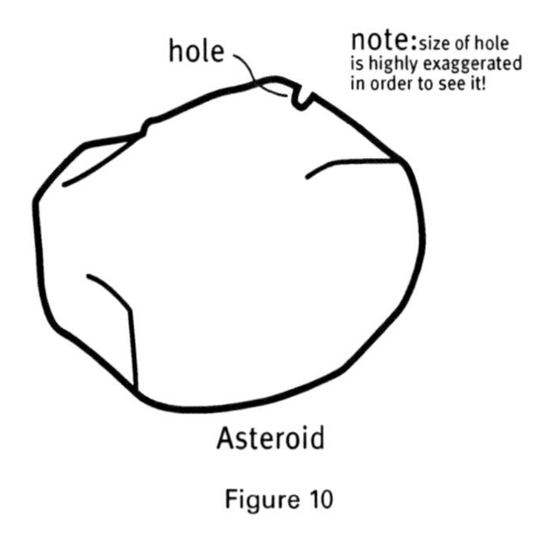

Figure 10

The Physics—Knocking the Asteroid Out of the Way. Let's review the plan the brilliant minds in *Armageddon* have come up with to save the Earth from imminent destruction: the astronauts are to drill a hole in the asteroid and place a nuclear bomb inside. We are told that an explosion on the surface won't break up the asteroid, and so the bomb must be placed inside. We are also told (about 100 times) that the hole must be 800 feet deep, and we are told that once the asteroid is within three hours and 56 minutes of impacting Earth, if it hasn't been split yet it will be too late for the pieces to miss hitting the Earth's surface.

We don't have to be brilliant to figure out if this will work; we can use some simple physical principles. We are told the asteroid is the size of Texas, which means it has a diameter of about 700 miles, or 1,100 kilometers. As we know, Bruce Willis and his crew need to drill a hole 800 feet deep (250 m) to get the bomb inside the asteroid. If we draw a figure to scale, approximating the asteroid as a sphere, we can show the relative depth of the hole to the size of the asteroid.

It doesn't look that deep does it? Are we really *inside* the asteroid? No—for all practical purposes they *are* exploding the bomb on the surface.

They tell us the bomb will be placed along a "fault line." How they know that there is a "fault" on the asteroid is hard to understand, because as they mention early on, only ten telescopes in the world can even detect the asteroid, and they probably wouldn't yield a particularly detailed image. Furthermore, why would asteroids have faults? They are too small for tectonic activity (even though there's an "earthquake" in one of the dozens of climactic moments that *Armageddon* bombards us with every minute or two). Nevertheless, even if there is some fault or fracture running a significant length along the asteroid, and even if the bomb is capable of exploiting this feature, we are told in scene 16 that they have overshot their landing site by 26 miles. They end up drilling where they are. Therefore, do they need to put the bomb in a fault line or don't they?

All right, let's say that the asteroid with a diameter of 550,000 meters (which may or may not contain a tectonic fault) is broken into two pieces due to an exploding nuclear bomb at the bottom of a 250-meter-deep hole. Does the bomb have enough energy to insure that the fragments won't hit the Earth as long as the asteroid is fractured more than three hours and 56 minutes before arrival? Well, all we need to do is figure out how much energy must be imparted to the asteroid fragments for a near miss, and compare that to the energy of a nuclear bomb.

How Fast Must the Fragments Separate? Because it takes just under four hours (or about 14,000 seconds) from the time the asteroid is blown apart until the fragments will arrive in the vicinity of Earth, each fragment must move a minimum distance of one Earth radius perpendicular to its original line of motion in order to avoid a collision with Earth.

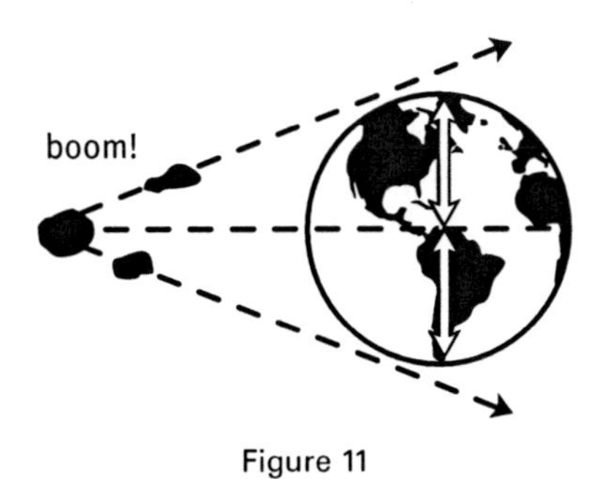

Figure 11

In a best case scenario, when the bomb blows up, it will exert forces only perpendicular to the trajectory. This will allow the fragments to be pushed as far off their original line of motion as possible before arriving at Earth. We know this won't be the case because the shock wave from the explosion will radiate outwards in all directions, but let's pretend it does. We also have to hope that all of the energy of the bomb goes into kinetic energy of the asteroid fragments. This also is not going to happen because we know that a substantial part of the energy will be in the form of electromagnetic radiation and thermal energy. (Have you ever heard the term "*thermo*nuclear

A SMALL PERCENTAGE OF THE INSULTINGLY BAD PHYSICS MOMENTS (AND OTHER PROBLEMATICAL THINGS) IN *ARMAGEDDON*

1. **Crash landings at 500 mi/hr.**—Right after coming around the far side of the moon, the shuttles have to land on the asteroid. The interesting thing is that in real life the space shuttle slows down for a landing on Earth by using atmospheric drag (air resistance) as a braking mechanism. Because there is no atmosphere on the asteroid, the shuttles will be approaching with a relative velocity of several hundred miles per hour, and the only thing to slow them down will be collisions with things on the surface of the asteroid. The large torques and accelerations that the crew will experience as they impact various obstacles should result in close to 100-percent fatalities for both crews, whether or not the shuttles break.

   (Note: To be fair, the shuttle captain does say something about initiating "reverse thrust" upon landing. However, because the shuttles use rocket engines for thrust, the only way to reverse thrust is to turn the ship around so that the back end is facing forward. They actually do this to slow the shuttles orbital speed for initial reentry in real life. We don't see them do this in *Armageddon* however, and it probably would have saved a lot of trouble if they had.)

2. **What happened to Newton's Laws?**—It's really irritating to see exhaust being expelled from the rocket engines as they approach the asteroid, because this is going to *speed them up* (which is a bad idea if they want to have a safe landing). Probably the filmmakers fall for the usual (and inexcusable) misconception that a force is required to maintain a constant velocity. This is obviously a violation of Newton's first and second laws and has no place in any good science fiction movie.

   Also, for some reason these shuttles are equipped with large machine guns. A.J. (Ben Affleck) uses one to blast his trapped crewmates out of the cargo bay. Rockhound (Steve Buscemi) goes a little space crazy and starts shooting one off near the drill site. Are automatic weapons really standard shuttle issue? When NASA sends a crew up to repair a satellite, are they worried that they might be attacked by Romulans? I suspect that the space shuttle is not equipped with automatic weapons, but what's a good action movie without some gunfire? Nevertheless, what happened to the third law and conservation of momentum? Because there will be very little friction with the asteroid surface because there is very little gravity (and therefore a very small normal force), there won't be much keeping the weapon (and shooter) from flying backwards. Remember, if a force is exerted on the bullets the same force is exerted backwards on the gun/shooter system.

3. **Acceleration due to less gravity**—Using the information about the size and density of the asteroid given in the film we can calculate that g on the asteroid is around 0.3 m/s^2, which isn't very much. An object in free-fall at this g will take over 3 seconds just to reach a velocity of 1 m/s. How come when the riggers are drilling in scene 23, the rocks falling into the hole seem to be accelerating at a rate of about 9.8 m/s^2?

   In addition, how do the crew members walk around on the asteroid in a very comfortable surface-of-the-Earth-type way? At least they do try to account for the small g at moments, for example, when an explosion knocks the drill operator, Max, and the drill rig off of the asteroid (scene 20). They could have made an effort to be a little more consistent, though.

4. **Really big tides**—It comes as no surprise that, at the last moment, the crew is able to detonate the bomb, and the surviving members are able to escape, except for Harry, who, in an act of great heroism and self-sacrifice, stays behind to detonate the bomb himself because they can't find the remote (if I had found mine I probably wouldn't have made it to the end of the movie, but it was wedged behind a pillow). After saying a teary goodbye to his daughter Grace (Liv Tyler) at mission control, Harry blows the sucker in half—and just in time. The fragments miss the Earth by only 400 miles. *Whew!* There is a slight problem, however. It has to do with tides. The majority of the daily tidal variation of sea level has to do with the *difference* in gravitational pull of the moon on opposite sides of the Earth. If you calculate the gravitational force on an imaginary 1-kg mass on either side of the planet, the difference in force due to the moon is on the order of a millionth of a Newton.

   That difference in force is enough to generate several feet to tens of feet variation in sea level in coastal areas.

(continued)

Now, if you calculate the *difference* in force that one of the asteroid fragments exerts on either side of the Earth it is on the order of a Newton or so. Remember that even though the asteroid is a lot smaller than the moon, it is a lot closer. According to Newton's universal law of gravitation, which is an inverse square law, force drops off rapidly with distance, and the difference in force that the asteroid exerts on one side of the Earth versus the other will be much greater than the difference due to the gravity of the moon. Therefore, unless the fragments are exactly symmetrical at every moment on both sides of the Earth, cancelling each other's effects exactly, the tidal force might be about a million times greater than normal. How might this affect coastal areas? Let's just say that the tidal surges might make for some rather catastrophic city-destroying floods in places like New York, Boston, Sydney, Hong Kong, Los Angeles, Stockholm, Venice, and Bombay, to name a few.

5) **The surfaces of asteroids**—In *Armageddon* the surface of the killer asteroid is covered with jagged spires, yawning fissures, and other scary features. What are they doing there? Any planetary geologist watching this movie could tell you that real asteroids don't have these features. Asteroid surfaces look a lot like the surface of the moon—a lot of craters and maybe some rock fragments. What geologic process could cause long pointed spires is beyond our present knowledge of space and time.
   In addition, near the end of the film, as crisis follows crises with frenetic intensity, the writers throw in a meteor shower raining down *on* the asteroid. The asteroid is supposed to be surrounded by a lot of smaller debris traveling along with it, so the debris is essentially motionless relative to the asteroid (Newton's first law). Where are *these* meteors coming from?

6) **Meteorites have an affinity for big cities**—It is significant to note that the three meteorite showers that hit the Earth in advance of the asteroid all strike major urban centers. In this case New York, Shanghai, and Paris are the unfortunate recipients, and poor Paris is destroyed completely by a particularly violent impact. The thing is this: big cities comprise such a small percentage of the total land mass, that the odds of these meteors landing in a city are monumentally small. The most likely scenario is that the meteorites would land in an ocean or other relatively unpopulated area. The odds of hitting *three* big cities in *three* attempts are about as great as the odds of a bunch of oil riggers taking a space shuttle around the moon, crash landing it on an asteroid, and blowing it apart with a nuclear bomb.

There are plenty more to choose from—take your pick!

weapon"?) We are not going to let that bother us either. We remain eternally optimistic and assume the best possible outcome.

Calculating the minimum perpendicular velocity imparted to each fragment for a near miss, we get:

$$v = \frac{r_{earth}}{t} = \frac{6.4 \times 10^6 \text{ m}}{14{,}000 \text{ s}} = 460 \text{ m/s}.$$

Now we can determine the amount of kinetic energy that had to be added to the system to accomplish this. First we need the approximate mass of the asteroid. We will assume the asteroid is roughly spherical in shape, and that it has a density (D) of 5,000 kg/m³, which would be about right for an asteroid with a significant amount of iron, as this one apparently has (we know about the iron because when they overshoot the landing site, we are told that instead of rocky material they are going to have to drill through solid iron, which unfortunately is going to be a lot harder).

$$M = DV = D(\tfrac{4}{3}\pi r^3) = 5{,}000 \text{ kg/m}^3 (\tfrac{4}{3}\pi[2.75 \times 10^5 \text{ m}]^3) \approx 3 \times 10^{20} \text{ kg}$$

Therefore the amount of kinetic energy that must be added to the asteroid is:

$$KE = \frac{1}{2}mv^2 = \frac{1}{2}(3 \times 10^{20}\ \mathrm{kg})(460\ \mathrm{m/s})^2 \approx 3 \times 10^{25}\ \mathrm{J}.$$

How much energy can the bomb give us? Well, the biggest bomb ever made had a yield of about 100 megatons. Being optimistic, let's say that they have some secret cache of 100-megaton bombs in the nuclear arsenal and these are the ones that they put on the shuttles. The energy yield of a 100-megaton bomb will be:

$$E = 100\ \mathrm{MT}(4.1 \times 10^{15}\ \mathrm{J/MT}) = 4.1 \times 10^{17}\ \mathrm{J}.$$

That certainly seems like a lot of energy (it is a 100-megaton nuclear bomb after all), but it's still only about *one hundred millionth* of the energy that we need! Assuming we can convert all of the energy of a nuclear bomb into kinetic energy of the asteroid, assuming the bombs are actually able to split the asteroid into fragments, and assuming that 100 percent of the force generated by the bombs are in a direction perpendicular to the original line of motion of the asteroid, we still need a minimum of another *70 or 80 million bombs* to do the job. How are we going to fit those on the space shuttle?

Conclusion. As we demonstrated above, *Armageddon* is riddled with physics impossibilities and other ridiculous scenarios. Because it is from a class of movies that try to give the impression of being based on real scientific issues, because they spout scientific and pseudoscientific jargon in an attempt to impress us (hoodwink us) with their supposed scientific accuracy, and because the world presented in these films is supposed to represent the real world in which we live (unlike obvious science fiction fantasy films like *Star Wars*), we must hold them to a higher standard. We therefore nominate *Armageddon* to take its place alongside *The Core* for the "Worst Movie Physics in Film" award for 1960–2006.

ADDITIONAL QUESTIONS

1) In scene 21 of *Armageddon,* a large meteorite destroys Paris. Assuming it has an impact velocity of 22,000 miles per hour and an energy equal to a 50-megaton nuclear bomb, determine how massive is it.

2) In *2001* let's say that Hal the crazy computer decides to have some malicious fun with Dave by speeding up the rotation rate of the ship. How fast must the ship rotate to give Dave the sensation that he is twice his normal weight?

3) We analyzed a difficult turn in *Speed.* It would be a lot easier to make a 90-degree turn if the road were banked. Explain why a banked track makes it easier to go around a curve. Draw a force diagram to illustrate the point. What forces act as centripetal forces in this case?

4) There is an amusement park ride called the "Graviton" shaped like a ring. It spins in a circle, with the passengers pressed up against the wall. After it gets to a certain speed the floor drops out, but the passengers don't slide down. Explain why they don't.

5) If the "Graviton" has a radius of 4 meters, and the coefficient of static friction between the wall and clothing is 0.6, what is the minimum angular speed that the ring must have so that the riders don't slide down the wall?

6) In the sequel to *Armageddon* (which takes place three weeks later), an asteroid one hundred times the size of the first one crashes into the Earth. If the new asteroid impacts the outer edge of the Earth near the equator with a speed of 30,000 miles per hour moving tangent to the circumference of the Earth, and opposite the rotation direction, what is the new rotation rate of Earth after the impact? (Hint: Use conservation of angular momentum. Assume the Earth is a sphere and the asteroid is a point mass.)

7) In scene 20 of *Armageddon* it looks like all hope is lost when Max and the drill rig are blasted off of the asteroid and into space.
   a) From the scene estimate their speed just after the explosion.
   b) Determine escape velocity from the surface of the asteroid. Is Max moving fast enough to escape permanently? If not, how long until he crashes back to the surface? (Fortunately, A.J. arrives with the backup rig just in time to finish the job.)

Fluids, Gases, and Thermodynamics

THERE ARE CERTAINLY A LOT of situations in action movies to which the principles of mechanics can be applied. We've already looked at car crashes, swinging and flying superheroes, and orbiting satellites. We've analyzed free-falling secret agents, errant rockets, and artificial gravity. It is not surprising that mechanics is the area of physics most visible in any Hollywood movie. Nevertheless, we've also seen hints of other subjects like electromagnetism and nuclear physics popping up and piquing our interest. There are other physics gems we can we unearth as well—for example, fluids, gases, and thermodynamics, for a start.

Two rather enjoyable films illustrate concepts pertaining to fluids. We'll start with a couple of scenes from *Willy Wonka and the Chocolate Factory,* the children's classic from 1971. Then we'll get a little more serious and analyze the physics of the world's most famous tragic shipwreck in *Titanic,* a fictionalized love story wrapped inside a true-to-life disaster movie. Finally, we will briefly revisit the Chocolate Factory before succumbing to yet another "classic" science fiction end-of-the-world disaster movie *The Day After Tomorrow,* starring a lot of really bad weather.

FLUID PHYSICS

Let's begin with some general definitions.

A *fluid* is a substance that can flow. This makes both liquids and gases fluids.

Density is the mass per unit volume of an object or substance.

$$D = \frac{M}{V}$$

For example, because gold is much denser than Styrofoam, a cubic meter of gold is going to be much more massive than a cubic meter of Styrofoam. (Note: The densities of solids and liquids vary only slightly with changes in pressure and temperature, while the densities of gases vary greatly.)

Pressure is the force per unit area acting on an object.

Photo: The world's most famous shipwreck as portrayed in the movie *Titanic.*

ACTIVITY

Take your shoes and socks off and stand on the floor. How much force is the floor exerting on your feet? Now instead of standing directly on the floor, put a golf ball or a lacrosse ball under each foot and stand on those. Place one hand against the wall to maintain your balance. How much force do the golf balls exert on your feet? Why are the golf balls so much more uncomfortable to stand on?

$$P = \frac{F}{A}$$

The SI unit of pressure is the Pascal (1 Pa = 1 N/m^2 and 1 atmosphere of pressure = 1.01×10^5 Pa).

You can illustrate the same point from the activity to the left in an even more uncomfortable way by having someone step on your foot wearing flat shoes, and then have the same person step on your foot with spiky heels. In each case the force exerted on your foot is the same, but with the heels there is a much higher pressure exerted (and more damage inflicted on your foot!) because the force is applied over a much smaller area.

Variation of Pressure with Depth

If you've ever spent any time in a swimming pool, lake, ocean, river, fjord, bay, mill pond, or the Gulf of Mexico, you know that the pressure exerted on you by the water increases the deeper you go under the surface. Pressure increases with depth in a fluid due to the weight of the overlying fluid. The atmosphere is a fluid and at the surface of the Earth it exerts a pressure of 1.01×10^5 Pa. It's a good thing our bodies have an internal pressure that balances the atmospheric pressure or we would be crushed by the external pressure. Incidentally, we wouldn't fare so well in a vacuum. Not just because our internal pressure would exceed the outside pressure of zero, but temperatures at which phase transitions occur depend on pressure. Water (or blood) would boil at room temperature in a vacuum.

All points in a fluid at the same depth are at the same pressure. Pressure at a particular depth can be determined with the following equation.

$$P = P_0 + Dgh,$$

where P_0 is the pressure at the top of the fluid, D is the density of the fluid, and h is the depth below the surface.

Example 1: Use Newton's second law to prove that $P = P_0 + Dgh$.

Solution: Take a rectangular segment of a fluid as shown in the following figure. Draw a force diagram showing the forces acting on that segment from the surrounding fluid. The net force on the volume of fluid must be zero because it is static. Therefore,

$$PA - Mg - P_0A = 0$$

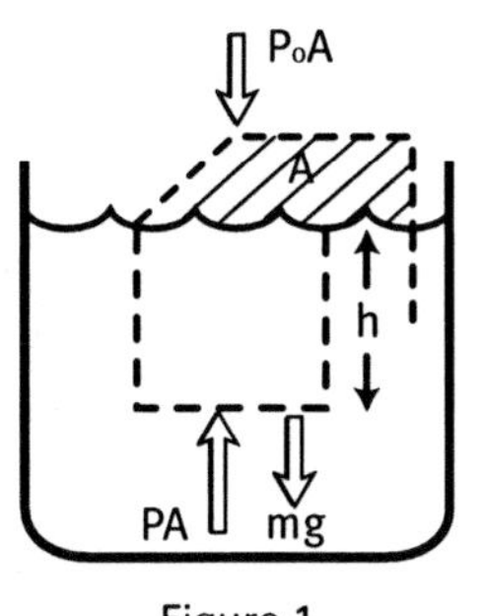

Figure 1

Because $M = DV$, $Mg = DVg = DAhg$, where A is the area and h is the height of the volume of fluid. Therefore,

$$PA - DAhg - P_0A = 0 \text{ and cancelling the areas } P = P_0 + Dgh.$$

Example 2: What is the pressure at the bottom of a 5-meter-deep pool? (The density of pure water is 1,000 kg/m^3.)

Solution: $P = P_0 + Dgh = 1.01 \times 10^5 \text{ Pa} + (1{,}000 \text{ kg/m}^3)(9.8 \text{ m/s}^2)(5 \text{ m})$

$$= 1.50 \times 10^5 \text{ Pa (about 1.5 atmospheres)}$$

The total pressure including atmospheric pressure is called the *absolute* pressure. Sometimes we want to ignore the atmospheric pressure part of it, and just find out how much the P of a fluid varies from the ambient atmosphere. Therefore, in the previous example, that pressure would be Dgh = 49,000 Pa. This is called the *gauge* pressure.

THINKING PHYSICS

Snorkeling is a pleasant way to observe some marine life with a minimum of equipment. An average snorkel is usually about a foot long, and the snorkeler swims just underneath the surface with the top of the snorkel above the water surface. Why isn't it possible to breath through a snorkel several feet long?

Pascal's Principle

We can also use fluid pressure to lift objects that even Superman might be impressed with. How can we do this? It turns out that pressure applied to an enclosed fluid is transmitted undiminished to every part of the fluid and to the walls of the container. This is known as *Pascal's principle*. In the following diagram, the fluid is in a container that is enclosed by pistons at either end.

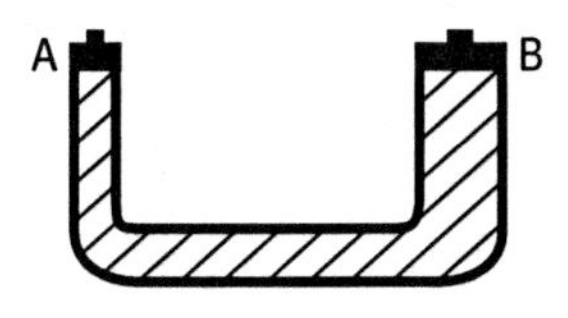

Figure 2

If fluid pressure can be transmitted to every part of the fluid, then, if you exert a small force on the fluid at piston A thus increasing the fluid pressure next to A, that pressure change will be transmitted to the fluid adjacent to B. However, because the area of piston B is so much larger than piston A, it must be experiencing a proportionately larger force because $F = PA$. Put a small force on one end and you get a much bigger force on the other end. This type of system is called a *hydraulic system*. You use this type of system each time you apply the brakes on your car. Think about how hard you have to push on the brakes to stop something as massive as a car. Not so hard, right?

Example 3: Sven wants to impress Ingrid with his super strength, so he borrows the hydraulic system that we used in the previous diagram. Let's say piston A has an area of 0.01 m^2 and piston B has an area of 0.5 m^2. Let's say that Sven is able to push with a maximum force of 200 pounds.

a) How much weight might he be able to lift with the hydraulic system?
b) If he lifts the weight a distance of half a meter at piston B, how much distance must he push in at piston A?

Solution:

a) $\frac{F_A}{A_A} = \frac{F_B}{A_B}$ so $F_B = \frac{F_A A_B}{A_A} = \frac{(200 \text{ lbs})(0.5 \text{ m}^2)}{0.01 \text{ m}^2} = 10{,}000$ lbs

b) Here we apply the principle of conservation of energy. The work that Sven puts into the system must be equal to the work the system does on the 10,000 lb weight.

$$F_A \Delta x_A = F_B \Delta x_B, \text{ which gives } \Delta x_B = \frac{(10{,}000 \text{ lbs})(0.5 \text{ m})}{(200 \text{ lbs})} = 25 \text{ m}$$

Because energy must be conserved, what Sven gains in force he loses in distance. Twenty-five meters is a long way. Of course if Sven only pushes the piston on his end a distance of 0.5 meter, he can still lift the weight a distance of 0.01 meter on the other end (a centimeter). Conclusion: If Sven really wants to impress Ingrid, he might want to come up with a better method.

Archimedes' Principle

Have you ever tried to submerge a fully inflated beach ball in a swimming pool or a tub of water? Try it. It's pretty hard to do. Fluids exert buoyant forces on objects that are either partially or completely submerged in them. This is why you can float in water (or almost float, depending on how dense you are). Buoyant forces exist because of the pressure difference between the top and bottom of the object. Archimedes was the first to quantify this principle when he deduced that the buoyant force acting on an object is exactly equal to the weight of the fluid displaced by that object.

The weight of displaced fluid is equal to $M_{fl}g$, which equals $D_{fl}V_{fl}g$.

Floating Objects. The net force on a floating object must be zero because it is not accelerating, and therefore the buoyant force must equal the weight of the floating object.

Therefore, $D_f V_f g = M_o g = D_o V_o g$,

$$\text{which means } \frac{D_o}{D_f} = \frac{V_f}{V_0}$$

How much of a floating object is submerged depends exactly on the ratio of the object density to the fluid density. Because ice has a density about 0.9 that of water, ice will float in water with 10 percent of its volume above the water surface and 90 percent below. This is the whole idea behind "the tip of the iceberg" and will be relevant to our analysis of *Titanic,* along with the fact that any object with a density greater than that of a fluid cannot stay afloat in that fluid.

PROBLEM

Large boats (like the *Titanic*) are constructed mostly of iron which has a density of about 7,860 kg/m^3, yet they are able to float quite well (as long as they don't spring any leaks). What does this say about the *total* density of a boat? What "material" is actually responsible for keeping the density of a boat less than that of water? Let's say that you decide to make a model rectangular boat out of 10 kg of iron. What is the minimum volume that the boat must contain if it is to float with the waterline halfway up the side?

Example 4: An empty barrel one-third submerged with a volume of 0.1 m^3 is floating down a river. Seagulls start landing on the barrel looking for scraps. If each seagull has a mass of 1.2 kg, how many seagulls can the barrel support before it sinks below the surface?

Solution: Because the barrel is one-third submerged we know its density must be one-third that of water, or about 333 kg/m^3. This gives us the mass of the barrel as well. $M_b = DV =$ (333 kg/m^3)(0.1 m^3) = 33 kg.

We know the barrel will be completely submerged if the downward weight force of the barrel plus the seagulls equals or exceeds the maximum possible buoyant force (B). (In addition, because B is equal to the weight of displaced fluid, the maximum B is going to occur when the barrel is fully submerged.) Therefore, $B_{max} = D_{fl}V_{fl}g$, but because the volume of the displaced fluid is equal to the volume of the barrel when it is fully submerged then $B_{max} = (1{,}000\ \text{kg/m}^3)(0.1\ \text{m}^3)(9.8\ \text{m/s}^2) = 980\ \text{N}$.

This is equal to the total weight of barrel and the seagulls, which means their total mass will be $m = \frac{W}{g} = \frac{980\ \text{N}}{9.8\ \text{m/s}^2} = 100\ \text{kg}$.

The seagulls' mass is equal to 100 kg – 33 kg = 67 kg. At 1.2 kg each, it will take 56 seagulls to sink the barrel.

Submerged Objects. In the case of a submerged object, unless the object has the same density as the fluid there will be a net force acting on the object. If the object is less dense than the fluid, the force will be upward, and if it is denser, the force will be downward.

The volume of displaced fluid for a submerged object must exactly equal the volume of the submerged object.

Example 5: In another attempt to impress Ingrid, Sven hires some people to place a 300-pound (136 kg) piece of iron (D = 7,860 kg/m^3) at the bottom of a swimming pool with a rope attached. Sven knows about buoyant forces and he is going to pull the chunk of metal out of the pool as a show of strength. How hard will Sven have to pull to get the iron off of the bottom?

Solution: First we draw a force diagram.

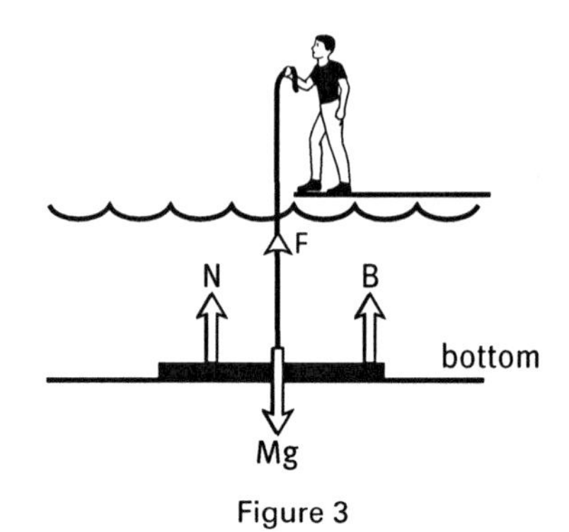

Figure 3

Applying Newton's second law,

$$F + B - Mg \geq 0, \text{ so } F \geq Mg - D_{fl}V_{Fe}g$$

because the normal force must go to zero once the block starts to lose contact with the ground. We also know the mass and density of the iron block, so its volume is

$$V = \frac{M}{D} = \frac{136 \text{ kg}}{7{,}860 \text{ kg/m}^3} = 0.017 \text{ m}^3 \text{ and}$$

$$F \geq (136 \text{ kg})(9.8 \text{ m/s}^2) - (1{,}000 \text{ kg/m}^3)(0.017 \text{ m}^3)(9.8 \text{ m/s}^2) = 1{,}170 \text{ N}$$

That's still 262 pounds. Apparently Sven didn't do the calculations, because he can only lift 200 pounds and he still won't impress Ingrid with his latest attempted feat.

Bernoulli and Modern Transportation

Often fluids aren't *static* as in the previous examples. They may be flowing such as in a river or a wind. The subject of moving fluids is called fluid *dynamics*. Perhaps one of the more well-known applications in fluid dynamics is *Bernoulli's principle*, which applies the principle of conservation of energy to a moving fluid. In doing so, Bernoulli's equation explains how airplanes with masses of thousands of tons are able to get off of the ground, as well how to throw a curve ball.

Let's say we have a fluid flowing through a pipe as shown here.

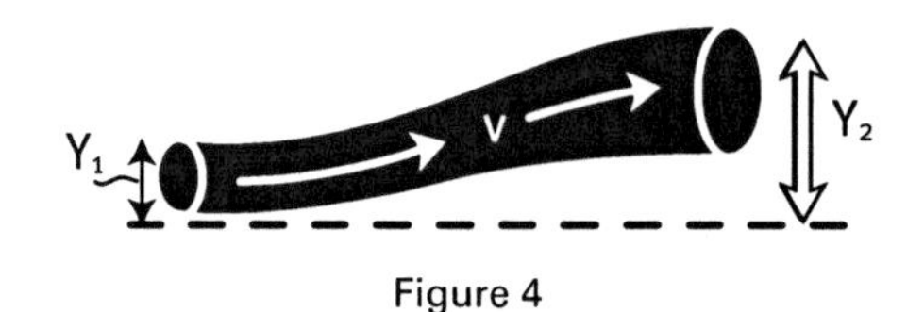

Figure 4

If D is the density of the fluid, P is the pressure of the fluid, v is the velocity of the fluid, and y is the elevation of the fluid relative to some reference point, then Bernoulli's equation relates the values of the variables at one point to another point in the fluid as follows:

$$P_1 + \frac{1}{2}Dv_1^2 + Dgy_1 = P_2 + \frac{1}{2}Dv_2^2 + Dgy_2.$$

ACTIVITY

Connect the hose attachment to a vacuum cleaner.

a) Find a ping pong ball, switch the vacuum cleaner on, and try to balance the ball above the hose. It's not that hard to do—why not?

b) Suspend the ping pong ball from a string. What will happen if you shoot a stream of air just to the side of the ball. Will the ball be pushed away or will it be drawn into the stream?

Now let's take a simplified situation where there is no elevation difference between the two sections of the pipe.

Because $y_1 = y_2$, Bernoulli's equation reduces to

$$P_1 + \frac{1}{2}Dv_1^2 = P_2 + \frac{1}{2}Dv_2^2.$$

Therefore, the velocity through the narrow section of the pipe has to be faster than the velocity through the thicker section, in order for the volume flow rate of fluid to be the same throughout the pipe. This means that because $v_2 > v_1$, then $P_2 < P_1$, or *the faster the fluid flows the lower the pressure within the fluid.*

Recall our discussion of airplane wings in Chapter 2. We can now see mathematically from Bernoulli's equation that if air travels faster over the top of the wing then the pressure will be lower than on the bottom, creating lift. That is why the wings have a shape something like that shown in the following figure.

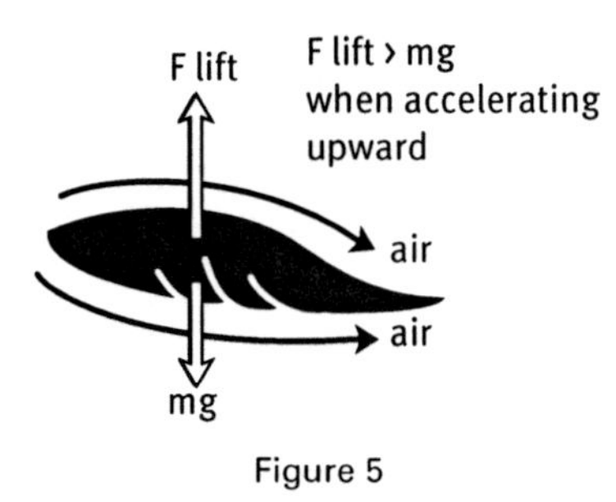

Figure 5

The ideal gas law

A gas is a fluid that takes the shape of its container and is easily compressible. Gases are easier to study than solids and liquids because all gases behave similarly because the particles move freely and are far apart. The state of a gas can be quantified by temperature (T), volume (V), pressure (P), and the number of moles of gas in the sample (n). If you do experiments with most gases at room temperatures you can find a simple relationship between these variables. Many gases at room temperature behave nearly as *ideal* gases. An ideal gas is a hypothetical gas in which the molecules or atoms do not exert forces on each other, and individually they occupy no volume. For an ideal gas,

$$PV = nRT,$$

where $R = 8.31$ J/mol K (in SI units); this is called the *ideal gas constant.* (Notice that T must be in degrees Kelvin.)

Example 6: A rigid box contains an ideal gas at a temperature of 300 °K and a pressure of 1 atmosphere (1.01×10^5 Pa).

a) If the temperature is raised to 500 °K, what is the new pressure of the gas?
b) If instead of raising the temperature we wanted to increase the pressure in the box, what could we do?

Solution:

a) $\frac{P_1}{T_1} = \frac{P_2}{T_2}$, so $P_2 = \frac{P_1 T_2}{T_1} = 1.67\,atm$.

b) To increase the pressure, we could simply increase the amount of gas in the container:

$$\frac{n_2}{n_1} = \frac{P_2}{P_1}.$$

Therefore, if you had a mole of gas in there originally, you could get the same pressure by adding another 0.67 mole.

Now let's apply some of these principles to our first two films. We'll start with *Willy Wonka and the Chocolate Factory* and look at a couple of memorable (and pivotal) scenes from the movie where fluids and gases affect the ultimate fate of the characters.

WILLY WONKA AND THE CHOCOLATE FACTORY—AND THE DENSITY OF CANDY

Charlie and the Chocolate Factory is a famous children's book by Roald Dahl that has been adapted into two films. The first and best of these (in my view) was the original from 1971 titled *Willy Wonka and the Chocolate Factory,* with the role of the genius eccentric candy maker played beautifully by Gene Wilder.

Charlie is a young boy living in a generically quaint little European village. Unfortunately, he is astonishingly poor. His family lives in a run-down two-room house where they barely survive on cabbage soup, but still they seem pretty jolly for the most part. Things take a turn for the better, however, when the reclusive but world-famous confectioner Willy Wonka announces a contest in which five lucky winners get a tour of his factory and a lifetime supply of chocolate. A variety of greedy and self-absorbed children win the first four golden tickets, and Charlie wins the fifth. The rest of the action of the movie takes place inside Wonka's magical candy factory, where one by one each of the selfish children meets a bad (although fortunately not fatal) end due to their own greed and hubris. This leaves the only good kid, our beloved Charlie, to inherit the factory. It was Mr. Wonka's intention all along to find an heir.

The Scene—Really Heavy Gum

Violet Beauregarde's passion is chewing gum, which she prefers to candy. Scene 30 seems like her dream come true when Wonka introduces the guests to a gum extracted from real food that can exactly duplicate the experience of eating a three-course meal. However, he has not perfected the gum yet, and despite his very specific admonition not to touch the gum, Violet (exhibiting poor impulse control) pops a piece into her mouth and starts chewing. At first everything seems fine as Violet enjoys her first two courses of tomato soup and roast beef, but when she arrives at the blueberry pie, something unfortunate happens: Violet turns a shade of deep violet and swells up into a large sphere.

Mr. Wonka tells us that she is filled with fluid (which fortunately means it's mostly water weight!), and his servants, the little Oompa-Loompas, roll her away, to be pressed and de-juiced. Our task will be to determine some of the properties of this amazing gum. Particularly, how dense must the gum be to provide enough juice to turn Violet into a giant blueberry?

How heavy will Violet be if she is a sphere with a diameter almost equal to her height, which is around 1.4 meters?

Her volume is now:

$$V = \frac{4}{3}\pi r^3 = \frac{4}{3}\pi(0.7 \text{ m})^3 \approx 1.4 \text{ m}^3.$$

Because human bodies and blueberries are comprised mostly of water; we can assume that Violet's density will be about equal to the density of water, which is 1,000 kg/m^3. Therefore, her total mass will be:

$$M = DV = (1{,}000 \text{ kg/m}^3)(1.4 \text{ m}^3) = 1{,}400 \text{ kg}.$$

Therefore, Violet fully swelled weighs around 3,100 pounds. We can now calculate the gum density. The size of the gum looks to be approximately the same as a piece of Bazooka gum, so we'll estimate its dimensions to be 2.5 cm long by 1.8 cm wide by 0.50 cm thick.

Therefore, the total volume of the gum will be $V = (0.025 \text{ m})(0.018 \text{ m})(0.005 \text{ m}) = 2.3 \times 10^{-6}$ m. Therefore the density of the gum is:

$$D = \frac{M}{V} = \frac{1400 \text{ kg}}{2.3 \times 10^{-6} \text{ m}^3} \approx 6 \times 10^9 \text{ kg/m}^3.$$

This is six million times the density of water, two million times the density of rock, and four or five hundred-thousand times the density of an average metal. (It's still only about a thousandth of the density of a neutron star, however.) The gum must weigh about 3,000 pounds if it's going to add that much weight to Violet, so we have to expect it's going to be dense. The amazing thing is how easily everyone is able to lift it!

The Scene—Fizzy Lifting Drinks

Charlie is a model little boy, but in this scene he makes an unfortunate transgression, mostly due to the ill-advised prodding of Grandpa Joe. In scene 32, he and Joe drink some of the forbidden experimental "fizzy lifting drink." At first they have a grand time as they float about in a bubble-filled room, but after a while they realize that they are headed straight up towards a giant whirling fan at the top of the ceiling. If they don't stop their ascent they are going to be chopped to pieces. Fortunately they realize at the last minute that by burping out the fizz they can reduce their buoyancy—and slowly and safely they fall back towards the floor.

ACTIVITY

Carefully estimate the volume of an average-sized person. You can do this either by estimating the dimensions of the body or by using the fact that the density of the human body is just a little higher than the density of water.

We can reasonably assume that the incredible buoyancy Charlie and Joe have is due to the volume of gas that they ingest. Let's find out what volume that would need to be to get them off of the ground. We know that the minimum buoyant force acting on each of them must be at least equal to their weight.

We also know from Archimedes' principle that the buoyant force acting on a submerged object is equal to the weight of the volume of fluid that is displaced by the object. In this case the object is Grandpa Joe, and the fluid displaced is air. Therefore,

$$D_{air} V_{\text{air disp.}} g = M_{Joe} g, \text{ which gives}$$

$$V = \frac{M_{Joe}}{D_{air}} = \frac{70 \text{ kg}}{1.3 \text{ kg/m}^3} \approx 54 \text{ m}^3$$

assuming Grandpa Joe has a mass of around 70 kg. So how big is a volume of 54 m^3? Well, if Grandpa Joe were spherical in shape he would need to have a diameter of almost 5 meters. The only way Grandpa Joe and Charlie would actually be able to float is if they expand to at least twice the diameter that Violet attains in the previous scene.

At that size, Joe and Charlie might not even fit in the room. Nevertheless, how much gas do they have to ingest to achieve this volume? We know the gas will expand until it equalizes with the external pressure (as long as the container is able to expand), which in this case is one atmosphere or 1.01×10^5 Pa. We will assume normal room temperature (about 293 °K), and that the volume of Grandpa Joe without the gas is only a fraction of a cubic meter.

Now we can use the ideal gas law to calculate the number of moles of gas that we have:

$$n = \frac{PV}{RT} = \frac{(1.01 \times 10^5 \text{ Pa})(54 \text{ m}^3)}{(8.31 \text{ J/molK})(293 \text{ K})} = 2{,}200 \text{ moles.}$$

That seems like a lot of gas! However, stomachs can't expand that much. Therefore, the fizzy lifting drink is probably just going to cause a lot of discomfort rather than increase buoyancy.

Because the stomach volume can't increase much, the pressure is going to go up. How much? If we account for just a little bit of expansion and approximate the average distended stomach to be a sphere with a radius of about 8 cm, we get the volume of the stomach to be:

$$V = \frac{4}{3}\pi(0.08 \text{ m})^3 = 2 \times 10^{-3} \text{ m}^3.$$

Using the gas law again we'll determine the amount of gas pressure in Grandpa Joe's stomach if he can't balloon out:

$$P = \frac{nRT}{V} = \frac{(2{,}200 \text{ moles})(8.31 \text{ J/molK})(293 \text{ K})}{2 \times 10^{-3} \text{ m}^3} = 2.7 \times 10^9 \text{ Pa}.$$

That is about 26,000 atmospheres, which is enough to crush a tank.

Now let's delve a little deeper into Archimedes principle applying it to one of the most famous maritime disasters of all time: the sinking of the *Titanic.*

TITANIC—ICEBERGS NEAR, FAR, WHEREVER THEY ARE . . .

Director James Cameron and his crew went to great lengths to accurately recreate the events leading up to the iceberg collision, as well as the physical details of the sinking ship. We are given a stunning, and horrifyingly visceral sense, of what it might have been like to be a passenger on the *Titanic* as the boat went down. The love story draws you in and the accurate reenactment keeps you riveted.

As to the physics involved, we recall from section one that the density of icebergs is just a little less than water, so we will be able to determine the threat posed by the submerged ice. Using Archimedes' principle, we will discover why the *Titanic* must sink when we determine the extent of the damage. Because holes are punctured in the *Titanic's* hull beneath the waterline, we can use Bernoulli's principle to estimate the rate at which water will inundate the ship. Let's have a look at each of these issues.

The Movie

In probably the world's most famous maritime disaster, the luxury ocean liner *Titanic* struck an iceberg in the Atlantic Ocean four days out from Southampton, England, on its maiden voyage and sank to the bottom. Out of over 2,200 people on board, only about 700 survived. The film goes to great lengths to reenact accurately the events that led up to, and occurred during, the disaster. Many of the characters, including the ship's captain, designer, first and second officers, telegraph operators, and so on are depictions of real people. Inside this compelling story of the fate of the *Titanic* a fictional love story is created, starring Kate Winslet as Rose, a reluctant debutante socialite, who finds true love and an escape from the straightjacket of Edwardian-Era aristocratic society when she meets indigent free-spirited artist Jack Dawson (Leonardo DiCaprio). Jack saves her life both metaphorically and literally on multiple occasions, during the final horrific hour on the sinking *Titanic.*

Probably one of the greatest ironies in the tragedy of the *Titanic,* and one of the greatest examples of tragic hubris in modern times, was that the ship received a lot of publicity as the first "unsinkable" ship. The fact that it went down on its maiden crossing stands out in sharp relief to its prevoyage hype. However, the *Titanic* did contain design innovations that theoretically should have made it safer in the event of a collision. In particular, the hull was divided into 16 separate compartments, separated by watertight doors that could be closed in the event of a puncture. Therefore, if one compartment was punctured it could be sealed off, preventing water from entering the rest of the ship and keeping the vessel afloat. However, in the collision with the iceberg six separate compartments were punctured. The *Titanic* could stay afloat with four compartments breached, but no more. Moreover, even though the actual holes were no more than an inch wide, and the total amount of area ripped open was only around 12 square feet (1.1 m^2); it was enough to do the ship in. Unfortunately, the bulkheads (walls) that separated the compartments only came up to 10 feet above the waterline. If water got above the level of a bulkhead it would start flooding the adjacent compartments even if they were intact. If the compartments had been *completely* watertight, that is if water could not spill over the top of the bulkheads, then *Titanic* would not have sunk—as we will see. It's too bad the design innovations didn't go that far.

Titanic Facts

We'll start by using some data about the *Titanic* to calculate the weight, mass, and overall density of the ship.

Weight. In the seafaring world, the weight of a ship is quoted as the amount (weight) of water displaced by the ship when it is operating as intended. Because the ship is floating, the weight of displaced water must equal the weight of the ship.

The maximum displacement of the *Titanic* is quoted as being 52,000 long tons (a long ton is equal to 2,240 pounds). However, during the voyage as fuel is consumed the displacement should decrease, and by the time of the accident, the *Titanic's* displacement was probably around 48,000 tons. Given this information, we can easily determine the total mass of the ship:

$$m = \frac{W}{g} = \frac{(48{,}000 \text{ tons})(2{,}240 \text{ lbs/ton})(4.45 \text{ N/lb})}{9.8 \text{ m/s}^2} \approx 4.9 \times 10^7 \text{ kg}.$$

Volume and Density. Here's an interesting fact: When discussing ships the word "ton" is also used to mean something completely different from weight. The term "gross tonnage" actually refers to the total *volume* contained by the ship. A ton in this context is equal to 2.83 cubic meters. The gross tonnage of the *Titanic* was about 46,000 tons, which is equal to 131,000 m^3.

From this information we can determine the actual density of the *Titanic*. In a problem earlier in the chapter you were asked to think about what factors determine the overall density of a ship. The point is, although a ship may be constructed primarily of iron, it is the large volume of air contained within the structure that makes the overall density less than that of water. The density must be less than water or the boat will sink.

If the *Titanic* displaces 48,000 tons (back to weight here), then the volume of displaced water must be:

$$V = \frac{M}{D} = \frac{4.9 \times 10^7 \text{ kg}}{1{,}027 \text{ kg/m}^3} \approx 48{,}000 \text{ m}^3,$$

where we have used the average density of seawater at the surface of the ocean (1,027 kg/m^3). This volume must be equal to the volume of the part of the ship that lies below the waterline. We'll need to use this quantity in a minute.

Because the weight of the ship must equal the weight of displaced water, and because the weight of displaced water is contained in a volume of 48,000 m^3 (while the weight of the ship is contained in a volume of 131,000 m^3), then the density of the *Titanic* must be

$$D_{ship} = D_{water} \frac{V_w}{V_s} = 1{,}027 \text{ kg/m}^3 \frac{48{,}000}{131{,}000} = 380 \text{ kg/m}^3.$$

The Scene—The Accident

At 11:40 PM on April 14, 1912, the *Titanic* collided with an iceberg on its port side in the waters off of Newfoundland. The collision punctured six compartments near the bow. In less than three hours the ship sank to the bottom of the Atlantic. The collision is dramatically reenacted in scene 16 of the film.

The Iceberg. According to eyewitnesses the iceberg was somewhere between 50 to 100 feet high and 200 to 400 feet long. From this information we can estimate the total volume and mass of the iceberg. Let's approximate the shape of the berg to be roughly that of a four-sided pyramid with volume = $1/3Ah$.

We can then determine the volume of pyramid above the water surface from the dimensions quoted above. We'll take the height to be 75 feet (23 m) and the length of the base at the waterline to be 300 feet (91 m). Therefore the volume of ice above the water is:

$$V = \frac{1}{3}(91 \text{ m})^2 (23 \text{ m}) = 64{,}000 \text{ m}^3.$$

Because the density of ice is 90 percent that of seawater (920 kg/m^3), this means that 90 percent of the volume of the iceberg is submerged, and therefore the total volume of the iceberg is 640,000 m^3, and the total mass of the iceberg is approximately

$$M = DV = (900 \text{ kg/m}^3)(640{,}000 \text{ m}^3) = 5.8 \times 10^8 \text{ kg}.$$

This is ten times the mass of the *Titanic,* so in a collision the iceberg wins. Our estimate may be a little high because our assumption of a pyramid shape is an oversimplification. Published estimates of the mass of the iceberg are closer to five times the mass of the ship, but remember our program throughout is to make reasoned assumptions, and along with the fundamental principles we are using, these can be pretty powerful in giving us an accurate feel for a situation.

In addition, as we all know, the submerged segment, which extends outward from the exposed base, is where you're most likely to collide. The *Titanic* was struck approximately 4 meters above the keel.

ACTIVITY

Take a small plastic food container with shallow sides about an inch deep. Float it in a large pot of water. From this, roughly estimate the density of the little "boat."

Now take two of these containers and tape or glue them together so that the connected sides are flush with each other. Add some coins to each side to get the boat to float as flat as possible, and then fill one compartment with water. Does the "boat" stay afloat?

Next take the boat out of the water, and cut a groove three-sixteenths of an inch wide or so between the two containers, extending most of the way down. Fill one compartment with water. Does it stay afloat this time?

Note: You might have to tinker with the experiment a little by adjusting the number and location of the coins, but if you play with it you should find that with one compartment filled the boat floats, but if there is leakage into the second compartment the boat will sink. It is also interesting to wedge some paper towels in the groove. It takes a while for the water to seep in, causing the boat to sink gradually at first—more like a real shipwreck.

Why Did the Titanic Sink?

The previous activity is an attempt to approximate the *Titanic's* compartments. In our model the boat doesn't sink with one compartment flooded, but if water gets into the second compartment it does. As we mentioned previously, the *Titanic* would have stayed afloat if only four compartments had been breached, but not any more than that—and the *Titanic* sustained damage to six compartments.

Once a fifth compartment started to fill, the ship would have floated low enough for water to spill over the top of the compartments because the compartments were only watertight to 3 meters above the waterline. Therefore, if water got above that level it would have started filling additional compartments that hadn't been breached, continuously increasing the mass within the ship and continuously decreasing its buoyancy, thus causing the boat to float lower. Then it would take on more water, causing more loss of buoyancy and finally causing the *Titanic* to plunge beneath the surface of the ocean and descend to the bottom.

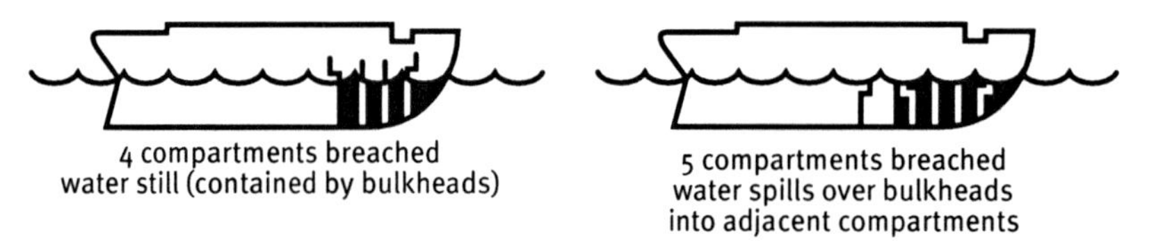

Figure 6

There are some great moments in scene 17 when Mr. Andrews (Victor Garber) explains this to the ship's officers and cruise line executive Mr. Ismay (Jonathan Hyde). Ismay, in another tragic, ironic twist, is partly to blame for the accident because he had been pushing the captain (on his final transatlantic crossing prior to retirement) to drive the ship faster in order to triumphantly enter New York Harbor a day early and "make headlines." When Ismay exclaims, "This ship can't sink!" Andrews silences the room with a dose of grim reality. "She's made of iron, sir, I assure you she can, and she will. It is a mathematical certainty."

Let's do some rough calculations to support Mr. Andrews's point.

We will assume for the sake of simplicity that the ship's compartments have equal volumes, even though some of the forward compartments were actually a little smaller. Therefore, if the *Titanic* floats optimally with 48,000 m^3 of its volume submerged, then dividing this volume among 16 compartments gives 3,000 m^3 per compartment. (Because the top of the compartments are actually another 10 feet above the waterline this would tend to make our estimate too low, but the volume of the hull and interior walls would take away from the compartment volume that can fill with water so we'll stick with the 3,000 m^3 as a reasonable estimate.) The ship can drop 3 meters deeper before water can spill over the top of the bulkheads into the other compartments. Another important piece of information is that the distance from waterline to boat deck was 60 feet under normal conditions.

Because the *Titanic's* total volume is 131,000 m^3, then under normal conditions 83,000 m^3 lies above the waterline. If the ship drops 10 feet, then approximately one-sixth of that volume will be submerged, or about 14,000 m^3. Each compartment has a volume of 3,000 m^3, so if you fill four compartments you only submerge 12,000 additional cubic meters, which means the water level inside the compartments will not rise above 10 feet. However, fill one more compartment and . . . disaster.

How Fast Is Water Getting In? We can conduct another rough approximation that will give us a feel for how fast the water would have flooded the *Titanic*. The punctures on the hull were around 4 meters above the keel, or 6 meters below the waterline.

We can use Bernoulli's principle to estimate the velocity at which water would be entering the ship through the various punctures. In addition, from the area of the punctures we can then estimate the volume of water per time entering the ship. According to Bernoulli's equation:

$$P_1 + \frac{1}{2}Dv_1^2 + Dgy_1 = P_2 + \frac{1}{2}Dv_2^2 + Dgy_2.$$

In this case we are looking at the water at the ocean's surface (which is at atmospheric pressure, and essentially has a velocity of zero) versus the water rushing in at the punctures (6 meters below the level of the ocean's surface, but also at atmospheric pressure when it enters the ship). Therefore the equation reduces to:

$$Dgy_1 = \frac{1}{2}Dv_2^2 + Dgy_2, \text{ and setting } y_2 = 0 \text{ we get}$$

$$gy_1 = \frac{1}{2}v_2^2, \text{ or } v_2 = \sqrt{2gy_1} = \sqrt{2(9.8\ \text{m/s}^2)(6\ \text{m})} \approx 11\ \text{m/s}$$

Notice how the previous equation to determine velocity is exactly the same as what we would use to calculate the velocity of an object starting from rest and falling a distance y using conservation of energy. Because Bernoulli's equation is, in fact, a statement of conservation of energy, that shouldn't be a surprise.

We stated that the total puncture area of the hull was only 1.1 m^2, so at a speed of 11 m/s, we get a volume of water entering the ship of:

$$V = Av = (1.1\ \text{m}^2)(11\ \text{m/s}) = 12\ \text{m}^3/\text{s},$$

which is about 700 m^3 per minute (corresponding to approximately 700 tons of water entering the *Titanic* every minute), or 42,000 m^3 per hour. The ship's pumps could not handle anything near that rate of flooding.

Conclusion. Like *Apollo 13*, *Titanic* is a great example of a film based on an historical event in which principles of physics play a leading role. Because the makers of *Titanic* attempted to recreate the accident as accurately as possible it gives us an excellent opportunity to apply our knowledge of physics to analyze and understand the disaster.

THE DAY AFTER TOMORROW—WHAT'S THERMODYNAMICS GOT TO DO WITH IT?

In this section we are going to introduce *The Day After Tomorrow,* yet another end-of-the-world disaster movie, and this time it's all about bad weather. The premise of this cinematic gem is based loosely on a controversial theory that global warming could actually end up triggering a global deep freeze (otherwise known as an "ice age"). Any discussion of weather clearly calls for a little foray into the principles of thermodynamics.

The Deep Freeze: Ice Ages and Inclement Weather

While the causes of ice ages are not completely understood, scientists think that ice ages may be triggered in part by a kind of positive feedback loop. (A positive feedback loop describes a situation where a small change in conditions alters a system in such a way that it reinforces the potential for the system to keep changing in the same direction.) For the case of a worldwide ice age the theory suggests that the process starts with a period of time where the average temperatures in the Northern Hemisphere decrease a few degrees. While the temperature decrease is not worldwide (in fact other parts of the Earth could actually get warmer), this local cooling could lead to increased glaciation in the upper latitudes. Once glaciers start forming, because they are good reflectors, the more glaciers there are, the more incoming solar energy is reflected back into space. The more energy reflected into space, the cooler the Earth, and the cooler the Earth, the more glaciers can form, and so on. This effect could perhaps lead to a worldwide temperature decrease and an associated ice age over a relatively short period of time.

In scene 14 of *The Day After Tomorrow*, there is one big storm surge. The biggest storm surges ever recorded were on the order of 13 meters high. Storm surges that high are generated by tropical cyclones with wind speeds exceeding 130 miles per hour or more. The storm surge in *The Day After Tomorrow* rises to the shoulders of the Statue of Liberty. The top of the Statue of Liberty is about 100 meters above the water, which means that this storm surge is at least 80 meters (260 feet) high—six times as high as the biggest storm surges ever recorded. What kind of energy/wind would it take to generate a surge this big? What kind of destructive capacity would the water have when it hits the statue and the buildings in its path? Would they still be standing? It's actually kind of surprising all things considered that they don't show buildings being knocked over. It is a disaster movie after all. What a missed opportunity. They did go to the trouble of making a wave that big; why not get all they can get out of it? Come on! Furthermore, after the surge hits New York, why doesn't the water recede? Have you ever seen a wave wash up on the beach? Does it stay up there? Why not? Are we to believe that the wave freezes solid before it has a chance to wash back out?

Now the disaster in *The Day After Tomorrow* is based on the following controversial theory: Because global warming causes melting of glacial ice, including continental glaciers like those on Greenland, the influx of fresh water reduces the salinity of the oceans in those areas. This change in salinity could affect something called *thermohaline circulation*, which affects ocean convection. Some models predict that thermohaline circulation could be affected such that it would reduce the amount of warm water flowing into the North Atlantic with the Gulf Stream, thus leading to locally cooler temperatures in parts of the Northern Hemisphere and subsequently runaway glaciation according to the process just discussed. Although some earlier models from a decade ago indicated that these cooler temperatures in the Northern Hemisphere might lead to an ice age, most climatologists today do *not* believe that this would happen.

Nevertheless, *The Day After Tomorrow* takes the idea of a disruption in the thermohaline circulation and turns local cooling in the Northern Hemisphere into the apocalyptic scenario of an "instant ice age." In climatological terms, *extremely* rapid climate change means about a decade. A movie in which the weather gets gradually worse over a 12-year period might not be that dramatic, so *The Day After Tomorrow* takes global climate change and slams it into overdrive. Things happen so fast in this movie that it should be called "The Millisecond After Tomorrow."

The Movie

Dennis Quaid plays Jack Hall, the usual, brilliant, work-driven, socially awkward, misunderstood, scientist "type" whose work on ancient climates (*paleoclimates*) suggests that due to the effects of global warming an ice age is imminent in the next several centuries. However, it turns out when Professor Hall's paleoclimate models based on ice core data from ancient climates are somehow adapted and run on current ocean temperature, they can be used as a forecasting tool. They predict that the new ice age will be here in a *few weeks!*

However, the time scale of a *few weeks* turns out to be a gross *overestimate*.

The weather starts getting really bad, really fast, but despite an alarming sequence of worldwide meteorological disasters (including, in one of the better special effects moments in the movie, the impressive obliteration of downtown Los Angeles by dozens of incredibly violent tornadoes)

One of the most mind boggling plot devices in *The Day After Tomorrow* takes place when Professor Hall decides he is going to *walk* from Philadelphia to New York (a distance of 125 miles) in the middle of unarguably the worst blizzard in recorded history, and fetch his son Sam. He may not have been there for Sam in the past, but this time, by God, he's going to come through! We are supposed to believe that because of his cold weather experience and know-how gained by years of collecting ice cores in Antarctica, he can handle conditions that would freeze a polar bear in its tracks. He finally arrives just after the storm breaks, to find Sam alive and well in the library where he has made some nice friends. He then radios for a helicopter to come and pick them up. Couldn't they have saved some trouble, waited for the storm to break, and then gone up to New York with the helicopter?

somehow the Dick Cheney look-alike vice president refuses to take the situation seriously. A series of "super storms" then descend from the Arctic, one of these causes a storm surge almost as high as the Statue of Liberty, which inundates Manhattan, flooding the entire island up to the second floor of the New York City Public Library. Within about a day this water freezes solid as most of the Northern Hemisphere is headed for the deep freeze.

For dramatic effect these super storms look like giant (tropical) hurricanes, and in the "eye" of each storm "super-cooled air in the upper troposphere at a temperature of negative 150 degrees Fahrenheit is being forced to the surface" where anyone who survives the preceding blizzard will be flash frozen when the eye of the storm passes through. The government finally gets the message at this point but is it too late?

Interspersed with scenes of truly horrific weather, *The Day After Tomorrow* is full of earnestly acted, overly heartfelt dramatic moments and predictable dialogue. There really isn't any character development, so we rely on the fact that we're familiar with the stock relationships and character types to navigate the standard subplots successfully. Finally, there is some almost incomprehensible physics that we will try to deal with.

The major questions we will address are:

1) Is it remotely possible that climate change on the scale that we see in the movie could actually occur?
2) What would happen to our planet if extremely cold air from the upper troposphere descended onto the Earth's surface?

Because we are talking weather here, we are talking thermodynamics, because all weather is driven by heat. Let's look at the principles.

Some Thermodynamics

Remember the principle of the conservation of energy? Well here it comes again. Before we get to it, however, let's define a few terms.

Internal energy is the total energy contained *within* a system, including thermal, chemical, and nuclear energy.

Thermal energy is the part of the internal energy that is due to motion (kinetic energy) of the particles. The thermal energy will change when the temperature of the system changes because temperature is proportional to the average kinetic energy of the particles in a system.

We will use the term *heat* to refer to the *transfer* of thermal energy from one region to another due to a difference in temperature between the regions.

The First Law of Thermodynamics. We know from the principle of conservation of energy that for an isolated system, the total energy of the system must remain constant. However, if a system is not isolated, energy can be transferred between the surroundings and the system. This energy can be transferred as heat or by doing work. For example, say you have a pan of water. When you place the water on a hot stove heat is transferred into the water, increasing the thermal energy of the water. The amount of energy gained by the water would be equal to that lost by the heating coil system if none of the energy were lost to the surroundings. You could also accomplish the same thing (although much less efficiently) by stirring the water vigorously with a spoon or a paddle because the work done on the water by the spoon must go into increasing the energy of the water.

The first law of thermodynamics quantifies the change in energy of a system by looking at how much work is done on or by the system, and how much heat is added to, or removed from, the system. Mathematically,

$$\Delta U = Q + W,$$

where U is the internal energy of the system, Q is the heat added to or removed from the system (positive for added heat and negative for heat removed), and W is the work done on the system. If the system does work on the surroundings, W must be negative. (Sometimes the first law is given as $\Delta U = Q - W$, in which case W refers to the work that the system does on the surroundings.)

Example 7: 100 Joules of heat is added to a system and the system does 40 Joules of work on the surroundings. How much does the internal energy of the system change?

Solution: $\Delta U = Q + W = 100\text{ J} + (-40\text{ J}) = 60\text{ J}$

Internal Energy. In the next examples, the chemical and nuclear energies will remain constant, that is, no chemical and nuclear *reactions* will occur. Therefore, changes in internal energy will be equal to changes in thermal energy only. In cases where the temperature is constant there is no change in U. Conversely, any change in U will be reflected by a change in temperature. The total internal thermal energy will also depend on the total number of particles in the system. Mathematically,

$$U = \frac{3}{2}nRT.$$

Example 8: Using the system in the last example, if the system contains two moles of an ideal gas, and the specific heat (c) of the gas under these conditions *in terms of moles* is 20 J/molK, how much will the temperature of the gas increase?

Solution: In this case not all of the heat (Q) entering the system goes into raising the temperature; some goes into doing work. Therefore,

$$\Delta U = nc\Delta T \text{ so } \Delta T = \frac{\Delta U}{nc} = \frac{60 \text{ J}}{(2 \text{ moles})(20 \text{ J/molK})} = 1.5 \text{ K}$$

Isobaric and Isovolumic Processes. How does a system do work on the surroundings, or how do the surroundings do work on a system? Because work involves motion the system must either push the surroundings or vice versa. We can show this with the classic example of a gas contained within a cylinder with a movable piston at one end of the cylinder, as depicted in the following figure.

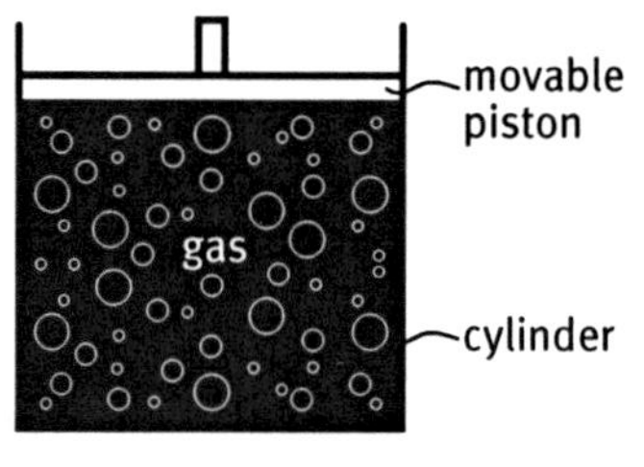

Figure 7

The gas occupies a volume V, at a pressure P, and a temperature T. If the gas expands slowly enough that P remains constant (an *isobaric* process), and the gas expands by a distance Δy, the work it does on its surroundings will be $W = F\Delta y$. However, because $F = PA$, where A is the area of the piston, then $W = PA\Delta y = P\Delta V$. We can graph this process of expansion on a graph of P as a function of V (a PV diagram).

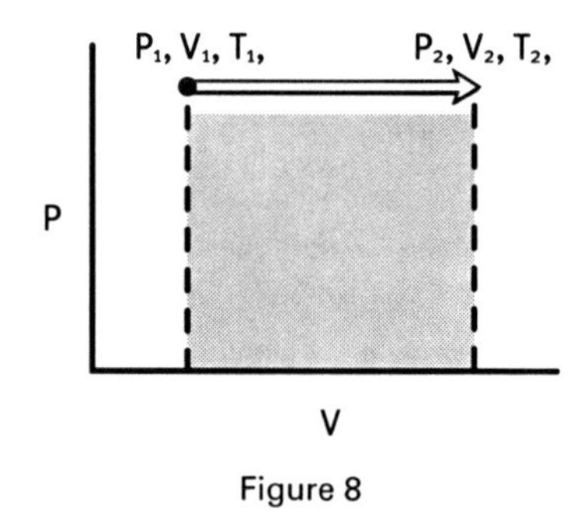

Figure 8

The shaded area underneath the line is actually equal to $P\Delta V$. Therefore, the area underneath is equal to the work done by the gas on the surroundings as it expands. Even if P is not constant,

SPECIFIC HEATS OF GASES

Specific heats for gases actually depend on the process. For example, the specific heat for an isobaric process will be more than that for an isovolumic process for the same gas. If you think in terms of energy it is easy to see why this is the case. In the case of an isovolumic process all of the heat that enters (or exits) the system goes into changing the thermal energy, and therefore the temperature of the system, because none goes into doing any work. However, for an isobaric process some of the energy goes into doing work because the gas expands, so you need more total energy to change the temperature of the gas a certain amount than for the isovolumic process.

In the first example we took a gas and put energy into the system in the form of heat. This energy accomplished two things: of the 100 Joules of heat put into the system, 60 Joules went into increasing the thermal/internal energy of the gas, and 40 Joules was given back to the surroundings in the form or work when the gas expanded. If the volume was forced to remain constant, then 100 Joules would have gone into increasing the internal energy, and the temperature would have gone up more.

For a gas it is more useful to put specific heat in terms of moles rather than mass. Therefore, we can define c_v as the molar specific heat at constant volume, and c_p as the molar specific heat at constant pressure ($c_p > c_v$). In examples 1 and 2 let's assume that the process was isobaric. In that case the specific heat we quoted was c_p.

however, the area under any curve representing a change from one state to another on a *PV* diagram is equal to the work done *by* the gas (if it expands) or *on* the gas (if it contracts).

Let's say we have a known amount of an ideal gas and its thermodynamic state at some instant is represented as a point on a PV diagram. Therefore, we know not only the pressure and volume of the gas, but its temperature as well, because $PV = nRT$.

Let's think about the first law to see what it can tell us. A lot of people are familiar with the ideal gas law from chemistry class, and recognize the mathematical relationship between *P*, *V*, and *T*, but don't consider how changes in those variables might be physically caused. For example it is true that if you hold the volume and number of moles of gas constant and *P* increases, then there must be a corresponding increase in *T*. However, what is the cause and effect relationship? An increase in *P* does not cause an increase in *T*, but rather the other way around. More specifically, to increase *T* energy must be added to the system, in this case in the form of heat, because if the volume is held constant (an *isovolumic* process) then no work is done on the gas. This added heat increases the thermal energy of the gas, increasing its kinetic energy and therefore the velocity with which the particles hit the walls of the container. The result is a greater force exerted on the walls due to the impacting particles or greater *pressure*.

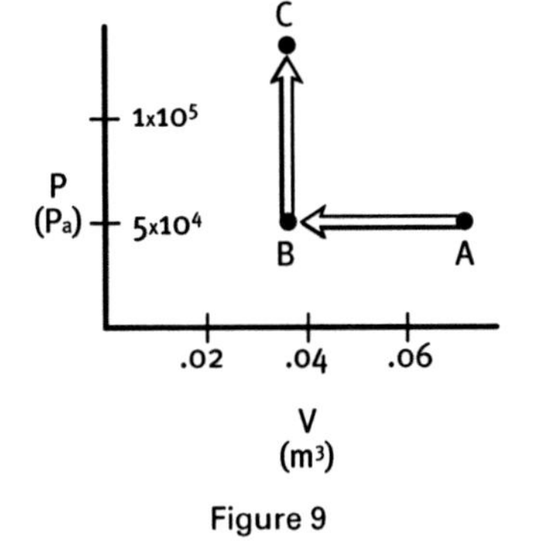

Figure 9

Example 9: An ideal gas in a piston-cylinder container goes from state A to B to C as represented by the accompanying *PV* diagram. At state A the temperature of the gas is 300 °K.

a) What is the temperature of the gas at points B and C?
b) How many moles of the gas are there?
c) What types of thermodynamic processes are represented by A to B, and by B to C?
d) How much work is done in process A to B, and in process B to C? Is the work done by the gas or on the gas in each case?

e) If ΔU from A to B is 200 J, what is ΔU from B to C?

Solution:

a) $\frac{P_A V_A}{T_A} = \frac{P_B V_B}{T_B} = \frac{P_C V_C}{T_C}$

From A to B the pressure is constant so

$$T_B = T_A \frac{V_B}{V_A} = 300\text{ K}\left(\frac{0.02\text{ m}^3}{0.04\text{ m}^3}\right) = 150\text{ K}$$

and B to C is isovolumic so

$$T_C = T_B \frac{P_C}{P_B} = 150\text{ K}\left(\frac{1 \times 10^5\text{ Pa}}{5 \times 10^4\text{ Pa}}\right) = 300\text{ K}$$

b) $n = \frac{PV}{RT} = \frac{P_A V_A}{R_A T_A} = \frac{(1.01 \times 10^5\text{ Pa})(0.02\text{ m}^3)}{(8.31\text{ J/molK})(300\text{ K})} = 0.81\text{ moles}$

c) A to B is a constant volume or *isovolumic process,* and B to C is a constant pressure or an *isobaric* process.

d) From A to B, $W = P\Delta V = (2.02 \times 10^5\text{ Pa})(0.01\text{ m}^3) = 2{,}020\text{ J}$ done on the gas. No work is done from B to C because volume is constant.

e) Because the temperature at C is the same as the temperature at A the total ΔU must be 0 from A to C. Therefore because $\Delta U_{A\text{ to }B} = 200$ J, $\Delta U_{B\text{ to }C} = -200$ J.

THINKING PHYSICS

When pressurized gas is released from its container it tends to get very cold as it expands into the air. Why does this happen? (Think in terms of work and energy!)

Isothermal and Adiabatic Processes. Isobaric and isovolumic processes represent simplified cases where the state of a system changes while pressure is held constant (isobaric) or while the volume is held constant (isovolumic) throughout the change. Applying the first law:

$$\text{Isobaric: } \Delta U = Q + P\Delta V$$
$$\text{Isovolumic: } \Delta U = Q$$

There are two additional easily definable processes that we can look at.

An *isothermal* process is one in which the *temperature is held constant.* This means that ΔU must be zero for the process.

$$\text{Isothermal: } Q = -W$$

This means if work is done on the system it must expel an equal amount of heat for U (and therefore T) to remain constant.

In example 9, states A and C were at the same T. To go from A to C at constant temperature you get the following curved path.

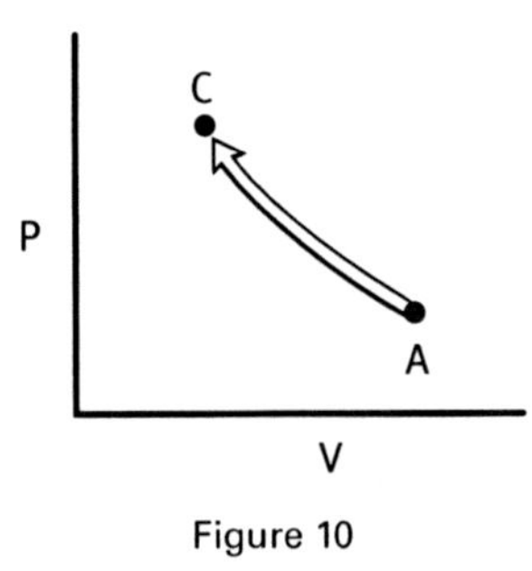

Figure 10

Lines of constant temperature on a PV diagram are called *isotherms.*

An ***adiabatic*** process is one in which no heat is exchanged between the system and the surroundings. This applies to insulated systems. This does not mean that the temperature remains constant, because if the system expands or contracts, work is being done, and the system will therefore change its internal energy. Remember, heat is not synonymous with temperature. Heat is the flow of thermal energy from one place to another. The heat energy might go into changing the temperature, but it could also go into doing work. Temperature is proportional to the average kinetic energy of the particles in a material, and is therefore also proportional to the total thermal energy. Temperature can be changed by an addition or subtraction of heat, but it can also change if work is done.

$$\text{Adiabatic: } \Delta U = W$$

In this case, if the system contracts, work is done on the system and all that work will go into increasing the energy of the system. Therefore, U and T will increase during an adiabatic contraction. If the system expands, then it does work on the surroundings and loses an amount of energy equal to that much work. U and T decrease during an adiabatic expansion.

Referring to the previous thought question: when pressurized gas is released from a container, (the same as what might happen if you were to open the valve on a scuba tank), the gas expands rapidly with very little heat exchange with the surroundings. The expansion is essentially adiabatic and you would notice the gas gets very cold as it expands. This is because the gas loses energy in the work that it does on the surrounding air.

Example 10: Let's take example 9 a step further, and return the gas to state A by an isothermal process. After all three steps the gas has returned exactly to the state where it started. This is called a cyclic process.

a) What is ΔU for one complete cycle?

b) Is work done on or by the gas during the isothermal expansion?

c) How much heat is added to or removed from the system from A to B, and from B to C?

d) *Estimate* the total work done during one complete cycle, and the total heat added or removed from the system during one complete cycle.

Solution:

a) $\Delta U = 0$, because the gas ends at the exact same thermodynamic state where it started. ($\Delta T = 0$)

b) Work is done *by* the gas, because it is expanding into/pushing on the surroundings.

c) $Q = \Delta U - W$

For A to B the process is isovolumic so $Q = \Delta U = 200$ J.

For B to C the process is isobaric $Q = \Delta U - P\Delta V = -200 \text{ J} - 2{,}020 \text{ J} = -2{,}220$ J.

Therefore, because work is being done on the gas and the volume is decreasing, the tendency is for both the temperature and pressure to increase. To keep the pressure constant at a smaller volume the temperature must actually drop, so heat must be expelled from the system.

d) We know $W_{A \to B} = 0$, and $W_{B \to C} = 2{,}020$ J and we can estimate $W_{C \to A}$, because we know it is equal to the area under the curve. Roughly we get $W \approx -1{,}400$ J. It is negative because the gas is doing work on the surroundings. Therefore, the net work is $W \approx 0 + 2{,}020 \text{ J} - 1{,}400 \text{ J} \approx 600$ J. Because ΔU for the complete cycle must be zero, $W_{net} = Q_{net} = 600$ J.

This means that while we had to remove heat from the system during the isobaric compression, we had to add heat (about 2,600 J) during the isothermal expansion to make up for the energy lost as the gas does work on the surroundings.

Example 11: One mole of an ideal gas goes through the cycle shown in the following figure.

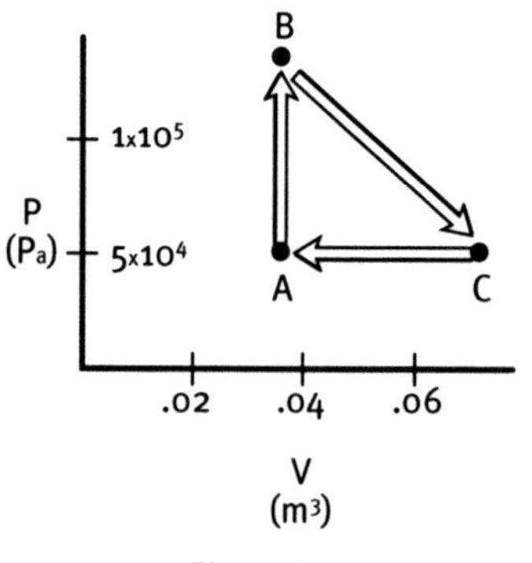

Figure 11

If $c_p = 20$ J/molK and $c_v = 12$ J/molK, find ΔU for each step of the cycle.

Solution: For A to B the process is isobaric and therefore $\Delta U = Q + P\Delta V$

Where $Q = nc_p\Delta T = nc_p(T_B - T_A)$

We can use the ideal gas law to find T_B and T_A, and then solve for ΔU.

$$T_B = \frac{P_B V_B}{nR} = \frac{(5 \times 10^4 \text{ Pa})(0.02 \text{ m}^3)}{(1 \text{ mole})(8.31 \text{ J/molK})} = 120 \text{ K}$$

$$T_A = \frac{(5 \times 10^4)(0.04 \text{ m}^3)}{(1 \text{ mole})(8.31 \text{ J/molK})} = 240 \text{ K}$$

$$\Delta U = (1 \text{ mole})(20 \text{ J/molK})(120 \text{ K} - 240 \text{ K}) + (5 \times 10^4 \text{ Pa})(0.02 \text{ m}^3 - 0.04 \text{ m}^3) = -3{,}400 \text{ J}$$

For B to C the process is isovolumic so $\Delta U = Q = nc_v\Delta T = nc_v\,(T_C - T_B)$

$$\Delta U = (1 \text{ mole})(12 \text{ J/molK})(240 \text{ K} - 120 \text{ K}) = 1{,}440 \text{ J}$$

For C to A because the total ΔU for one complete cycle must be zero then ΔU from C to A must be −3,400 J + 1,440 J = 1,960 J.

PROBLEM

For example 11 above, determine the total work done during one complete cycle, the amount of work done during each step of the cycle, the total heat either added to or removed from the gas during one complete cycle, and the heat added or removed during each step of the cycle.

The Scene—Grandpa Joe's Isobarically Expanding Stomach!

Let's apply some thermodynamics to the fizzy lifting drink scene from *Willy Wonka and the Chocolate Factory* before we march on to *The Day After Tomorrow.* We've already had some fun illustrating the volume that Charlie and Grandpa Joe would have to occupy before they could float, but we didn't discuss the specific thermodynamics of their situation. Let's assume for the moment that Grandpa Joe is able to expand easily to the final volume of 54 m^3 calculated previously. If he expands without any energy (heat) exchange between his stomach and his surroundings, what happens to the temperature in his stomach? What kind of thermodynamic process would this be?

Because no heat is exchanged, this would be an adiabatic process. The gas inside Joe's stomach would be doing work on the surroundings, thus losing energy and dropping temperature.

We can actually calculate how much the temperature in his stomach would drop. For an adiabatic process the following equations relate P, V, and T at some initial point in the process to some later point, where γ is the ratio of c_p to c_v. For diatomic gases like those found in the atmosphere, $c_p/c_v = 1.4$, and

$$PV^{\gamma} = P_0 V_0^{\gamma}$$

$$TV^{\gamma-1} = T_0 V_0^{\gamma-1}$$

$$\left(\frac{1}{P}\right)^{\gamma-1} T^{\gamma} = \left(\frac{1}{P_0}\right)^{\gamma-1} T_0^{\gamma}$$

Because we know the initial and final volume of Joe's stomach, and if we assume an initial body temperature of 98.6 °F (310 °K), then

$$T = \frac{T_0 V_0^{\gamma-1}}{V^{\gamma-1}} = \frac{300\ \text{K}(2 \times 10^{-3}\ \text{m}^3)^{0.4}}{(54\ \text{m}^3)^{0.4}} = 5\ \text{K}.$$

That's five degrees above *absolute zero.* Grandpa Joe's stomach would freeze solid!

Willy Wonka wouldn't design a candy drink capable of freezing people to death, would he? Perhaps the gas in the drink is highly exothermic when it reacts with stomach acid, and it generates heat, which prevents energy loss that would occur in the adiabatic case. So what if the expansion is isobaric instead? Now what's the final temperature of Joe's stomach? For the isobaric case:

$$\Delta U = 3/2\ nR\Delta T = Q - P\Delta V = nc_p\Delta T - P\Delta V$$

(the negative sign means the gas is doing work on the surroundings)

We're assuming that we start with the initial pressure of 26,000 atmospheres in Grandpa Joe's stomach that we calculated earlier and we maintain that pressure by adding energy to the system as Joe's stomach expands. Because the volume is expanding the temperature must go up in order to maintain pressure. Because pressure is caused by collisions of the gas particles with the container walls, if the container gets bigger, the gas particles must hit with higher energy to maintain the same average pressure. This is the essence of the isobaric expansion. Therefore,

$$\Delta T = \frac{P\Delta V}{(\frac{3}{2}nR + nc_p)}.$$

We will assume a specific heat of the fizzy lifting gas to be similar to that of the atmosphere (around 28 J/molK). Therefore,

$$\Delta T = \frac{(26{,}000\ \text{atm})(1.01 \times 10^5\ \text{Pa/atm})(54\ \text{m}^3)}{(\frac{3}{2}[2{,}200\ \text{mol}][8.31\ \text{J/molK}] + [2{,}200\ \text{mol}][28\ \text{J/molK}])} = 1.6 \times 10^6\ \text{K},$$

which is essentially the same as the temperature at the center of the sun. In this case Grandpa Joe is going to vaporize in a nuclear fusion reaction. I guess Wonka was right when he said his drink wasn't perfected yet!

The Scene—Global Freezing

In *The Day After Tomorrow* the world plunges into an ice age in a matter of a few days. If you look at glacial versus interglacial cycles (like the one that we're now living in) the temperature data show a drop of about 6 to 8 °C averaged globally during the glacial cycles. This roughly implies that during a glacial cycle, the net amount of thermal energy contained in the atmosphere (and hydrosphere) is less than during interglacial periods, but where does the energy go? In fact, where does the thermal energy come from in the first place, and what determines average global temperatures? This can be a simple or complex question depending on how you look at it, but we will go with the simple route. The energy comes from the sun. The temperature of the atmosphere depends on how much of that energy is trapped by the atmosphere and converted to molecular thermal energy, and how much is reflected back into space. The equilibrium between the two determines the average amount of thermal energy, and therefore the average temperature of the atmosphere.

We are dealing with the principle of the conservation of energy again. To get a sense of what the movie is suggesting, let's look at the situation very generally. Remember the positive feedback loop we discussed earlier? It depends on energy. If glaciers grow and therefore reflect more of the sun's incident radiation back into space, the total thermal energy contained in the atmosphere will decrease, leading to a net drop in temperature over a period of time. If we approximate the process to be roughly isovolumic, where V is the volume of the atmosphere, then the change in thermal energy depends only on the net heat flow into or out of the system:

> **PROBLEM**
>
> Estimate the volume and mass of the troposphere. If we assume an average specific heat for air of 1,000 J/kg, determine how much energy would have to exit the atmosphere to drop its temperature by 8 °C. If the temperature drops this much in one week, at what rate (power) is thermal energy leaving the atmosphere?

$$\Delta U = Q.$$

In an equilibrium state the net Q is zero, and U is constant. To drop the temperature, we are going to need a net Q out of the atmosphere. Something has to cause that. In glaciation models, it is due to the increased reflection of energy due to the build up of glacial ice. That would take time, at the extreme rapid end of the spectrum a decade or two. That is drastic or catastrophic climate change according to climatologists. In *The Day After Tomorrow* how in the world is the atmosphere getting rid of all of all that energy so fast?

Descending Air Contracts Adiabatically Baby! Probably the most dramatic and life-threatening meteorological phenomena (on a hemispheric scale) in *The Day After Tomorrow* are the famous –150 °F (–100 °C) masses of air descending to the surface from the upper troposphere, causing everything to freeze instantly. Temperatures in the upper troposphere range from around –45 °C to –75 °C. Where did this "super-cooled" –100 °C air come from? Even if the air somehow magically got that chilly at the top of the troposphere, would it really be that cold once it arrived on the ground? It seems that there might be a moment of physics clarity when Ian Holm as meteorologist Terry Rapson poses to Professor Hall, "Shouldn't the air warm up as it descends?" (Yes, it should!) However, that hope is dashed when Hall tells us that the air is descending too fast to warm up (also dashing our hopes that the movie might adhere in any way to known physical principles). They are suggesting that the reason the air mass would warm is because it would mix with the air lower down, but that's not the way it works.

Air is a very good insulator, and rising or falling air masses tend to do so adiabatically. There is very little heat transfer into or out of these air masses as they rise or fall. Consider also that the atmospheric pressure in the upper troposphere is about one-tenth of that on the ground. What does this suggest? If the surrounding pressure increases as the air mass descends what happens to the volume of the air mass? It decreases. If the volume decreases ***adiabatically*** what happens to the temperature of the air mass?

According to the first law of thermodynamics if the volume is decreasing then work is being done on the air mass, thus increasing its internal energy, and therefore its temperature. Remember, for an adiabatic process,

$$\Delta U = W.$$

We can do a rough calculation to estimate how much the super-cooled air will warm up before it gets to the ground. For an adiabatic process,

$$(\frac{1}{P})^{\gamma-1} T^{\gamma} = (\frac{1}{P_0})^{\gamma-1} T_0^{\gamma},$$

and

$$(\frac{T}{T_0})^{\gamma} = (\frac{P}{P_0})^{\gamma-1}.$$

If the pressure in the upper troposphere is one-tenth that at the surface, using $\gamma = 1.4$, and a temperature of 173 °K (–150 °F), we get:

$$T^{1.4} = (173\ \mathrm{K})^{1.4}(10)^{0.4}, \text{ which gives } T \approx 330\ \mathrm{K}.$$

Well, that's a balmy 57 °C (135 °F) (picture Death Valley in the summer). The deadly super-cooled air isn't so cold anymore is it? The point is, as the air compresses it heats up—a lot.

ACTIVITY

Put air into a bicycle tire with a manual pump. Pump it up and down rapidly several times. Then feel the temperature of the pump and tire. Why is it so warm? Now release some air from the tire. How does the air feel as it comes out of the tire? Explain it in terms of thermodynamics.

Admittedly, we are making a rough estimate using an equation that describes an *ideal* gas under perfectly adiabatic conditions. It turns out generally in real life the temperature of a rising or falling air mass changes by about 10 °C per kilometer. The troposphere is around 10 kilometers thick at upper northern latitudes, so possibly the air would only heat up by 100 °C to a final temperature of 0 °C (32 °F) at the ground. In addition, once the descending air reaches the same temperature as the surroundings, it will become neutrally buoyant, and stop descending. Therefore, if the temperature equalizes before the air mass gets to the ground, it won't ever get there.

Conclusion. As is often the case in an "action sci-fi disaster movie," some potentially realistic physical phenomenon is distorted out of recognition for the purposes of dramatic effect, and to give the special effects guys something fun to work on. *The Day After Tomorrow* is no exception. Joining *The Core* and *Armageddon,* these three notable films form a kind of action sci-fi disaster bad physics trifecta.

ADDITIONAL QUESTIONS

1) With a depth of about 7 miles beneath the surface, the Marianas Trench is the deepest part of the ocean. What is the pressure at the bottom of the trench?

2) Air has a density of 1.26 kg/m^3 and helium has a density of 0.179 kg/m^3. What is the buoyant force acting on a spherical helium balloon, with a diameter of 1 meter at the surface of the Earth? What is the net force acting on the helium balloon assuming the balloon material has a mass of 10 g?

3) In scene 25 of *Willy Wonka and the Chocolate Factory* the gluttonous Augustus Gloop falls into a chocolate river because he leans over too far while taking a long drink. The chocolate flows into a pipe that rises above the level of the river. What must be true about the pressure inside the pipe in order for the chocolate to rise above the level of the river? What is the maximum possible height to which the chocolate can rise? Remember the pressure in a fluid is the same everywhere at the same depth, so the pressure inside the pipe at river level must be equal to the pressure at the surface of the river. (Assume chocolate has the same density as water.)

4) In the same scene described in problem 3 Augustus gets wedged up in the pipe when it sucks him in. Wonka says the pressure in the pipe is building up underneath Augustus, and eventually Augustus is explosively dislodged. Is it possible for the pressure to build up this way if the river is exposed to the air? Why or why not?

5) Take a hypothetical ship with the same mass and volume as the *Titanic,* but comprised only of 16 completely enclosed watertight compartments. Assuming each compartment has the same volume, what is the maximum number of compartments that could be punctured before the ship would sink?

6) If an iceberg punched a hole the size of a quarter in one of the compartments of the ship from problem 5 at a depth 10 meters below the waterline, what would be the initial rate at which water would enter the ship?

7) Estimate how much energy would be given off in freezing a 20-meter deep system of canals filling the streets of Manhattan. How would water turning to ice affect the temperature of the adjacent air? At what rate would heat need to exit the water in order for it to freeze solid in less than a day?

8) If the pressure in the upper troposphere is one-tenth that at the surface, what happens to the volume of an air mass as it descends? Calculate the ratio of the volume at the surface to the volume at the top of the troposphere.

9) Will it take more energy to cause something to expand isothermally or isobarically from the same initial to the same final volume? Explain.

10) If you blow on the back of your hand with your mouth open wide, the air feels warm. If you do the same thing making your lips into a small opening it will feel cool. Explain.

Electrostatics, Electricity, and Magnetism

HAVE YOU EVER HEARD A movie reviewer describe a film as "electrifying" or even "magnetic"? Well these one-word reviews would work literally for the movies we are going to discuss in this chapter—two of our favorite films, *Apollo 13* and *The* (infamous) *Core*. We will revisit these films in order to explore electric circuitry and expose ourselves to some electromagnetism. We will also take a scene from *An Enemy of the State* to examine the interesting phenomenon of electromagnetic shielding.

APOLLO 13 AND ENERGY-EFFICIENT CIRCUITRY

Chapter 3 discussed some of the problems that befell the *Apollo 13* mission crew, after an explosion in one of the oxygen tanks ripped a hole in the service module. One of their major problems was energy-related.

The explosion knocked out the fuel cells that were the prime source of electrical energy used to power the command module (CM). The crew had no choice but to use the lunar module (LM) as a "lifeboat" until reentry. However, upon reentry they were going to have to jettison the LM, return to the CM, and use battery power to run the CM systems. Unfortunately, as the circuits were configured mission control knew that there just wasn't enough energy to do the job. They had to find a way to reconfigure the CM circuits and resequence the start-up such that the battery could provide sufficient energy to run the systems for reentry. If they could not find a way to do this then they would not make it back alive. In scenes 11 and 15 director Ron Howard recreates mission control's intense efforts to find a solution to the circuit problem.

Although the actual circuitry in the command module is more complicated than we want to deal with here, we can make some simpler models and go through the same exercise that mission control had to go through to make the most of the energy they had. With a basic understanding of simple electrical circuits, we can get a feel for how we can affect the rate at which a circuit uses energy. First, let's lay out some basic principles of electrostatics and electric circuits.

Photo: A high voltage electrostatic discharge.

Electrostatics

In order to analyze the behavior of electric circuits, which involve moving electric charge, it is helpful first to understand how static electric charges affect each other. We have spent some time in earlier chapters discussing one of the fundamental forces of nature—the gravitational force. Here we will introduce another fundamental force—the electrostatic force—which exists between all electric charges.

Charge. Just as mass is the fundamental property of matter responsible for the gravitational force, the fundamental property of matter that is responsible for all electrical forces is called *charge*. There are two types of charge arbitrarily designated either "positive" or "negative." The smallest possible charge is that carried by the electron—the particle-like entity found outside the nucleus of an atom. The charge of an electron is equal to -1.6×10^{-19} Coulomb (C), which is equal in magnitude to the charge on a proton but of the opposite sign. The proton, which is one of the constituents of the nucleus of the atom, has a charge equal to $+1.6 \times 10^{-19}$ C. The other constituent of the nucleus is called the *neutron,* which has no net charge. The behavior of electric charge is such that *like charges repel each other, and unlike charges attract.*

Example 1: Sven places three charges, A, B, and C, in a region of space in an attempt to impress Ingrid. Charge A attracts charge B, but repels charge C. How will charge B affect charge C?

Solution: Because A attracts B they must be oppositely charged. Because A repels C, C must be the same type charge as A. Therefore, B is opposite in sign to C and will attract it. The question is will Sven attract Ingrid?

Charge may be transferred from one object to another; however, because the electron resides on the outside of the atom, it is much more mobile than the proton and is the particle that is actually able to move from one atom to another. Therefore, charge transfer is accomplished primarily through the motion of electrons.

The Law of Conservation of Charge. While charge may be transferred from one location to another, the total amount of charge is constant. For example, if an object gains 1 C of charge, this amount of charge must have come from another object, or objects, that has lost the exact same amount.

Coulomb's Law. Coulomb's law relates the electrostatic force between any two charges. If, for example, there are two charges, q_1, and q_2, whose centers are separated by a distance r, then the force that each exerts on the other is given by:

$$F = \frac{kq_1q_2}{r^2},$$

where k is the Coulomb's law constant and is equal to 8.99×10^9 Nm^2/C^2. Like charges repel each other, while unlike charges attract.

Example 2: Two charges are separated by a distance r and exert forces on each other equal to F. If Sven doubles the distance between the charges, how much is the force between the charges reduced?

Solution: The new force is $\frac{1}{4}$ the original force. Because $F \propto \frac{1}{r^2}$, if r is doubled the force is reduced by $\frac{1}{4}$.

$$F_1 = \frac{kq_1q_2}{r^2} \text{ and } F_2 = \frac{kq_1q_2}{(2r)^2} = \frac{kq_1q_2}{4r^2} = \frac{1}{4}F_1$$

Notice how the form of Coulomb's law is identical to the universal law of gravitation. Both are *inverse square laws*, where the force falls off rapidly as a function of distance.

While the gravitational force depends on the product of the masses of two objects, the electrostatic force depends on the product of the charges. Notice also that k is 20 orders of magnitude greater than G, the gravitation constant! This means that the electrostatic force is actually trillions of times stronger than the gravitational force.

There may be more than two charges in a region in space. In this case, to calculate the total or net force acting on a given charge in the region, calculate the force on that charge due to each of the other charges separately, and then take the vector sum of these forces to determine the net force.

THINKING PHYSICS

If the electrostatic force is so much greater than the gravitational force why doesn't the electrostatic force dominate over the gravitational force in everyday life? (Hint: Consider that there are two types of charges, but only one type of mass.)

Example 3: Three positively charged particles each with a magnitude of 1×10^{-6} C float by the *Apollo 13* spacecraft in a line such as the one in the following figure.

If $r = 0.1$ m, determine the net electrostatic force on q_1.

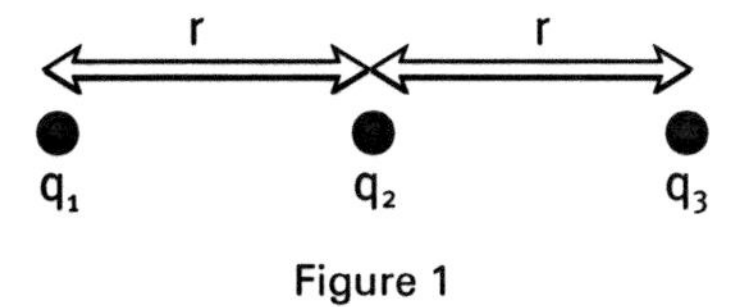

Figure 1

Solution: Calculate the forces that q_2 and q_3 exert on q_1 separately:

$$F_{q_2 \text{ on } q_1} = \frac{kq_1q_2}{r^2} = \frac{(9 \times 10^9 \text{ Nm}^2/\text{C}^2)(1 \times 10^{-6} \text{ C})(1 \times 10^{-6} \text{ C})}{(0.10 \text{ m})^2} = 0.90 \text{ m to the left}$$

$$F_{q_3 \text{ on } q_1} = \frac{kq_1q_3}{r^2} = \frac{(9 \times 10^9 \text{ Nm}^2/\text{C}^2)(1 \times 10^{-6}\text{ C})(1 \times 10^{-6}\text{ C})}{(0.20 \text{ m})^2} = 0.23 \text{ m also to the left}$$

The total force is the sum of these two forces = 1.1 N to the left.

ACTIVITY

a) Take two 6- to 8-inch pieces of Scotch tape and press them onto a flat table with a bit of one side hanging over the edge so that each piece can be pulled off easily. Then pull each piece of tape off of the table, keeping them separated from each other, holding them so that each piece hangs down vertically. Now move each piece of tape slowly towards each other. What happens? Explain?

b) This time take a single piece of tape and charge it as we did in the first activity. Now slowly move it towards the wall, the back of a chair, or some other neutrally charged object. What happens?

Transfer of Charge. Charge transfer is accomplished primarily by motion of electrons. In the accompanying activity, electrons are either stripped off of the tape onto the table, or stripped off of the table onto the tape such that each piece of tape (and the table) is left with a net charge. We can't tell what has which charge just from the experiment, because we don't have an object of known charge to compare it to. We do know, however, that each piece of tape must have the same sign of net charge because we treated them identically. If you did the experiment carefully you will have seen the two pieces of tape move away from each other due to their like charges repelling.

It turns out if we take the charged piece of tape and place it in the vicinity of another material that has no net charge, without touching it, there can still be a force exerted between the neutral material and the charged tape. This is because the electrons in the neutral material redistribute themselves. If we assume the tape is negatively charged, then the tape will repulse the electrons in the chair resulting in a region of net positive charge closer to the tape, and net negative charge further away from the tape (this separation of charge is called *polarization*). The tape and the chair will therefore exert attractive forces on each other due to this temporarily *induced* charge in the chair.

The Electric Field

How is it possible that fundamental forces like the gravitational force and the electrostatic force can be felt across empty space? In order to explain this behavior, physicists model the empty space around a mass or charge as having something called a *field*. A field is something that exists whether or not there is anything there to be affected by it. However, if you place a mass in the gravitational field caused by another mass it will feel a gravitational force. An *electric field* defines the region in space around a charge or charges in which another charge will feel a force if placed in that region. We define the direction of the electric field at a point in space to be in the direction of force that a very small positive "test charge" would feel if it happened to be placed at that point. Therefore, the direction of the electric field in the space around positive charges would be directed away from those charges, and the field around negative charges would be directed towards them. The strength of the electric field is given by $E = \frac{F}{q}$ where E is the field strength in

Newtons per Coulomb of charge, and F is the force that a small positive test charge q would feel if it happened to be placed in the region.

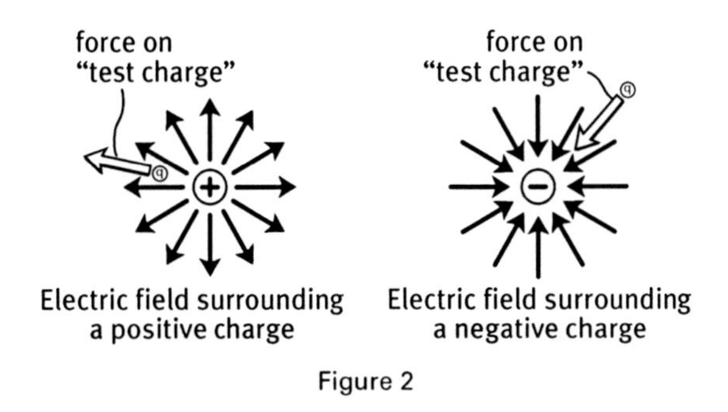

Figure 2

Example 4: At a point in space near a charged metal pie plate Sven measures the electric field strength to be 500 N/C pointing in a direction due east. If Ingrid comes over carrying a small piece of copper with 0.001 C net charge and places it at that point, what magnitude and direction of force will the piece of copper experience?

Solution: $F = qE$ = (0.001 C)(500 N/C) = 5 N due east because a positive charge will experience a force in the same direction as the field.

Electric Potential/Voltage

The *electric potential (V)* is proportional to the amount of energy that is available to do work on electric charges in an electric field. The potential difference between two points is the amount of work it would take to move a charge q between those two points against the electric field, divided by the amount of the charge:

$$\Delta V = \frac{Work}{q}$$

Note: The units of potential and potential difference are Joules/Coulomb (1 J/C = 1 Volt *(V)*.

ΔV is therefore proportional to the amount of energy that the charge has gained by having been moved a distance against the force of the electric field. The idea is similar to lifting a mass up off of the ground against the Earth's gravitational field, thus giving the object gravitational potential energy relative to the ground. The *gravitational potential* would be the change in potential energy per mass. Electric potential is also called *voltage*, and is an essential concept in analyzing electric circuits. Because voltage is proportional to electrical potential energy, we often talk of a *voltage drop* if the potential level goes from a higher to a lower value.

In the case of a uniform electric field in a region of space we can relate E to V as follows. The work to move a charge q through a potential difference ΔV is $W = q\Delta V$, but for a constant E the amount of work is also $W = Fd$ where F is the force acting on the charge q, and d is the distance that the charge is moved through the field.

Because $F = qE$, $W = qEd = q\Delta V$, so $\Delta V = Ed$.

Electric Circuits

We can utilize the energy contained in moving electric charge to do work like toasting some bread, operating a drill, or starting a bus. An *electric circuit* consists of a complete path of conducting material through which electric charges can flow. They require some means by which energy is supplied to the circuit (for example, a battery) so that the charges will be able to flow. Usually there are various electrical components within the circuit, such as resistors, capacitors, diodes, and so on, that are designed to do work or perform a specific task using the flow of electric charge as an energy source.

Conductors and Insulators. *Conductors* are materials through which electric charge can flow easily. Metals are particularly good conductors due to the fact that electrons in metals are extremely mobile and move from atom to atom quite easily. Most wiring in electrical circuits is made of a metal core (copper is very common) surrounded by a thin layer of *insulating* material.

Insulators are materials that do not conduct electric charge easily due to the relative immobility of their electrons. Good insulators include rubber, many plastics, and Styrofoam.

Current, Voltage, and Resistance

Current (I) is defined as the rate at which charge flows through a conductor. The units of current are Coulombs/second. 1 C/s = 1 Ampere (A). Although for the most part it is electrons that are actually in motion in a circuit, by convention the direction of I is said to be in the direction that positive charge would flow through the circuit. When analyzing the behavior of circuits this presents no problem, as a positive charge flow in one direction is mathematically (and conceptually) equivalent to a negative charge flow in the other direction.

As we discussed, *voltage* or electrical potential is proportional to how much energy is available to move charge through a circuit. Therefore a 9-Volt battery provides more energy to the circuit than a 1.5-Volt battery.

As charge moves through a circuit it encounters resistance to flow. *Resistance* is a measure of how severely the charge flow is impeded in its motion. It's analogous somewhat to friction in mechanical systems. The wires and all of the elements and devices in the circuit have resistance, although we often neglect the resistance of the wires when there are other elements in the circuit because the wire resistance is usually very small relative to the other resistances. Many devices convert electrical energy to some other form of energy. For example, a lightbulb converts electrical energy to heat and light, and an electric fan converts electrical energy to mechanical energy. These devices create resistance to the current that flows through it and are called *resistors*. The unit of resistance is the ohm (Ω). $1\ \Omega = 1$ J/C.

Ohm's Law

As the voltage is increased in a circuit the resulting current also becomes larger. In *many* circuits there is a linear relationship between the voltage and the current.

This linear relationship implies that current is directly proportional to voltage, or that the ratio of V to I is a constant. This ratio is equal to R, the resistance of the circuit, expressed by *Ohm's law:*

$$V = IR$$

Although this is not a fundamental principle (like Newton's second law, for example) and there are many circuits that do not obey this relationship, many simple circuits can be analyzed using Ohm's law.

Example 5: In an attempt to study electric circuits before the filming of *Apollo 13* Gary Sinise puts together the following circuit from odds and ends in his garage. The battery has a voltage equal to 3 Volts, and the resistor has a resistance of 1 Ω.

a) What is the current in the circuit?

b) With what value resistor would you have to replace the 1 Ω resistor to increase the current in the circuit to 12 A?

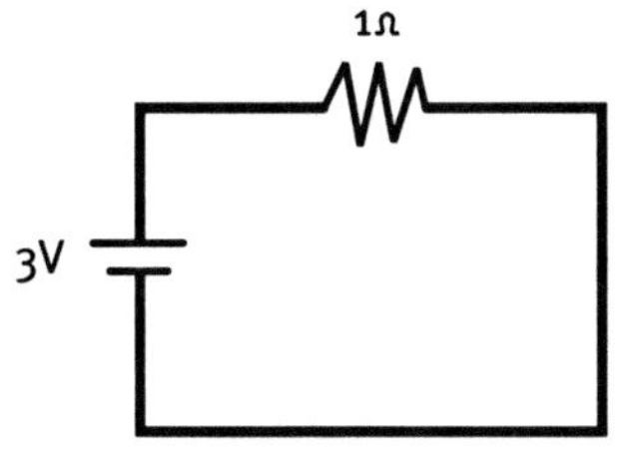

Figure 3

Solution: a) $I = \frac{V}{R} = \frac{3\text{ V}}{1\ \Omega} = 3\text{ A}$ b) $R = \frac{V}{I} = \frac{3\text{ V}}{12\text{ A}} = 0.25\ \Omega$

Electrical Power in Circuits

Power is defined as the *rate* at which work is done, or the rate at which energy is transferred from one form to another: $P = W/t$ or $P = E/t$. In an electric circuit we can show that power is related to voltage, current, and resistance as follows:

$$P = IV = \frac{V^2}{R} = I^2 R$$

The units of power are Joules/second (J/s), and I J/s = 1 Watt (W). A 60-Watt lightbulb means that the filament of the bulb converts 60 Joules of electrical energy to heat and light energy every second.

Example 6: For the circuit in the previous example determine the power dissipated by the circuit for the situations described in (a) and (b).

Solution:

a) $P = IV = (3\text{ A})(3\ V) = 9\text{ W}$ or, $P = I^2R = (3\text{ A})^2(1\ \Omega) = 9\text{ W}$
b) $P = IV = (12\text{ A})(3\ V) = 36\text{ W}$

Note that the circuit with lower resistance and more current uses energy at a much greater rate.

The Kilowatt-Hour. Despite the way it sounds, *the kilowatt-hour (KWh)* is actually a unit of energy, not power. One KWh is the energy used by a circuit, or electrical device, operating at a power of 1 KW (= 1,000 W = 1,000 J/s), for a time of 1 hour. In Joules, 1 KWh = 1,000 J/s(1 hr)(3,600 s/hr) = 3.6 million Joules. The electrical meter outside a home measures the amount of energy used in units of KWh.

Example 7: Over a one-month period of time, Ingrid's household electrical meter reads a total energy use of 120 KWh.

a) How much energy in Joules has she used in the month in question?
b) The electric company charges a rate of 20 cents per KWh. What will Ingrid's electric bill be for that month?

Solution:

a) $E = 120\text{ KWh}(1{,}000\text{ W/KW})(1\text{hr})(3{,}600\text{ s/hr}) = 4.3 \times 10^8\text{J}$
b) Total cost = 120 KWh(20 cents/KWh) = 2,400 cents = $24

Circuits with Multiple Resistors

Resistors in a circuit can be placed in *series*, in *parallel,* or in a combination of series and parallel arrangements.

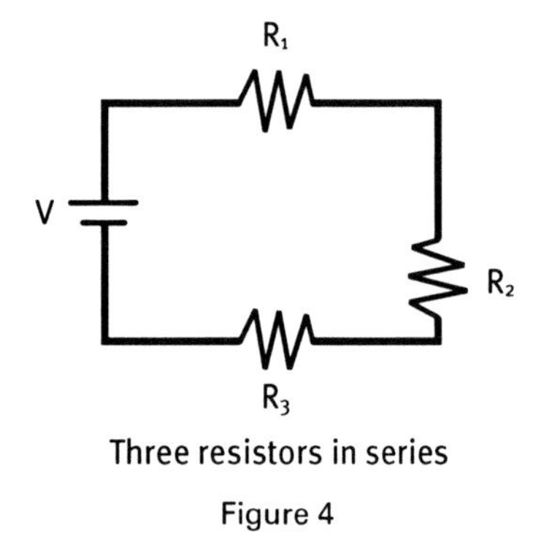

Three resistors in series

Figure 4

Series Circuits. Resistors in series are all in the same path in the circuit (as shown in the accompanying figure).

The current must therefore be the same through each resistor in the series arrangement.

The total resistance (also called *equivalent resistance*) in a series circuit is the sum of the individual resistances:

$$R_{\text{equivalent}} = R_1 + R_2 + R_3 + \ldots$$

Example 8: Tom Hanks is at Gary Sinise's house playing with circuits. They have three resistors $R_1 = 3\Omega$, $R_2 = 5\Omega$, and $R_3 = 7\Omega$. They want to get a feel for resistors in series so they put together a circuit like the one shown in figure 4. Determine

a) the total resistance of the circuit.
b) the total current in the circuit.
c) the amount of voltage dropped across each of the resistors.
d) the power dissipated by the circuit.

ACTIVITY

With an inexpensive circuit kit including a couple of "D" batteries, some wires, and small bulbs (resistors) you can show the effect of adding resistors in series. First make a complete circuit using a single bulb. Notice the brightness of the bulb. Next add a second bulb, and compare their brightness to the single bulb (and to each other). Repeat for three bulbs. Assuming the brightness is proportional to the current through the bulb, what happens to the total current each time you add a bulb? What does this suggest is happening to the total resistance each time you add a bulb?

Solution:

a) $R_{total} = 3\ \Omega + 5\ \Omega + 7\ \Omega = 15\ \Omega$

b) $I = \frac{V}{R_{eq}} = \frac{5\ V}{15\ \Omega} = \frac{1}{3}A$

c) Use Ohm's law for the individual resistors in the circuit

V across the 3Ω resistor = $IR = \frac{1}{3}A(3\ \Omega) = 1\ V$

$V_{5\Omega} = \frac{1}{3}A(5\ \Omega) = \frac{5}{3}V$

$V_{7\Omega} = \frac{1}{3}A(7\ \Omega) = \frac{7}{3}V$

The total of the individual voltage drops add up to the 5-Volt of the battery.

d) $P = IV = \frac{1}{3}A(5\ V) = \frac{5}{3}W$

Parallel Circuits. Resistors in parallel are arranged in separate branches in the circuit as shown in the following figure.

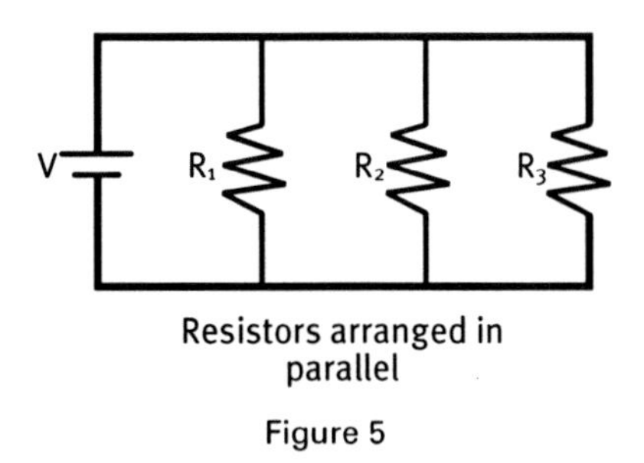

Resistors arranged in parallel

Figure 5

Because they are in separate branches, the current through each resistor may be different; however, the voltage drop across each resistor in parallel must be the same. When placed in parallel the total resistance of the circuit is actually less than any of the individual resistors, and the more resistors you add in parallel the lower the total resistance becomes, and the higher the total current.

ACTIVITY

Using the same materials from the previous activity, make a circuit with one bulb and make a note of its brightness. Next, make a circuit with two bulbs *in parallel.* Compare the brightness of the two bulbs to each other, and to the brightness of the single bulb circuit. Repeat with three bulbs in parallel. Rank the circuits in terms of total current and total resistance. (There may be small differences in brightness due to slight differences between filaments. Try to look for obvious differences.)

$$\frac{1}{R_{eq}} = \frac{1}{R_1} + \frac{1}{R_2} + \frac{1}{R_3} + \ldots$$

If you did the last activity you probably found that the brightness of all of the bulbs was essentially the same. If the brightness depends on current, this means that each of these bulbs is receiving the same current. However, if branches are in parallel then the current must divide into each branch, so the total current must go up the more branches you add in parallel, and therefore the total resistance must go down.

Example 9: Tom and Gary have been playing around with the circuit equipment again, and they put together the circuit shown in figure 6. Determine

a) the total resistance.
b) the voltage drop across each resistor.
c) the current through each resistor.
d) the power dissipated by the circuit.

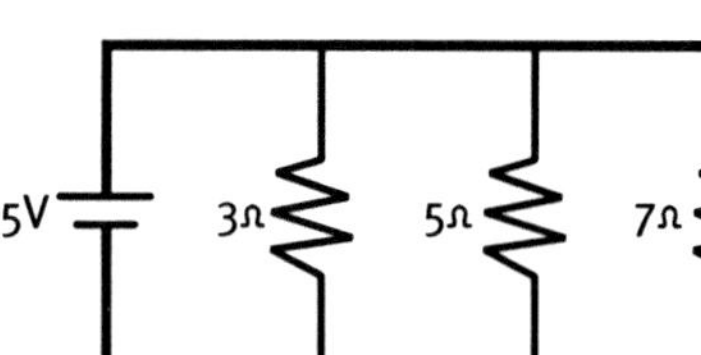

Figure 6

Solution:

a) $\frac{1}{R_{eq}} = \frac{1}{3\ \Omega} + \frac{1}{5\ \Omega} + \frac{1}{7\ \Omega} = \frac{0.68}{\Omega}$, therefore $R_{eq} = \frac{1}{0.68}\ \Omega = 1.5\ \Omega$

$I_{total} = \frac{V_{battery}}{R_{eq}} = \frac{5\ V}{1.5\ \Omega} = 3.3\ A$

b) Each resistor must have the same voltage drop as the battery, therefore $V_1 = V_2 = V_3 = 5V$

c) Use Ohm's law individually for each R, such that

$I_1 = \frac{V_1}{R_1} = \frac{5\ V}{3\ \Omega} = \frac{5}{3} A$, $I_2 = \frac{5\ V}{5\ \Omega} = 1\ A$, $I_3 = \frac{5\ V}{7\ \Omega} = \frac{5}{7} A$, and these three currents add together to give the total current of 3.3 A.

d) $P = IV = 3.3\ A(5\ V) = 16.5\ W$

Notice that for the same three resistors, the parallel arrangement gives a much lower resistance and a higher total current and power for the circuit. Both Gary and Tom find this illuminating.

It is easy to see that for resistors arranged in parallel, burning out one element in the circuit leaves the others unaffected. In a series arrangement, if one element in the line breaks the entire circuit is disrupted. This is one of the reasons that household wiring is arranged in a parallel system. Another, more important reason is that each device that you plug in is designed to have a 120-V drop across it. If you were to place electrical appliances in series with each other the 120-V drop would be divided up among the appliances.

Circuits with Combined Series and Parallel Arrangements. Many circuits have combinations of series and parallel arrangements. For example, in the following circuit, the 8-Ω, 6-Ω, and 12-Ω resistors are in parallel with each other, but they are in a branch that is parallel with the 6-Ω and 12-Ω resistors. The entire parallel system in turn is in series with the 2-Ω resistor.

To get the equivalent resistance for the entire circuit just work from the most detailed to the most general. First reduce the three-way parallel to a single resistor. Then add the equivalent of the parallel system to the 2-Ω resistor as shown.

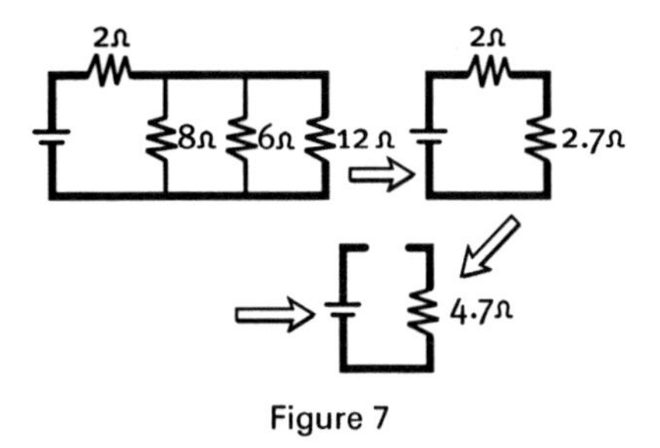

Figure 7

We can determine the current through each part of the circuit, the voltage drop across each resistor, and the total power dissipated by the circuit as follows:

$$I_{tot} = \frac{V_{tot}}{R_{eq}} = \frac{30\text{ V}}{4.7\ \Omega} = 6.4\text{ A}$$

$$V_{2\Omega} = I_{tot} R_{2\Omega} = (6.4\text{ A})(2\ \Omega) = 12.8\text{ V}$$

Which leaves 17.2 V dropped across each of the parallel branches. From this we can find the current through each one of these branches.

$$I_{8\Omega} = \frac{V_{8\Omega}}{R} = \frac{17.2\text{ V}}{8\ \Omega} = 2.2\text{ A}$$

$$I_{6\Omega} = \frac{V_{6\Omega}}{R} = \frac{17.2\text{ V}}{6\ \Omega} = 2.9\text{ A}$$

$$I_{12\Omega} = \frac{V_{12\Omega}}{R} = \frac{17.2\text{ V}}{12\ \Omega} = 1.4\text{ A}$$

Finally, the total power is given by:

$$P_{tot} = I_{tot} V_{tot} = (6.4\text{ A})(30\text{ V}) = 192\text{ W}$$

OTHER COMMON CIRCUIT ELEMENTS

Capacitors are devices that store charge and electric field in a circuit that can be used at a later time.

Diodes are devices that are used to regulate the voltage in circuits, and to make logic gates in computers. A diode has a very high resistance in one direction, and therefore current can only flow the other way. LEDs, or light emitting diodes, are commonly used for indicator lights in computers or televisions.

Thermistors are devices used as temperature sensors. Their resistance decreases with increasing temperature, so that the higher the temperature the more current will flow through them.

Light dependent resistors (LDRs) and are used to detect light levels. Their resistance decreases as light intensity increases.

Note: Sometimes resistors can be arranged in such a way that they cannot be reduced to a system of series and parallel. There could also be more than one battery, and these could be arranged in different branches in the circuit. Circuits with multiple batteries cannot be analyzed as we did with our previous example.

To analyze these types of circuits we can apply fundamental principles of conservation of energy and conservation of charge through the use of Kirchhoff's rules, which we won't address here. (Incorporating other types of circuit elements makes analyzing circuits more complicated as well.)

Measuring Voltage, and Current in Circuits. You can use an *ammeter* to measure current through a resistor. This must be placed in series with the resistor so that it receives the same current as that passing through the resistor. However, so as not to affect the current, the ammeter must have a very low resistance.

A *voltmeter* is used to measure voltage drop across a resistor. The voltmeter must be placed in parallel with the resistor so that it drops the same voltage as the resistor. However, it must have a very high resistance relative to the resistor so that very little current will flow through it.

Direct Current and Alternating Current

Direct current (DC) occurs when the current in the circuit flows in only one direction. Circuits powered by batteries are DC circuits. In a DC circuit the voltage is a constant value, resulting in a constant value of current in the circuit. The systems aboard *Apollo 13* ran on batteries, and therefore were DC circuits.

Alternating current (AC) is a current that constantly switches directions. This is the type of current supplied by the generators that supply power to large systems, such as a house, city, and so on. This current oscillates very rapidly (60 times a second, or 60 Hz (hertz) in the United States). The alternating current is due to a rapidly changing voltage source. Therefore, the voltage and current in this type of circuit are constantly changing, however; the oscillation is so rapid that circuit elements operate continuously.

Household Circuits and Wiring

Household circuits are wired in parallel. Separate branches in a parallel circuit remain unaffected by the other branches. Therefore, if an electrical appliance or device in the house breaks, for example, if the filament of a lightbulb burns out, then the other appliances will operate normally. In addition, appliances are designed to operate at specific voltages and currents. The only way to insure the proper voltage is for each appliance to be in its own separate branch in a parallel circuit.

Fuses and Circuit Breakers. As we saw, the more branches that are connected in parallel the smaller the total resistance and, therefore, the larger the total current in a circuit. So when a lot of appliances are turned on simultaneously in a house that drives the total current in the circuit up. When the current gets too high the wires may heat up to temperatures large enough to melt the surrounding insulation. This is a dangerous situation that can lead to an electrical fire. For this reason circuits are often protected from overheating. The idea is to break the circuit once the temperature gets beyond a certain level. One way to do this is to place a *fuse* in the circuit. A fuse contains a piece of wire or metal that melts easily. If the temperature gets too high the fuse melts, thus breaking the circuit. Another way to protect the circuit is to use a *circuit breaker.* With a circuit breaker a spring-loaded switch is held closed by a spring-loaded iron bolt. An electromagnet is set up so that if it receives sufficient current it can pull the bolt away from the switch, because the higher the current the stronger the electromagnet. A circuit breaker does not have to be replaced like a fuse; it simply needs to be reset once the wires cool down.

A *short circuit* occurs when for some reason the circuit is able to bypass the resistor. For example, if a cord becomes frayed, the exposed wires can touch and the current bypasses the resistor. Because the total resistance of the circuit becomes greatly reduced the current becomes very high. A fuse or circuit breaker could protect the system if a short circuit occurred. This is in fact what they believe occurred within one of the oxygen tanks aboard *Apollo 13.* A short circuit caused a spark that ignited the highly volatile oxygen. This resulted in the explosion that disabled the ship.

Grounding and Electrical Safety. Electrical circuits can also pose an electrocution danger in certain circumstances. Current above 100 mA driven through a person is often fatal. The amount of current will depend on how high the voltage is, and the resistance of both the person and the device or appliance to which the person is in contact. To receive an electric shock two parts of the body must be in contact with conductors at different potentials, because potential difference is a property of two points. In typical wiring one of the two wires is connected physically to the ground. This wire is said to be *grounded.* This arrangement prevents the wiring from reaching excessively high potentials; however this means that if a person contacts the high-voltage wire and any grounded conductor, including the ground itself, they will receive a shock. For example, if a short circuit occurs within an appliance the exterior of the appliance may be at high potential, and if a person were in contact with the appliance and the ground a current will flow through them. To avoid this many appliances and power tools are equipped with three wire cords. The third wire runs from the exterior of the appliance to a grounded wire in the outlet. Normally no current flows through this wire; however, if a short circuit should occur this third wire provides a very low resistance path for the current to travel to ground, bypassing the person.

Apollo 13 Short-Circuits

Now let's get back on board *Apollo 13* just before the accident depicted in scene 5. The crew is on the journey out to the moon and is just settling back in to the routine after doing a broadcast from space. Mission control gives Jack Swigert (Kevin Bacon) some standard "housekeeping" procedures, one of which is to "stir the oxygen tanks" to keep the oxygen slush from stratifying in the tanks. This is the moment when the explosion becomes imminent. As we segue into scene 6, a computer-generated sequence takes us through the wiring and into the interior of the defective oxygen tank. Inside is some damaged wire. It looks like the insulation has deteriorated and frayed, and the high and low voltage wires come into contact with each other causing sparking and heating. Because oxygen is highly explosive, if overheated, an explosion is inevitable. (It turns out that *Apollo 13*'s problem with the oxygen tank was due to an engineering oversight. The tank was refitted from a previous mission, and was designed to operate at 28 V. When it was installed, it had to be adjusted to handle 65 V. However, the thermostat was never altered and it malfunctioned at the higher voltage. The insulation probably melted when the tanks were tested before launch. The temperature inside the tank was much higher than it should have been, but no one noticed because the thermostat was stuck. Once the mission was underway, excessive heating and a short circuit was inevitable when they "stirred the tanks," leading to the explosion.)

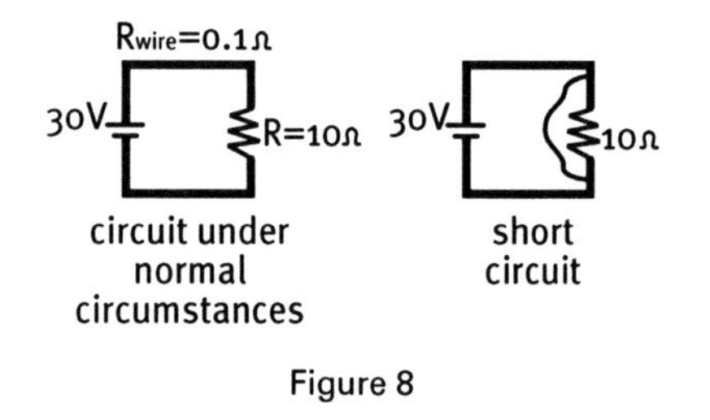

Figure 8

Let's look at how a short circuit can lead to sparking, and overheating. So far we have assumed that the resistance of the wires in a circuit is negligible compared to that of the resistors when applying Ohm's law. As an example, in the circuit in figure 8 let's say the resistance of the wire is 0.1 Ω.

Because the resistor has so much more resistance than the wire the total current in the circuit is approximately

$$I = \frac{V}{R} \approx \frac{30 \text{ V}}{10 \ \Omega} = 3.0 \text{ A}$$

However, what happens if there is a short circuit? In essence, in this situation a piece of wire makes a parallel branch with the resistor. Now the branch with the resistor has so much more resistance than the branch with the wire, *almost* all of the current will flow through the branch of least resistance and the total resistance of the circuit drops drastically. In this case:

$$\frac{1}{R_{eq}} = \frac{1}{10 \ \Omega} + \frac{1}{0.1 \ \Omega} \approx 0.1 \ \Omega \quad \text{and} \quad I = \frac{V}{R} \approx \frac{30 \text{ V}}{0.1 \ \Omega} = 300 \text{ A}$$

This is a huge amount of current, and without a circuit breaker the wire is going to get very hot, melt, and possibly start an electrical fire.

However, even if the shorting wire doesn't completely close the short circuit, if it is close enough there could still be a problem.

What causes a spark to jump between objects? Well, it is essentially the same thing that causes current to flow in a circuit. When you scrape your feet on the carpet you gain an excess charge that distributes itself around the exterior of your body, meaning there is a *potential difference* between your body and surrounding objects (and an *electric field* between your body and the objects). If you get close to a piece of metal then you can induce an opposite charge in the part of the metal closest to you. This means that electrons will feel a force compelling them to accelerate. The spark that you see, as well as the unpleasant tactile sensation that you feel, are the effects of a bunch of electrons jumping between the metal and your finger.

ACTIVITY

On a very dry day, walk on a carpet by shuffling your feet across it, trying to generate as much friction as possible. Now slowly move your finger towards a metal object like a doorknob and see what happens. (This works best inside the house on cold winter days in northern climates.)

Frayed wires can lead to sparking, fires, severe overheating, and in the case of *Apollo 13*, a life and death situation.

Saving Energy

To save electrical energy and get *Apollo 13* back home, mission control had to find a way to rearrange the circuitry. They had to find a startup sequence for the systems that would drastically reduce the power rate, but such that each system necessary for reentry would still operate. In fact, to have sufficient energy they had to reduce the total current of the circuit from 60 A to 12 A, or reduce the power by a factor of five. No wonder it wasn't easy. (In scene 11, mission control states that they have to reduce the current to 12 A, but in scene 15 we see the meter at 20 A during the successful simulation. If that reading was accurate, they would have had to reduce the power by a factor of three.)

Let's examine a hypothetical circuit and go through a power-saving exercise. Let's say on our space craft we have the following circuit running on a 60-V battery.

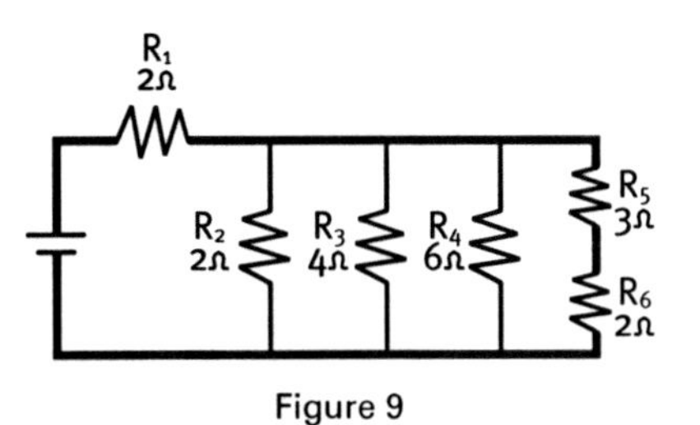

Figure 9

We need to calculate the normal operating voltages and currents for each resistor in the circuit. First we can get the equivalent resistance of the circuit and the total current. Because R_5

and R_6 are in series, they act as a single 5-Ω resistor. Now we have four branches in parallel, which can be reduced to one.

$$\frac{1}{R_{parallel}} = \frac{1}{2\ \Omega} + \frac{1}{4\ \Omega} + \frac{1}{6\ \Omega} + \frac{1}{5\ \Omega} = \frac{1.12}{\Omega} \text{ so } R_{parallel} = \frac{1}{1.12}\ \Omega = 0.89\ \Omega$$

Now we can add the equivalent resistance of the parallel branches to R_1 to get the equivalent resistance of the entire circuit. $R_{eq} = 0.89\ \Omega + 2\ \Omega = 2.89\ \Omega$. Then we can use Ohm's law to get the total current.

$$I = \frac{V_{battery}}{R_{eq}} = \frac{60\ \text{V}}{2.89\ \Omega} = 20.8\ \text{A}$$

From this we know that the total power for the circuit is $P = IV = (20.8\ \text{A})(60\ \text{V}) = 1{,}250\ \text{W}$.

To get the current and voltage drops through each resistor we can apply Ohm's law individually to each one. Because R_1 is in the "main branch" of the circuit with the battery before/after the current is divided between the parallel branches, it receives the total 20.8 A. The voltage drop across R_1 is then $V_1 = IR_1 = (20.8\ \text{A})(2\ \Omega) = 41.6\ \text{V}$. This leaves a voltage drop across each parallel branch of 18.4 V. Now we can determine the current through each of these branches.

$$I_2 = \frac{V_{parallel}}{R_2} = \frac{18.4\ \text{V}}{2\ \Omega} = 9.2\ \text{A}$$

$$I_3 = \frac{18.4\ \text{V}}{4\ \Omega} = 4.6\ \text{A}$$

$$I_4 = \frac{18.4\ \text{V}}{6\ \Omega} = 3.1\ \text{A}$$

$$I_{5 and 6} = \frac{18.4\ \text{V}}{5\ \Omega} = 3.7\ \text{A}$$

Finally, because R_5 and R_6 drop a total of 18.4 V together, we can find their individual drops.

$V_5 = I_{5,6}\ R_5 = (3.7\ \text{A})(3\ \Omega) = 11\ \text{V}$, and similarly $V_6 = 7.4\ \text{V}$.

Now let's say we have an emergency situation where the battery is damaged, and we determine that it can only last half as long as it should. Because it was designed to last for the duration of the mission, the only way to keep things running for the entire time is to cut the power dissipation rate of the circuit in half. Because $P = IV$ that means that the total current must be cut in half to no more than 10.4 A.

In our hypothetical spacecraft manual it says that each device will still be functional if it has *at least* half of its optimal operating current running through it. However, they will not function if they have more than their optimal current *except* R_6, which can take up to *three times* its normal operating current and still function properly.

Here is our task: How can we rearrange the circuit so that each device will still function for long enough to get us back to Earth?

> **PROBLEM**
>
> Do the appropriate calculations to illustrate that our reconfigured circuit for our hypothetical spacecraft will meet the power and current requirements. Find the total current in the circuit and the current through each resistor/device in the circuit.

Keep in mind that the current must not exceed 10.4 A, which means that the equivalent resistance must not be less than 5.8 Ω. It's easy to make R_{eq} greater than this by putting a lot of devices in series but it's no good if we don't have enough current in each device (at least half the normal operating currents we calculated). On the other hand we can't have too much current through any resistor. We can rearrange the devices any way we want in the circuit. We have to find the right combination of devices in series and parallel to meet the above requirements.

The following circuit can do the job and run the ship until we make it back to Earth as you can confirm with the accompanying boxed problem!

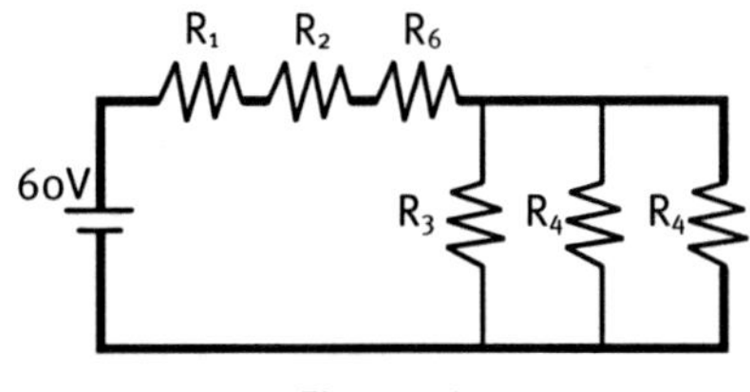

Figure 10

THE CORE REVISITED, RELOADED, REMAGNETIZED

We spent a lot of time talking about *The Core* in Chapter 3, and yet we still have more to say. In that chapter, we introduced some concepts pertaining to the Earth's magnetic field to discuss the fundamental premise of the movie—that the Earth's core has "stopped circulating," causing the magnetic field to disappear. As a result (according to the scientists portrayed in the film), without the protection of the field, incoming radiation will destroy all life on the planet. We discussed why this would be highly unlikely. In this section we are going to look at the physics behind the Earth's magnetism and at electromagnetism in general in more detail. In particular: What causes magnetic fields and how is the Earth's magnetic field created? How do magnetic fields affect charged particles? How does the Earth's magnetic field protect the Earth from incoming radiation? Why do they claim in the movie that life on Earth will be destroyed if the magnetic field disappears?

MAGNETISM

Eerily similar in nature to the electrostatic force is the ever popular magnetic force. While these two fundamental forces seem distinct at first inspection, it turns out that the two are actually *unified* in one fundamental force—the electromagnetic force. We will see some evidence for this interrelationship

ACTIVITY

Find a bar magnet, horseshoe magnet, or one of those really strong circular neodymium magnets. Lay the magnet flat on a piece of paper, and place a compass at several different locations around the magnet. Draw arrows representing the direction of the magnetic field at each of these locations.

between electricity and magnetism after we discuss some of the basics of magnets, magnetic poles, magnetic fields, and magnetic forces.

The Magnetic Field

We can define a *magnetic field* (B) as a region in space in which a magnet will experience a force, such that the *north* side of the magnet will feel a force in the direction of the field and the *south* side of the magnet will experience a force opposite to the field direction.

Magnetic fields exist around permanent magnets like those from the accompanying activity. Thanks to *The Core* we also know that Earth itself has a magnetic field, and a magnet (such as the needle on a compass) will feel a force in the direction of the Earth's field as long as no other stronger fields wash out the effect of that of the Earth. The unit of magnetic field is the *tesla* (T). $1\ T = N/C(m/s)$.

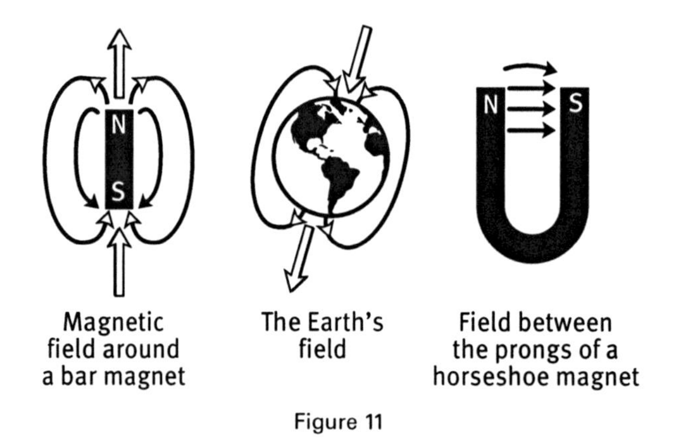

Figure 11

Permanent magnets such as the bar magnet and horseshoe magnet are created when certain susceptible materials called *ferromagnetic* materials are placed in an external magnetic field. The ferromagnetic material then aligns itself at the atomic level such that it produces a net magnetic field in the same direction as the external field. The term "ferromagnetic" is derived from ferrous (iron bearing) because the element iron is the most highly susceptible material to being magnetized.

Magnetic Poles

We know from the behavior of magnets that magnetic forces can be either attractive or repulsive. Therefore, similar to "positive" and "negative" electric charge, we can define a "north" and "south" pole of a permanent magnet. (Magnetic poles are not the same thing as electric charges, although people sometimes confuse them. We can see that they differ because magnets have no effect on static electric charges.)

ACTIVITY

(To do this one we might have to acquire a few items that aren't lying around the house, although if you're taking a physics class you'll probably find what you need in the classroom.)

Find a hemispherical object (like a plastic ball cut in half, or a "watch glass" from a chemistry lab) on top of which you can balance a meter stick. Place the meter stick carefully so that it is stable but can also swing in a circle without too much friction. Rub an acrylic rod with a piece of felt to give it a static electric charge, and hold it close to the meter stick without touching it. It should attract the meter stick, and you can have fun starting and stopping it with the electrostatic force. The normally neutral meter stick is attracted due to the induced charge polarization caused by the proximity of the rod.

Now take the strongest magnet you can find and hold it close to the meter stick. What happens? (Probably not much!) Therefore, we see that there is no force interaction between static charges and magnetic poles. In addition, unlike with electric charges, magnetic poles *always* come in pairs. (Break a bar magnet in half. What happens?) There's something different about magnetic poles compared to the physical properties responsible for gravity (mass) and the electrostatic force (charge). However, it turns out that magnetic poles, while not the same thing as electric charge, are actually *related* in an interesting way, as we shall discuss next.

We can see that a compass needle is deflected when the compass is placed in the proximity of an electric current, which means that there must be a magnetic field around the wire when the current is flowing. Therefore, while we don't detect a magnetic field surrounding stationary charges we do detect one surrounding *moving charges. Ultimately, the source of all magnetic fields is moving charges.*

What does this say about permanent magnets? What is the source of their fields? All materials have moving charge because electrons are in constant motion around the nuclei of atoms. Therefore each atom has its own magnetic field. The reason that most materials do not exhibit macroscopically observable magnetic fields is because the fields of the individual atoms are oriented randomly such that on average they cancel out. When a ferromagnetic material is placed inside an external field, the electron motions within the material can align themselves in the same direction, giving a net magnetic field in the material in the same direction as the external field. When the material is then removed from the external field it retains this magnetic alignment, and becomes a permanent magnet.

ACTIVITY

Make a simple circuit with a couple of batteries, wires, and a lightbulb. Keep the circuit open while you place a compass underneath one of the wires. Align the wire with the direction of the needle, and then connect the circuit. What happens to the compass needle?

Magnetic Field Due to a Current Carrying Wire. Because a current consists of moving charge, a current carrying wire has a magnetic field around it, and if you do a careful experiment you can show that the field forms a circular pattern around the wire. The direction of the field can be determined by *the first right-hand rule.* According to this rule, if you place your thumb of your right hand in the direction of current flow (using the convention that current is in the direction of positive charge flow), then your fingers will curl around your thumb in the direction of the field. The x's in the following figure represent B coming into the page, and the dots represent B coming out of the page.

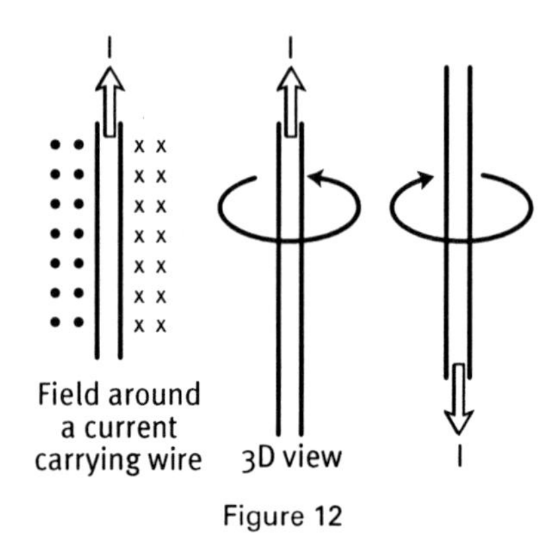

Figure 12

Not surprisingly, the strength of a magnetic field around a straight wire increases with increasing current and decreases with distance from the wire. The field strength a distance r from a wire, which contains current I, is

$$B = \frac{\mu_0 I}{2\pi r}$$

Where the constant $\mu_0 = 4\pi \times 10^{-7}$ Tm/A.

Example 10: Two wires on the space shuttle are parallel to each other and 6 centimeters apart. They each conduct a current of 5 Amperes in the same direction.

a) What is the magnitude and direction of the magnetic field in the middle of the wires 3 centimeters from each one?
b) What is the magnitude and direction of the field 5 centimeters outside the rightmost wire?

Solution:

a) Using the first right-hand rule we can see that in the plane of the page the wire on the left has a field pointing into the page, and the one on the right has a field pointing out. Because each wire has the same current, and the distance to each wire is the same, the net field is zero.
b) In this case both fields point into the plane of the page.

$$B_{total} = B_1 + B_2 = \frac{\mu_0 I_1}{2\pi r_1} + \frac{\mu_0 I_2}{2\pi r_2} = \frac{(2 \times 10^{-7}\ \text{Tm/A})(5\ \text{A})}{0.11\ \text{m}} + \frac{(2 \times 10^{-7}\ \text{Tm/A})(5\ \text{A})}{0.05\ \text{m}} = 2.9 \times 10^{-5}\ \text{T}$$

directed into the plane of the page.

The *solenoid* is a common electrical element in which the wire is wrapped into a stacked coil. If you were to bend a straight wire into a loop, the circular field would be such that the field inside the loop would be in one direction and the field outside the loop in the opposite direction. With several stacked loops, like in a solenoid, it can be seen that *the net field of the solenoid looks exactly like the field due to a bar magnet.*

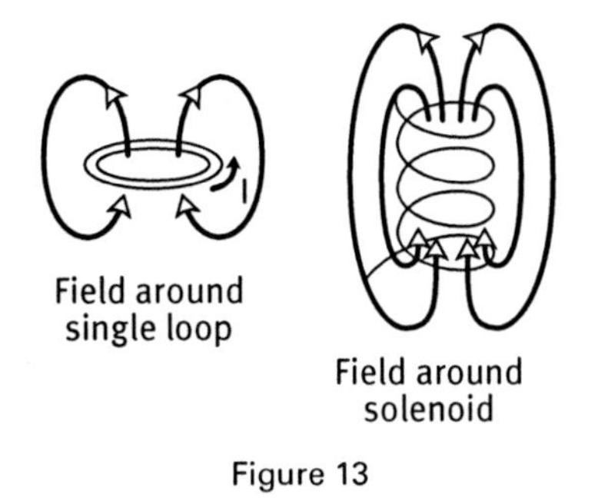

Figure 13

If you place an iron core inside the coils of a solenoid, the field lines are intensified, due to the fact that field inside the iron core aligns itself in the direction of the solenoids field. Such a device is called *an electromagnet.*

Moving Charges in Magnetic Fields. Not only do moving charges create magnetic fields, but also a charge moving through an external field *may* experience a force as it moves through the external field. The way this happens is not as simple as an electric force acting on a charge in an electric field, where the force is always parallel to the field lines and is independent of whether the charge is moving or not ($F = qE$). For there to be a magnetic force on an electric charge: (1) The charge must be in motion or there will be no force. (2) The charge must have at least some component of velocity *perpendicular* to the field lines. If the charge is in motion parallel to the magnetic field there will be no force. Mathematically, the magnetic force on a charge q moving through a magnetic field B with velocity v is given by:

$$F = qvB \sin \theta$$

where θ is the angle between the field lines and the direction of the velocity as shown.

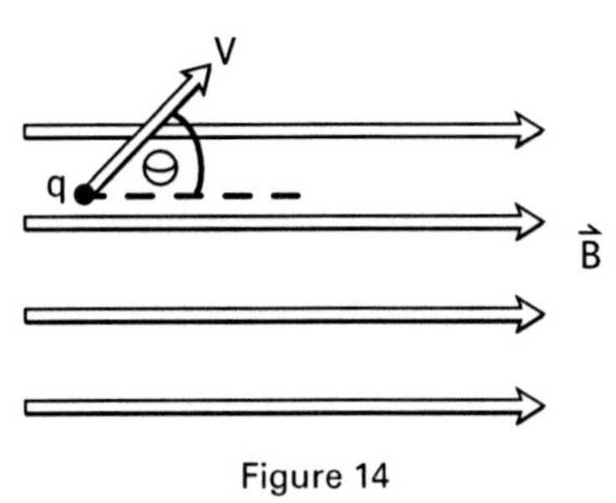

Figure 14

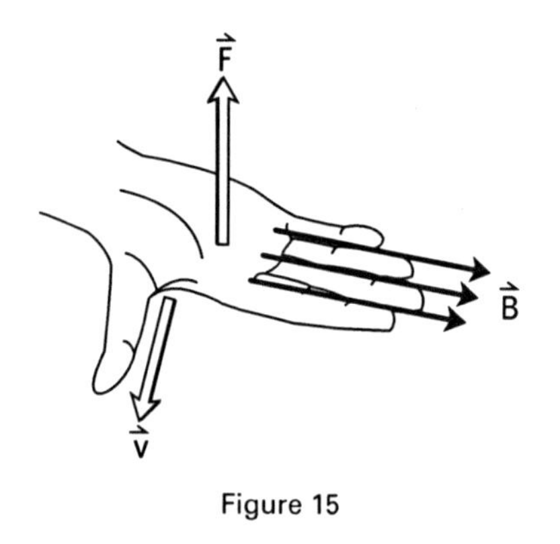

Figure 15

The direction of the force is neither in the direction of the field, nor in the direction of the velocity. It is perpendicular to both of these directions.

However, there are two possible directions perpendicular to the plane of the velocity vector and the field vector.

You can conveniently determine which of these two directions the force will be exerted on the moving charge using *the second right-hand rule.* For this rule, place the thumb of your right hand in the direction of the motion of the charge, and your fingers in the direction of the magnetic field. The force on a positive charge will be in a direction coming out of your palm. The force on a negative charge will be exactly the opposite (out of the back of your hand).

Alternatively, to determine the magnetic force on a negative charge you can use your left hand (*the left-hand rule)*, such that the direction of force comes out of the palm of your left hand. The charges shown in the following diagrams are all moving up the plane of the page. In the first example the field is oriented directly to the right, so the positive charge will experience a magnetic force into the page and the negative charge out of the page. In the second example B is out of the plane of the page resulting in a force to the right on the positive charge and to the left on the negative charge.

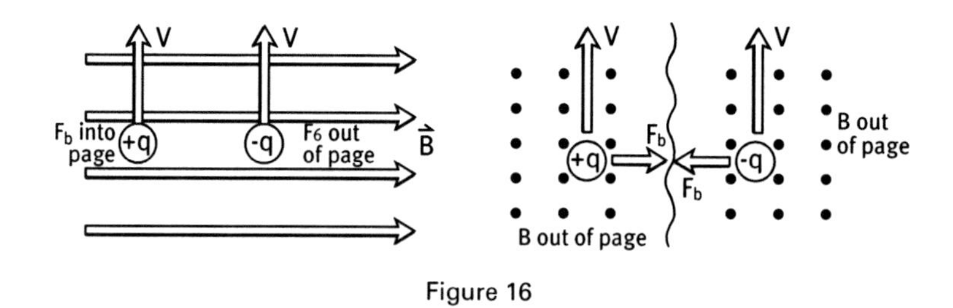

Figure 16

Example 11: For each of the following charges determine the magnitude and direction of the magnetic force acting on it. In each case B = 0.50 T, the velocity of the charge = 500 m/s, and the magnitude of the charge = 1×10^{-3} C (although in part b) the charge is negative).

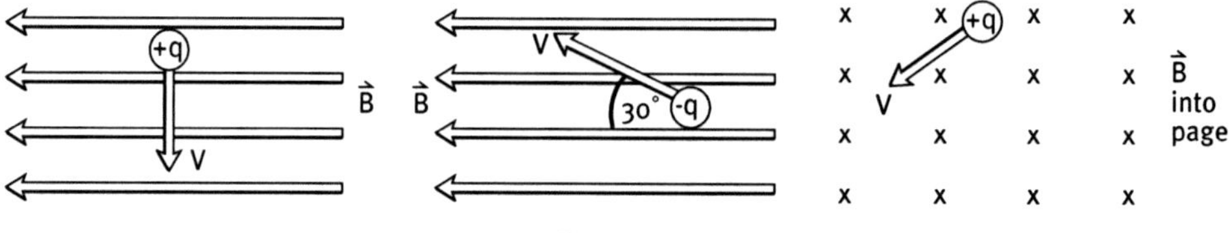

Figure 17

Solution:

a) $F = qvB \sin \theta = (1 \times 10^{-3}$ C) (500 m/s)(0.5T sin 90) = 0.25 N in a direction into the page according to the right hand rule.

b) $F = (-1 \times 10^{-3}$ C)(500 m/s)(0.5T sin 30) = 0.125 N into the page.

c) $F = (1 \times 10^{-3}$ C)(500 m/s)(0.5T sin 90) = 0.25 N down and to the right. (perpendicular to both v and B)

Force on a Current Carrying Wire. Because a current in a wire consists of moving charge, these charges and therefore the wire itself may experience a force under the same conditions as a single charge if the wire is placed inside a magnetic field. In this case, the force can be expressed in terms of current I, and length L of the wire that is immersed in the field.

$$F = ILB \sin \theta$$

The direction of force is given by the same right-hand rule as for a single charge with the velocity of the charge replaced by the direction of the current.

The Electric Motor. An *electric motor* (such as the kind you can get at a hobby shop to run a model car) is a device that operates because of the force that a current carrying wire experiences when immersed in a magnetic field. A motor consists of a coil of wire placed inside a magnetic field. If you run a current through the coil (by connecting it to a battery, for example) the wires experience forces that cause the coil to spin.

Electromagnetic Induction

We have shown in the preceding figure that a current produces a magnetic field, but the reverse is also possible—a magnetic field can produce a current. This phenomenon is called *electromagnetic induction*. You can *induce* an electric current by changing the field strength within a loop of wire. For example, if you move a magnet in the vicinity of a loop of wire, current will start to flow without any apparent voltage source such as a battery. An *induced voltage* is created in the wire, causing an *induced current* to flow. You can also induce a current by rotating a conducting loop in a constant field, or even by changing the area of the loop.

The key point is that in each of these cases the "amount" of field *within* the coil is varying over time, and it is this variation over time that creates the induced voltage, and current. If the field within the coil is constant there will be no induced current. The amount of voltage and current produced depends on the rate at which the field within the coil varies. *The faster the rate of change, the greater the induced voltage (and current) will be.* In order to measure the "amount" of field we can use the concept of magnetic *flux*. We define magnetic flux (Φ) as follows:

$$\Phi = BA \cos \theta$$

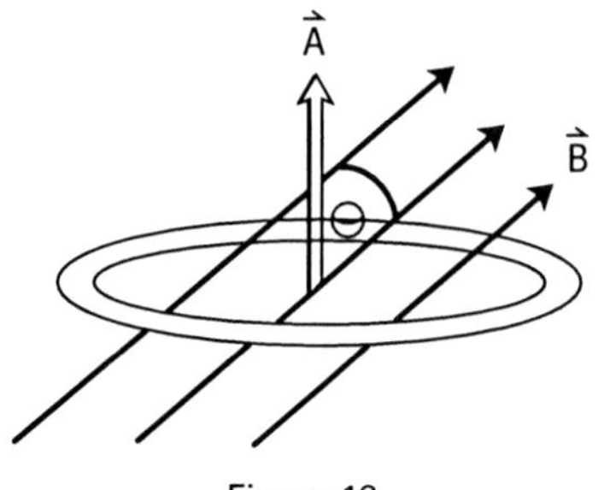

Figure 18

Where B is the strength of the magnetic field inside the loop, A is the area of the loop, and θ is the angle between the field lines and the area vector.

Unlike the way it sounds, the term "flux" does not indicate change. It is just an amount. If the flux is constant within a loop of wire, there will be no induction. However, it turns out that *the amount of induced voltage in a single loop of wire is equal to the rate of change of flux within the loop*. This is known as Faraday's law.

$$V = \frac{\Delta\phi}{\Delta t}$$

For multiple loops,

$$V = N\frac{\Delta\phi}{\Delta t}$$

where N is the number of loops in the coil.

Example 12: According to *The Core's* Dr. Keyes and Dr. Zimsky, the Earth is doomed due to the rapidly vanishing magnetic field. Because we now understand electromagnetic induction, we know that we can use the changing field to generate some power, so we decide to take advantage of the situation. We place a 100-turn coil of wire in the backyard, such that the Earth's field lines pass through the inside of the coil perpendicular to the plane of the coil (or parallel to the coil's area vector). The present magnitude of the field is about 5×10^{-4} T, and let's say before it fades out altogether it fluctuates between this value and 0 every tenth of a second. How much voltage and current can we generate in our coil during the period of oscillation if it has a total resistance of 0.5 Ω?

Solution: Because the coil is stationary and its area is constant, the changing magnetic flux is due solely to the changing field strength.

$$V = N\frac{\Delta\phi}{\Delta t} = NA\cos 0\,\frac{\Delta B}{\Delta t} = 100(0.1\text{ m}^2)(1)\left(\frac{5\times10^{-4}\text{ T}}{0.1\text{ s}}\right) = 5\times10^{-2}\text{ V}$$

$$\text{and } I = \frac{V}{R} = \frac{0.05\text{ V}}{0.5\ \Omega} = 0.1\text{ A}$$

This isn't that much current. If you really wanted to generate some power for free you could take your coil near a power transmission line. The fields close to the wires are much stronger than the Earth's field, and they oscillate at 60 Hz. However, because this is highly illegal *do not* try this at home! (Bonus question: If you were to place a coil near a power line in order to steal electrical energy, in what orientation should you place it relative to the line to get the maximum result?)

THE ELECTROMAGNETIC GENERATOR

The *electromagnetic generator* uses the principle of electromagnetic induction to produce a current. A generator in essence is an electric motor run in reverse. In an electric motor, electrical energy is converted to mechanical energy by running a current through a wire that is immersed in a magnetic field. In a generator mechanical, energy is converted to electrical energy. With a generator you apply a force to spin a coil of wire, which is within a magnetic field. Because the area within the loop of wire will therefore experience a changing amount of field over time, a voltage and current will be induced in the wire. Electromagnetic generators produce the electrical energy that powers cities. For example, a hydroelectric plant uses the energy of falling water to spin a turbine inside a magnetic field, thus converting mechanical to electrical energy. The energy to spin the turbine could also be produced by the burning of fossil fuels or in nuclear reactions. In nuclear power plants, energy is released in nuclear fission reactions, and is used to spin a turbine and generate electricity. Most generators produce AC current because as the coil spins inside the field the current is constantly reversing direction. In fact, the frequency of the AC current depends directly on the rate at which the turbine spins. A coil spinning at a rate of 50 Hz will generate a 50-Hz AC current.

TRANSFORMERS

A *transformer* is a device that changes the voltage from an AC power supply using the principle of electromagnetic induction. A transformer can change a high-voltage supply into a lower voltage (a step-down transformer), or vice versa (a step-up transformer.) Because more energy is lost to heat when current is higher, it is more economical to transmit electricity from the power stations at a high voltage, and a low current. For this reason step-up transformers are used at the power station. However, these voltages are much too dangerous to use in the home so step-down transformers are used locally to reduce voltages to safe levels. Transformers are essentially comprised of two solenoids placed in close proximity to each other. The primary coil is connected to an AC power supply. Because the current is constantly changing in the primary coil it produces a constantly changing magnetic field. This changing field in turn induces an AC current in the secondary coil. The primary coil and secondary coil contain a different number of loops, and the difference in voltage between the two coils is proportional to the difference in the number of loops. A step-down transformer is when the primary coil has more loops than the secondary coil, and vice versa for a step-up transformer.

The Earth's Magnetic Core, Cosmic Radiation, (and a Disaster of a Movie)

Why does the Earth have a magnetic field? As they suggest in the film (and as we discussed in Chapter 3), geophysicists believe that the field is due to moving iron ions in the liquid outer core. Why do they think this? Well, we know that magnetic fields are produced by moving electric charge and that in the outer core temperatures are sufficiently high to ionize some of the liquid iron. Therefore, if these charged particles are in motion they should produce a magnetic field.

We also know that the magnetic field at the Earth's surface looks similar to the field of a magnetic dipole (like a bar magnet). However, the Earth's magnetic field is more complicated. In *The Core* Dr. Zimsky is trying to convince us that the field is due to the core's rotational velocity. The general idea is not totally unreasonable. We've discussed that a current loop, such as in a solenoid, produces a dipolar magnetic field. It seems possible that a circular current in the core would produce something similar.

Nevertheless, Dr. Zimsky should know better, being a world-famous geophysicist. Most actual geophysicists believe that the field is due to convection currents produced by a process called a *self-sustaining geodynamo.* In a geodynamo an existing magnetic field is reinforced and sustained when currents in a conducting fluid flow in such a way that they create their own fields, which line up with the initial field.

Did you know that the Earth's "north magnetic pole" is actually a south magnetic pole? Because we know that opposite poles attract, and the north pole of a compass points towards magnetic north, then this must correspond to the south end of a magnet. The fact that the NMP is located close to (but not at) the geographic North Pole might be a reason for the misnomer. A geomagnetic reversal would straighten out the confusion because the south end of a compass needle would then point north.

Therefore, while the idea that the rotation of the outer core can somehow be brought to an abrupt halt is ridiculous, the idea that the Earth's field can fluctuate due to changes in the geodynamo is not only reasonable, but also accurate. In addition to continuous short-term variations, we know that the Earth's magnetic field reverses polarity frequently (geologically speaking) on the order of every 700,000 years or so. During reversals, geophysicists think that the field weakens considerably as it fluctuates and that the dipolar component of the field may fade out. It is uncertain whether the field disappears altogether during these transitions.

Magnetic Fields and Charged Particles. Dr. Zimsky and Dr. Keyes tell us that the magnetic field protects us from incoming radiation including charged particles and electromagnetic radiation. Let's look at how it might or might not do that.

As we discussed previously, if a *charged particle* is in motion in a magnetic field it can experience a force due to the field, with the direction of the force given by the right-hand (or left-hand) rule. Now consider a positive charge entering a magnetic field with an initial velocity in the direction shown in figure 19a below.

Because the force is perpendicular to the velocity of the charge it will start to deflect the charge to the left. However, as soon as it starts to turn, the direction of force changes so that it will still be perpendicular to the velocity. We can see that in fact the magnetic force acts as a centripetal force in this situation and steers the charge into a circle. Therefore, any charge entering a magnetic field in this orientation is going to start spinning in a circle, and as long as the field region is bigger than the circle the particle will be trapped.

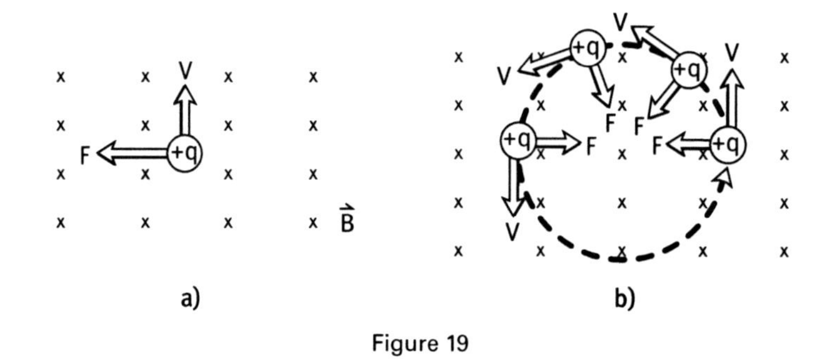

Figure 19

Now remember in the case of a charge moving parallel to the field lines that no magnetic force will act on the charge and it will pass through the field undeflected.

What if the charge enters the field at an angle to the field lines; what does the motion look like in this case? This will result in a combination of the two motions described. The velocity component perpendicular to the field will result in a circular motion, while the component parallel to the field means that the charge will continue to travel along the field direction. Therefore, the motion will be a spiral.

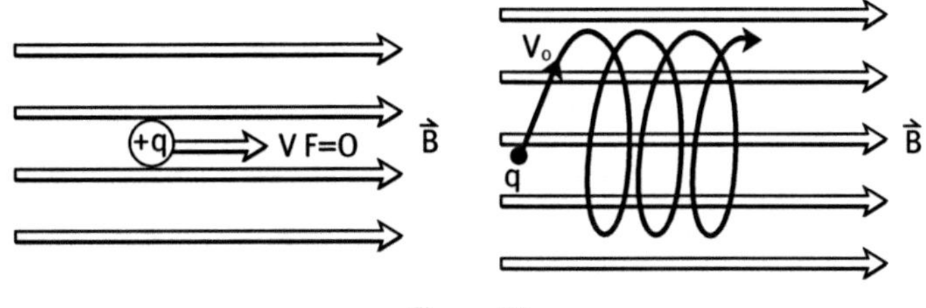

Figure 20

Now let's see how the Earth's field would affect incoming charged particles. Any charged particles (as long as their energy isn't too high) that have a component of their velocity perpendicular to the field lines, will be trapped by the field such that either they will be constrained to spin in a circle around the field lines, if their initial velocity is exactly at 90 degrees to the field, or more likely they will spiral along the field towards the magnetic poles.

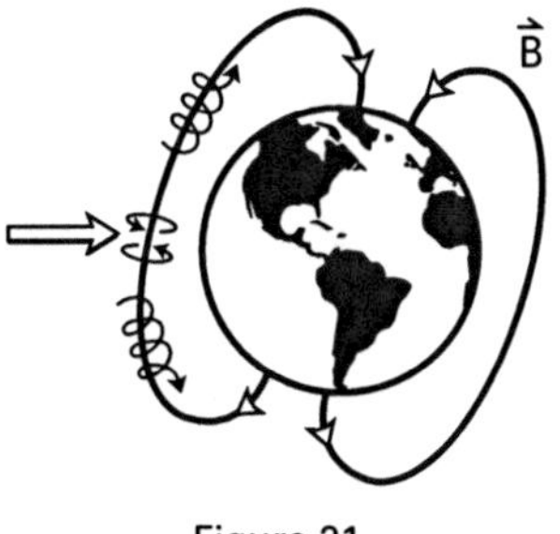

Figure 21

If the particles are moving parallel to the field, such as those heading straight towards the magnetic poles, they will remain undeflected.

Example 13: A proton is traveling at a velocity of 400 km/s when it encounters the Earth's magnetic field in a region where $B = 5 \times 10^{-5}$T. If it is initially moving at a 30-degree angle to the field lines describe the subsequent motion of the proton.

Solution: If we break the protons initial velocity into a parallel and perpendicular component we get $v_x = v_0 \cos 30 = 3.5 \times 10^5$ m/s and $v_y = v_0 \sin 30 = 2 \times 10^5$ m/s.

The field will not affect the parallel component so the proton will move in the direction of the field with a speed of 3.5×10^5 m/s.

Now the perpendicular force will act as a centripetal force pulling the proton in a circle and $F = qv_yB = \frac{mv_y^2}{r}$. Therefore, plugging in for the mass and charge of a proton we get

$$r = \frac{qB}{mv_y} = \frac{(1.6 \times 10^{-19}\ \text{C})(5 \times 10^{-4}\ \text{T})}{(1.67 \times 10^{-27}\ \text{kg})(2 \times 10^{5}\ \text{m/s})} = 0.24\ \text{m}$$

This means the proton is traveling along the field lines with a speed of 3.5×10^5 m/s while simultaneously circling with a radius 0.24m and a constant tangential speed of 2×10^5 m/s. The combined motion is a spiral like that shown in figure 20.

What this tells us is that the magnetic field can trap charged particles way above the atmosphere in what are called the *Van Allen Radiation Belts,* except at high northern and southern latitudes where they can pass "Go" without paying two hundred dollars. This is good news for astronauts on the Space Shuttle, and low Earth-orbiting satellites, because a high enough flux (incident particles) of charged particles can be deadly to humans, and destructive to electronics. However, during solar storms charged particles emitted by the sun can increase by as much as five orders of magnitude (that's one hundred thousand times). The energy of many of these particles is high enough that they can pass *through* the field, posing a threat to people outside of the atmosphere (astronauts hopefully).

However, here underneath the atmosphere we don't really notice this increased flux too much. This is because, as we discussed in Chapter 3, *the atmosphere blocks most of these particles before they ever reach the surface!* Remember that there is no magnetic shielding near the poles, yet the amount of charged particles reaching the surface isn't drastically more than at the equator. Therefore, if we are to believe what they say in *The Core*, everyone and everything in the high northern and southern latitudes should already have been scorched. Yet amazingly this hasn't happened. Why? Because of the atmosphere, that beautiful envelope of nitrogen and oxygen (and ever-increasing amounts of sulfur dioxide, carbon monoxide, nitrogen dioxide, benzene, and so on) which makes this all possible.

ACTIVITY

Find a really strong horseshoe magnet and a laser pointer (or flashlight). Shine the light onto a wall so that the beam passes close to the magnet. Is the beam bent? (Is there any deflection of the light on the wall?)

Now let's talk about *electromagnetic* radiation, as described by those idiots Keyes and Zimsky. We haven't discussed e-m radiation yet because static magnetic fields like the Earth's don't affect e-m radiation. It comes right on through to the Earth's surface, which is a good thing, if we like being able to see things, and having the temperature higher than –200 °C. The "professors" try to tell us the Earth's magnetic field protects us from incoming microwaves. It is true that the *atmosphere* (not the magnetic field) does filter out many frequencies of e-m radiation, but it is transparent to most frequencies of microwaves. Microwaves from the sun are coming right on through your living room window right now as you read this (if it's daytime). The magnetic field isn't stopping them.

So far we have uncovered a fair amount of incomprehensible physics in *The Core.* We have determined that the whole premise of drilling into the Earth's core with a "magic" ship and using nuclear bombs to restart the outer core, to be highly suspect. At a slightly less egregious level, in dealing with some aspects of electromagnetism, while the writers of the film seem to have a sense that the magnetic field can do something to shield the Earth from something (charged particles) we have seen that they completely misunderstand the fact that magnetic fields have no effect on e-m radiation. We have also determined that while they correctly assert the idea that moving charge creates magnetic fields, the specifics of how the Earth might produce its field is grossly misrepresented.

That being said let's take a few more minutes to talk about a couple more scenes in *The Core* that bring up some interesting electromagnetic issues.

"High-Level Static Discharge" and the Aurora Borealis

As we discussed, if the Earth's magnetic field disappeared, then more charged particles would impact the atmosphere than occurs at present. What effects might this increased flux of charged particles produce? Well, in *The Core* one of the effects that we see is displays of the northern lights (the aurora borealis) at low latitudes where they are not normally observed. It turns out this may be quite possible, when we consider what causes auroras in the first place and why we see these only at high latitudes. (Incidentally, this phenomenon also occurs at high southern latitudes where they are called aurora australis.) Although our understanding of auroras is incomplete and the details may be somewhat complex, the general explanation is that magnetic field lines funnel charged particles (probably electrons it turns out) toward the magnetic poles. (Recall the magnetic spirals we discussed.) As they enter the atmosphere these electrons collide with oxygen and nitrogen molecules. The molecules absorb some of this energy, which causes *their* electrons to increase in energy level. When the molecular electrons return to the "ground" state they give off energy in the form of visible light.

We can understand why auroras are only visible near the magnetic poles when we consider that large fluxes of charged particles will only impact the atmosphere at those latitudes. Therefore, it certainly seems feasible that without a magnetic field, because charged particles would be impacting the atmosphere all over the Earth, aurora-like phenomena might occur at any latitude. (Note that auroras don't occur continuously but show up mainly during periods of intense solar activity.) If you're going to do a movie about a disappearing magnetic field, showing aurora occurring in the tropics is a reasonable way to illustrate the point, so we can give the filmmakers a "thumbs-up" at least for this one little part of an otherwise very silly film.

It's a little harder to suspend our disbelief in scene 9 when we are treated to the most violent onscreen lightning storm in cinematic history. According to Dr. Keyes, "high-level static discharge" is responsible. It's a truly colossal storm, occurring on a global scale and focused particularly on Rome where thousands of lightning bolts converge and destroy the entire city. While most lightning as we know it usually lasts for about one-fifth of a second or so, with most of the energy discharged in a few milliseconds, the bolts in this scene discharge over a period of up to several seconds (meaning they last hundreds of times longer). They look more like weapons from an enemy spacecraft than real lightning the way they target the Coliseum and blow it into little bits, and the way one of the bolts travels along the street in an almost malevolent fashion. Normal lightning bolts have energies on the order of 500 million Joules, and the most energetic lightning (positive lightning) can have energies upwards of 250 billion Joules. The amount of energy released by this storm must be staggering, hundreds of thousands to millions of times that. (Interestingly, while there is a lot of sizzling and the sound of explosions, we never hear any thunder.)

What might cause such a vicious storm? To answer this, let's look at what causes lightning. Lightning is an electrical discharge, the same as when you get a shock touching the door handle, but on a larger scale. To elicit an electrical discharge, you need an electric field and a potential difference between two points. For example, consider two charged metal plates with an air gap in between (a capacitor). If charge continually builds up on the plates, the field between the plates gets continually stronger. If enough charge can build up, eventually the field will be so strong that the charge will be able to "jump" the gap between the plates, which you might see as a spark.

Example 14: A capacitor consists of two plates separated by a distance of 3 millimeters. If the potential difference between the plates is 100 V when the charge on each plate is 1,000 μC (+1,000 on one plate and -1,000 on the other), then

a) what is the electric field between the plates (assuming it is uniform)?
b) how much energy will be released if the capacitor is discharged?

Solution:

a) $E = \frac{\Delta V}{d} = \frac{100 \text{ V}}{0.003 \text{ m}} \approx 33{,}000 \text{ V/m}$

Note: 1 V/m is equivalent to 1 N/C.

b) The energy released is equal to the energy initially stored on the capacitor, and the energy stored is equal to the work it took to separate the charges. Because $W = q\Delta V$, for each charge the total W is the sum of the work done on all of the individual charges. However, ΔV between the plates is not constant as you build up the charge. We start with $\Delta V = 0$, and end up with $\Delta V = 100$ V. Taking an average ΔV we get:

$$W_{total} = \frac{q_{total}V_{final}}{2} = 1 \times 10^{-3} \text{ C}(50 \text{ V}) = 0.05 \text{ J}.$$

In a thunderstorm charge is separated within the clouds with the tops of the clouds becoming more positively charged and the bottoms more negatively charged. (Though there are a variety of theories, the exact process by which this charge separation occurs is still not agreed upon.) The bottoms of the negatively charged clouds can induce a positive charge on the ground, and eventually the field can become so strong that a discharge occurs between the cloud and ground or from cloud to cloud. (The discharge process is actually somewhat complicated and occurs in a number of discreet but very rapid steps.) The potential difference associated with a typical lightning discharge is hundreds of millions of volts.

Figure 22

"Positive lightning" is more rare, but much more powerful, and it can occur when the positive cloud tops are torn away from the negative bottoms.

We could estimate the maximum amount of current flowing in a typical lightning bolt as follows:

First we can find the maximum power of the discharge assuming that most of the energy is released in about 50 microseconds.

$$P = \frac{E}{t} = \frac{500 \times 10^6 \text{ J}}{50 \times 10^{-6} \text{ s}} = 1 \times 10^{13} \text{ W}$$

We can use this power then to determine the approximate current.

$$I = \frac{P}{V} = \frac{1 \times 10^{13} \text{ W}}{500 \times 10^6 \text{ V}} = 20{,}000 \text{ A}$$

Despite our rough calculation, the value of 20,000 A corresponds pretty well to experiments that have been done to determine lighting currents.

Can you think of an approach to measure experimentally the current generated by a lightning bolt? (Hint: Try applying Faraday's law.) Well, remember that moving charge generates a magnetic field, so a lightning bolt will have a large field surrounding it with the field proportional to the current and inversely proportional to the distance from the current. ($B = \mu_0 I/2\pi r$). The field will be changing very rapidly, and we know that according to Faraday's law a changing field within a conducting loop will induce a current in that loop ($V = \Delta\Phi/\Delta t = IR$). So let's say we set up a tower (lightning rod) in the middle of a large open field. We can surround the tower with coils of wire at known distances from the tower. When there is a lightning strike we could measure the currents induced in the coils, and theoretically deduce from these the lightning current.

Have you ever wondered why lightning tends to strike protruding and pointed objects like trees and tall buildings? It has to do with the electric field. For a charged conductor, E is going to be strongest where it's most curvy (or pointy) because the concentration of charge will be highest at those places. Because like charges repel each other, they want to get as far away from each other as possible. However, in the pointy sections of an object they are forced closer together, resulting in a stronger field outside of those areas. Therefore, an electrical discharge is more likely to occur there.

But what about the spectacular lightning in *The Core*? Could that really happen? The idea in the movie is that because charged particles are no longer being shielded by the magnetic field they are accumulating in the upper atmosphere, creating a much stronger than normal electric field between the atmosphere and ground. The field apparently gets so strong that it causes super lightning discharges. The general physical idea here makes some sense. The upper layer of the atmosphere, appropriately called the *ionosphere,* is charged (mostly due to ultraviolet e-m radiation ionizing the gas molecules rather than incident particles however). The field generated in the ionosphere may contribute to the charge separation process in storm clouds. It seems plausible that if the magnetic field were to disappear, then more charged particles might augment the ionosphere and other regions of the upper atmosphere, increasing the strength of the electric field and the potential difference between the upper atmosphere and the ground—but by how much? This is basically what happens near the magnetic poles. Therefore, while in a general way the idea that an increased potential difference would lead to discharges makes some sense, the ludicrously exaggerated portrayal of this phenomenon in *The Core*, while obviously highly dramatic, seems unlikely. According to the logic of the film, we should currently see violent lightning storms at the magnetic poles on a regular basis because there is no magnetic shielding there.

***Possible* Effects of a Disappearing Magnetic Field.** This isn't to say that if the Earth's magnetic field disappeared, unpleasant or inconvenient things wouldn't happen. For example, we know during solar storms the increased flux of charged particles can affect orbiting satellites and damage their hardware. Radio and television transmissions and other communications could be affected, especially during periods of intense solar activity. Space travel would be more dangerous than it already is due to the biological effects of high-energy particles. Power outages would probably occur more frequently due to the magnetic fields of the moving charged particles effecting generators, something that occasionally happens nearer the magnetic poles. On a grander (and more permanent) scale, over *millions* of years the increased magnetic flux could erode the atmosphere, *eventually* rendering Earth uninhabitable. (Scientists theorize that Mars lost most of its atmosphere this way after its own magnetic field disappeared.) Nevertheless, any large-scale immediate danger to the existence of life on planet Earth as portrayed in *The Core* falls within the realm of science-fiction action movie fantasy.

ELECTROMAGNETIC SHIELDING IN THE CINEMA

Let's close this chapter with one more important electromagnetic principle, illustrated in the political suspense thriller *Enemy of the State,* starring Will Smith and Gene Hackman.

The Movie

Powerful and vicious NSA administrator Thomas Reynolds (Jon Voight) has a congressman assassinated because he won't vote the way Reynolds wants him to on a security bill. This turns out not to be such a good idea because, unbeknownst to Reynolds, the entire incident is caught on tape by a counter-culture documentary waterfowl biologist. Reynolds sends his NSA henchmen after the biologist who, while being pursued, manages to slip the tape into Robert Dean's (Will Smith) shopping bag. An unsuspecting Dean becomes the target of the pursuit. Fortunately, he meets up with former NSA operative Brill (Gene Hackman) who has been living "underground" for the past 15 years in fear of his former employers. Brill reluctantly comes to his assistance, and after a sequence of nail-biting twists and turns, Dean is finally able to turn the tables on his tormentors, and exonerate himself.

The Electromagnetic Jar. Throughout the film we are wowed by the high-tech gadgetry of the NSA tech support people and their seemingly unlimited resources. They have the ability to tap electronically into every facet of Smith's life, freezing his assets and credit cards, monitoring his phone calls, and accurately pinpointing his whereabouts. However, because Brill is savvy about all of this gadgetry; in scene 20 he takes Smith to his "office" in an abandoned warehouse surrounded by copper mesh fencing. Brill explains (correctly) that the fencing prevents radio waves from penetrating into the warehouse, and along with the fact that the there are also no phone lines or wiring connected to the outside the NSA has no way to pierce the veil of Brill's sanctuary. Brill calls his office "the Jar." We could also call it a *Faraday cage.*

ACTIVITY

Find a small portable radio and some wire mesh. Make a pail out of the mesh big enough to fit over the radio. Place the radio on a piece of aluminum foil, and turn it on. While listening to your favorite station place the mesh pail over the radio so that the radio sits in the little wire cage. What happens to the reception?

(Either the radio goes dead or the reception gets really weak. The little container you made out of wire mesh is called a Faraday cage, and it can prevent static electric fields and electromagnetic radiation from either getting into or out of the cage.)

Faraday Cages, Electric Fields, and e-m Radiation. Let's look at how electric fields are shielded. Let's say you have a charged conductor in *electrostatic equilibrium,* meaning that the net force on the individual excess charges must be zero. If the conductor is in equilibrium two things must be true: (1) All of the excess charge must be on the surface, and (2) the electric field inside the conductor must be zero. These conditions must be met because if the electric field inside the conductor is not equal to zero then excess charges would experience forces and therefore not be in equilibrium. The excess charge therefore must be confined to the surface.

If you apply an external electric field to a piece of conducting material, the charges in the material will move as a result of the electrical forces acting on them until they achieve equilibrium. When this occurs the field created by the charge distribution must exactly cancel the external field such that the interior of the conductor has $E = 0$. This is the principle behind electric shielding. If a conducting material surrounds a volume it will cancel out any external electric fields.

In the case of radio waves and other electromagnetic radiation (which we will discuss in more detail in the next chapter), the principle is similar although the details are more complicated. E-m radiation consists of oscillating electric and magnetic fields. It can be shown that even if there are openings in the shielding conductor, as long as they are smaller than the wavelength of the incident e-m radiation, most of the e-m waves will be reflected, and the greater the size difference, the better the shielding. This is called *RF (radio frequency) shielding.* In the case of "the Jar" from the film, the wire mesh has openings of only a few centimeters, and because most radio and television waves have wavelengths of half a meter or more, Brill's office should be well-shielded from both incoming and outgoing signals. Kudos to *Enemy of the State* for that fine moment of movie physics!

ADDITIONAL QUESTIONS

1) In a television set electrons are accelerated from rest by an electric field that exists between the front and back of the picture tube. If the field region is 5 centimeters long and it has a strength of 25,000 N/C how fast is an electron traveling when it exits the field? (The mass of an electron is 9.11×10^{-31} kg.) What is the potential difference from one side of the field to the other?

2) Rank the following circuits from most total current to least total current. Some circuits may have the same current. The resistors are all identical, as are the batteries.

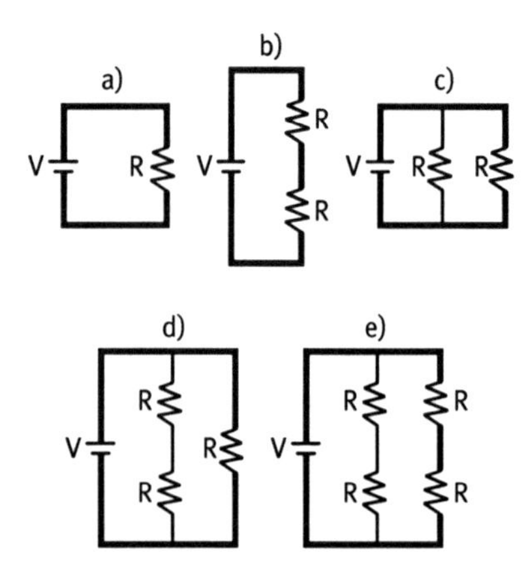

3) Two equal positive charges are a distance *d* apart. They are fixed in place so that they cannot move. Where could you place a third charge so that it experiences no electrostatic force? What if one charge is positive and the other is negative? What if they are oppositely charged and the positive charge is twice the magnitude of the negative charge?

4) Protons are fired horizontally exactly between two charged plates where there is a constant electric field of 600 N/C oriented upward. The protons have an initial velocity of 2×10^6 m/s and the plates are 4 centimeters apart.

   Describe qualitatively the motion of the protons. What is the shape of their trajectory? How far horizontally from where the protons enter the field do they strike one of the plates? Is gravity an important consideration in determining the trajectory? Why or why not?

5) The Earth's magnetic field is oriented due north at a certain location. At this location a high-energy electron is moving due east through the atmosphere. What direction of magnetic force will the electron experience?

6) Why do very high-energy cosmic rays pass through the magnetic field without being trapped and impact the atmosphere?

7) A long straight wire has a 3-A current running through it. A proton is fired parallel to the wire 5 centimeters away and opposite to the direction of the current. If the initial velocity of the proton is 1×10^7 m/s what is the initial magnetic force acting on the proton?

8) Remember the lightning storm we talked about in *The Core*? Roughly estimate from the information given in the previous discussion how much energy would be released in the storm. Compare this to the energy released in a 1-MT nuclear bomb

Waves, Light, Sound, and Sensurround

SOONER OR LATER IN ANY discussion of movie physics, you arrive at the inevitable—for many the apex of movie physics—where the frontiers of physics merge with the frontiers of imagination: science fiction. In these last two chapters we are going to deal almost exclusively with "hard-core" sci-fi, and we're going to spend some time in outer space. In this chapter we'll warm up by seeking out (in addition to new life and new civilizations) concepts of waves, sound, and light in some of these sci-fi masterpieces, and we'll also point out some common "science fiction physics film no-no's." Our format for this chapter will be slightly different, in that we won't spend as much time on individual movies, but we will weave segments of many films into our discussion to illustrate the principles.

THE OSCILLATING UNIVERSE

The universe as we know it is composed of matter and energy. It is in a continual state of change, in constant motion. One ubiquitous example of motion/change in the universe is the propagation of energy through space in the form of *waves.* A *mechanical wave* like a water wave, sound wave, or earthquake wave is one in which a disturbance or oscillation travels through a medium consisting of some kind of matter. *Electromagnetic waves* don't need matter to propagate. They can transfer energy through empty space. These consist of oscillating magnetic and electric fields and include visible light as well as a continuum of longer and shorter wavelengths such as radio waves, x-rays, gamma rays, and so on. which constitute the electromagnetic spectrum. All e-m waves travel through a vacuum at the same speed, which is the famous "speed of light": $c = 3 \times 10^8$ m/s.

Mechanical Waves

There are two basic types of mechanical waves: *transverse waves* and *longitudinal waves* (also called *compressional waves).* Each type is defined by the direction of the wave oscillation. In the case of transverse waves, the disturbance oscillates perpendicular to the direction of wave propagation. For longitudinal waves the oscillation is back and forth along the propagation direction.

Photo: Fighting giant insects in *Starship Troopers.*

ACTIVITY

a) Take a rope and fix it at one end. Take the other end and shake it up and down. Notice how while the wave (energy) travels along the rope, the matter (the rope) oscillates up and down perpendicular to the wave motion as the wave passes through. You can see that it is energy—not matter—that is transferred.

b) Find a Slinky and stretch it out flat on the floor. Have a friend hold one end while you hold the other. Now push your end of the Slinky back and forth. The compressions traveling through the Slinky are longitudinal waves where the oscillations vibrate back and forth along the direction of the propagating wave.

The accompanying activity is a demonstration of a transverse wave. Water waves are another example of transverse waves. As a wave travels through the water, the water rises and falls perpendicular to the direction of the wave motion.

General Properties of a Wave

The following graphs represent two views of a linear simple harmonic traveling wave. (A simple harmonic wave oscillates in a sinusoidal pattern.) Both graphs show the displacement of the medium from equilibrium on the *y*-axis. However, the first graph shows this displacement as a function of horizontal position along the medium at a specific time. It is like a photograph of the wave at one instant. The second graph shows the displacement at a specific horizontal point in space along the wave as a function of time. It is the position-time graph showing the motion of the oscillation of the medium at that point.

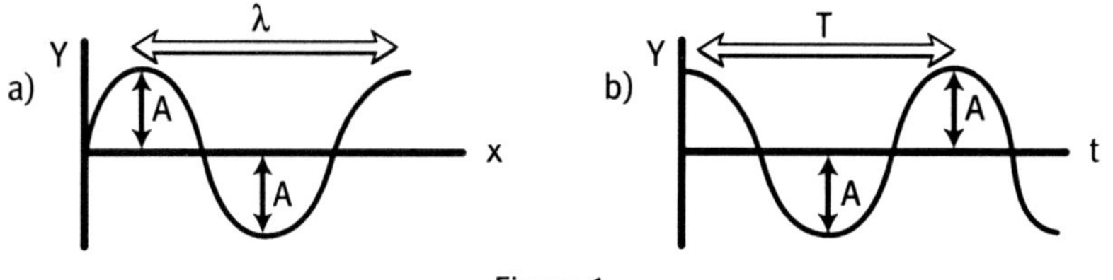

Figure 1

Amplitude. The *y*-axis on both graphs represents the displacement from the equilibrium position of the medium as it oscillates. We define the ***maximum*** displacement from equilibrium as the ***amplitude (A)*** of the wave. The amplitude represents the maximum displacement of the oscillations perpendicular to the direction of wave propagation in the case of a transverse wave, and the maximum displacement of the oscillations parallel to the wave direction for a longitudinal wave.

Wavelength, Period, and Frequency. On the *y-x* graph you can see that the wave travels through one complete cycle over a specific distance called the ***wavelength (λ)***. On the *y-t* graph the time for the wave to travel through one complete cycle is called the ***period (T)*** of the wave. The reciprocal of the period, called the ***frequency (f)***, is equal to the number of wave cycles per time. One cycle per second is also known as 1 Hertz (Hz).

Mathematically we can describe the displacement of the wave oscillations as a function of time, and as a function of position along the wave.

$$y(t) = A\cos(\omega t)$$

$$y(x) = A\cos(kx)$$

where $\omega = 2\pi f$ and is called the ***angular frequency***, and $k = 2\pi/\lambda$ and is called the ***wave number***.

Velocity of a Traveling Wave. The velocity of a wave can be related to the previous quantities as follows.

$$v = f\lambda \text{ or } v = \lambda/T$$

Example 1: In *StarTrek the Movie* (StarTrek 1) the crew receives an (e-m) signal from "V-ger," which, it turns out, is one of the old *Voyager* probes modified out of recognition into a gigantic super machine. Because V-ger destroys everything that doesn't answer its signal, obviously it's important to the crew that they figure out what it says. They're surprised that the signal has a frequency of "one million megahertz" (1×10^{12} Hz) instead of a radio frequency (around 1 megahertz). The point they are trying to convey is that "V-ger" is so advanced that it sends signals and information at an astonishing rate. What this really means, though, is that V-ger is inexplicably broadcasting e-m waves at microwave/infrared frequencies rather than at radio frequencies. Find

a) the speed of V-ger's e-m transmission.
b) the wavelength of V-ger's e-m transmission.
c) the period of V-ger's e-m transmission.

Solution:

a) This one is easy. The speed of all e-m waves is 3×10^8 m/s.

b) $\lambda = \dfrac{v}{f} = \dfrac{3 \times 10^8 \text{ m/s}}{1 \times 10^{12} \text{ Hz}} = 3 \times 10^{-4} \text{ m}$

c) $T = \dfrac{1}{f} = \dfrac{1}{1 \times 10^{12} \text{ Hz}} = 1 \times 10^{-12} \text{ s}$

Example 2: While floating on your surfboard you decide to count the time between passing waves. You find that that the wave peaks move through every 8 seconds, and the vertical distance between the trough and peak is about 3 meters. Approximating the wave pattern as sinusoidal/simple harmonic,

a) write an equation for the vertical position of the wave as a function of time at your location assuming you start timing at a peak.
b) If the waves are traveling at a speed of 4 m/s, write an equation for the vertical position as a function of horizontal position at an instant of time.

c) After 3 seconds, what is the vertical position of the wave?

Solution:

a) $y(t) = A \cos 2\pi ft = 1.5\text{m} \cos 2\pi[0.125 \text{ Hz}]t = 1.5\text{m} \cos \frac{\pi}{4}t$

where A is half the distance from peak to trough, and the frequency is the reciprocal of the period.

b) First we need to find the wavelength: $\lambda = \frac{v}{f} = \frac{4 \text{ m/s}}{0.125 \text{ Hz}} = 32 \text{ m}$.

Then we can use the wave equation:

$$y(x) = A \cos \frac{2\pi}{\lambda}x = 1.5\text{m} \cos \frac{\pi}{16}x .$$

c) $y(t) = 1.5\text{m} \cos \frac{\pi}{4}t = 1.5\text{m} \cos \frac{\pi}{4}(3s) = -1.1 \text{ m}$,

which means you are on the way down to the bottom of a trough.

ACTIVITY

a) Take a square pan and fill it with water to a depth of about one inch deep. Take a small block of wood or other small solid object a few inches wide and use it to generate a single wave pulse. Watch how the waves reflect off of the edge of the pan. Try creating wave-fronts at different angles and observe the reflections.

b) Put a pencil in a glass of water. Why does it look broken and bent?

Reflection, Refraction, and Diffraction of Waves. *Reflection* of a wave can occur when the wave encounters a barrier. The *law of reflection* states that the angle at which the wave is incident upon the barrier will be equal to the angle of reflection providing that the barrier surface is smooth. By convention, the angle of incidence and reflection are measured relative to a line perpendicular to the barrier called the *normal line.*

An echo is an example of a sound wave reflecting off of a barrier. Bats navigate by emitting high-frequency sound pulses, which reflect off of objects and back to the bat, which can then determine where the objects are (convenient for both avoiding walls and catching bugs).

Refraction occurs when the direction of a wave changes when it travels from one medium to another.

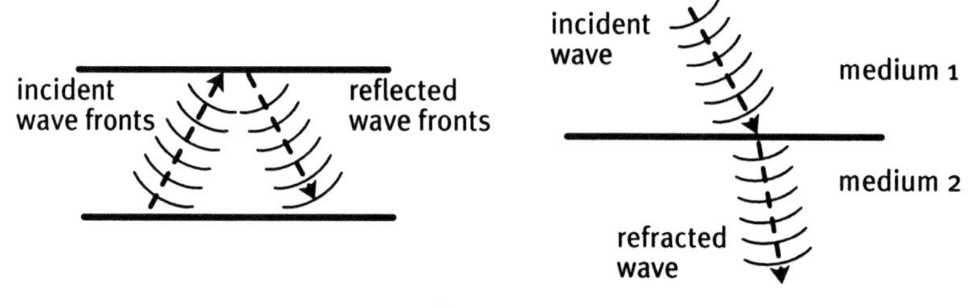

Figure 2

When a wave travels from one medium into another, the wavelength and speed of the wave will both change, although *the frequency will remain unchanged.*

In our discussion of waves, I'm afraid we have to bring up *The Core* yet again. Remember when Dr. Keyes is explaining wave motion to his Intro to Geophysics class? Well, he says wavelength and frequency *both* change when a wave travels from one medium into another. Wrong—the frequency does not change. Consider a series of wave-fronts crossing a boundary between two materials. If the frequency were to change across the boundary, there would be a mismatch between wave-fronts on one side of the boundary versus the other. The frequency of wave peaks crossing the boundary must be the same as the frequency coming out on the other side or something is really weird. So the change in velocity must be accommodated by a change in distance between the peaks (wavelength).

ACTIVITY

Make a barrier with an opening an inch or two wide in the middle, and place it in your water-filled pan. Generate plane waves on one side of the barrier using a flat piece of wood several inches long. Observe what happens to the plane waves as they pass through the opening.

Therefore, because $v = f\lambda$ if the velocity of the wave is reduced as it passes into a medium, the wavelength will also decrease, and vice versa. In addition, as a wave passes from one medium into another, usually not all of the energy will be transmitted. Some will be reflected back.

Diffraction is the bending of a wave that occurs as it passes around a barrier. For example, water waves passing through a small opening in a barrier will be slower at the edges due to their interaction with the barrier, while the waves are faster in the middle. Therefore, the waves will spread out as they pass through the barrier.

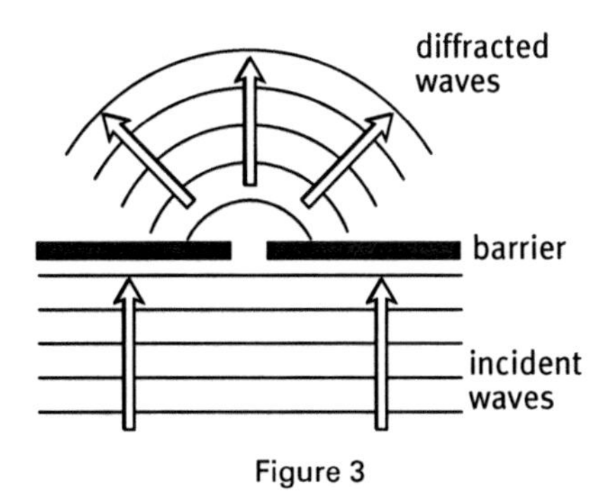

Figure 3

Interference of Waves. When two or more waves pass through the same location in space at the same time they will *interfere* with each other. The interference may be *constructive* or *destructive.* Constructive interference occurs when the waves are *in phase* with each other. This means that the displacement of each wave is in the same direction. When the displacements are in phase they will augment each other resulting in a larger wave than either of the individual waves during the period of time that they overlap. Once the waves move past each other they each return to their original form.

Destructive interference occurs when the waves are *out of phase*. This occurs when the displacements of each wave are in opposite directions. When waves interfere destructively, and are of equal magnitude, they will cancel each other out as they pass through each other. Examples of constructive and destructive interference are shown below.

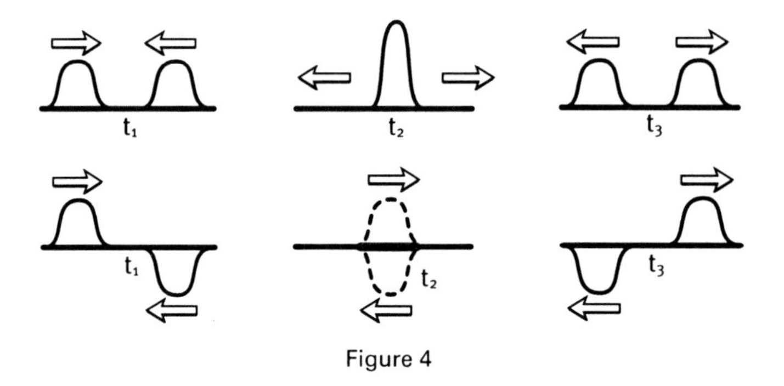

Figure 4

The examples shown above include only waves that contain a single pulse with a single phase direction. However, continuous or more complex wave patterns may interfere in such a way that part of the interference is constructive and part is destructive.

Standing Waves. *Standing waves* are interference phenomena that occur when two waves continuously interfere with each other in such a way that certain areas of the medium experience continuous constructive interference and other regions experience continuous destructive interference, forming a "standing" pattern.

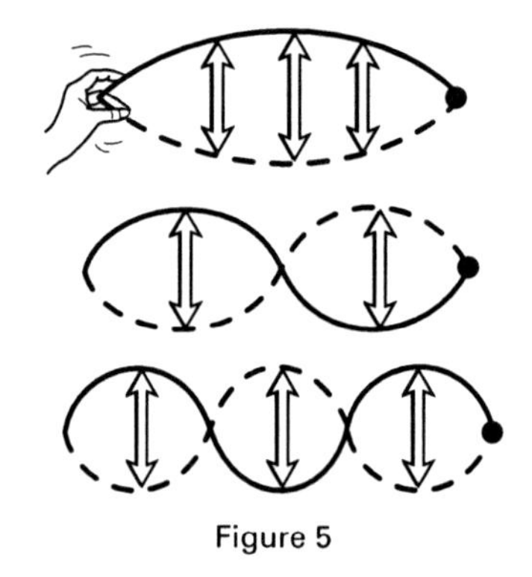

Figure 5

The areas of constructive interference where the rope has maximum amplitude are called *antinodes*, and those with destructive interference where there is no displacement are called *nodes*. The wave reflects off of the fixed end, and on the way back this reflected wave then passes through and interferes with the wave that is still traveling towards the wall. If the oscillation is at the right frequency then there will be points along the rope that will constantly be experiencing destructive interference (nodes), and other points repetitively experiencing constructive interference (antinodes).

A certain fundamental frequency of oscillation results in a single antinode in the middle of the wave, while integer multiples of this frequency will result in multiple antinodes. There will be no standing waves when the frequencies are not integer multiples of the fundamental frequency.

The frequencies at which standing waves form in the rope example depend on the velocity of the wave, and the length of the string.

$$f_n = n\frac{v}{2L}$$

Where n = some integer (1,2,3,4,etc.).

Sound Waves. *Sound waves* are longitudinal (compressional) waves. They are mechanical waves and require a medium through which they can travel, such as air or water. A sound wave is initiated when a disturbance in a medium creates regions of high and low pressure that propagate through the medium. As we have already emphasized at least twice, *sound cannot propagate through a vacuum and therefore even a supernova explosion in space would make no sound, never mind spaceships whizzing around.*

THE MOST COMMON CINEMATIC SCI-FI ERROR

Star Wars is perhaps the granddaddy of the modern science fiction explosion. In the original 1977 film, the evil Empire has developed a Death Star capable of destroying an entire planet with a single high-energy beam. The rebels, led by Princess Leia (Carrie Fisher) and Luke Skywalker (Mark Hamill), need to destroy the Death Star before the rebellion is crushed and all hope for a sequel is lost. Needless to say there are numerous exciting space battles between ships of all shapes, sizes, and political affiliations. Curiously, in every battle, the noise is deafening. The ships emit whizzing, screeching, and whirring sounds. Explosions are loud and dramatic. Even with increases in technical sophistication, and "cinematic scientific savvy" over the almost 30 years since *Star Wars* was released, we still, sadly, see the inevitable "sounds in space" phenomenon all the time in space movies like *Star Trek, Galaxy Quest,* and *Starship Troopers.* (*Star Wars* gets credit for being consistently loudest, and only the notable and beautiful exception of *2001* portrays the silence of space as it really is.) Although it may give more dramatic impact to motion, explosions, and weaponfire, nevertheless, this simple error constitutes probably the most obvious and common abrogation of physical principles in the sci-fi in space genre.

For those of us who don't like obvious physical principles offending their sense of fantasy and want to justify these noisy space scenes try the following hypothesis. During the construction and design phase of the spaceships, they are tested on the surface of a planet, and the sounds they make are recorded. The sounds are then broadcast via e-m signals (radio waves) to the surrounding vehicles so that they can have more fun during space battles by playing the whizzing sounds on their receivers inside the ship. As we watch the movie, we can pretend we are listening inside our own spaceship.

The Speed of Sound

The speed of sound in air is around 340 m/s, although it varies slightly with temperature and air pressure. Sound travels much faster in water or solid materials than it does in air.

Example 3: A Fangorian bloodworm emits a constant musical note, which has a frequency of 440 Hz. What is the wavelength of the sound wave?

Solution: Because $v = f\lambda$, $\lambda = v/f$ = 340 m/s/440 Hz = 0.77m.

Volume and Pitch. The ears perceive the amplitude of a sound wave as volume, or loudness of the sound, and the frequency of the sound as pitch. The higher the frequency of sound, the higher the pitch that we hear. Most sounds are combinations of several frequencies mixed together. Usually there is a fundamental frequency, which is the lowest of the frequencies, in conjunction with higher multiples of this frequency called *harmonics*. The pitch that we hear is the fundamental frequency, but the "color" of the sound depends on which harmonics are present. For example, a trumpet and a flute can play the same note, and our ear perceives that they are playing the same note, but we hear a difference in the quality or color of the sound due to each instrument having different harmonics.

Example 4: The following figure shows three different displacement time graphs of three different sound waves. In the case of sound waves the displacement is synonymous with pressure (P on the graphs). The scales on each graph are the same. Which sound is the loudest? Which sound has the highest pitch?

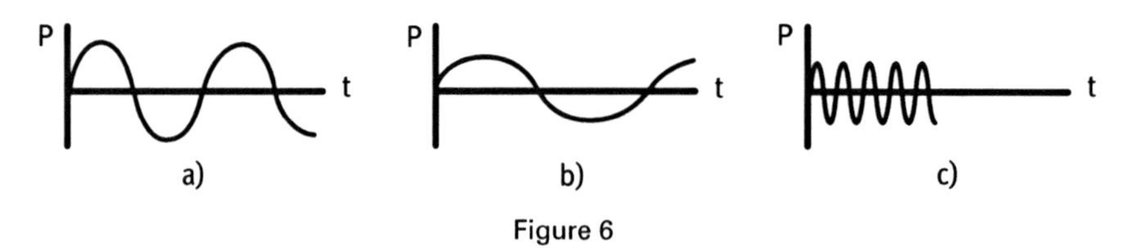

Figure 6

Solution: Graph (a) represents the loudest sound, because it has the largest amplitude. Graph (c) has the highest pitch, because it has the highest frequency.

Close Encounters with Friendly Aliens

In *Close Encounters of the Third Kind* it turns out those UFOs really aren't swamp gas, and they're not ball lightning either (natural phenomena commonly reported as UFOs). They're real live alien spaceships. Better yet, these ships are essentially a high-tech version of the old standard, that old favorite sci-fi standby: the flying saucer. They do, however, have a few more special effects bells and whistles than those low-budget photographs of plates being thrown into the air. Best of all, the aliens are friendly, and they have a nice meet and greet with government officials, scientists, and various other official riff-raff at the end of the movie in a secret location behind Devil's Tower, Wyoming.

Of course the general public is not allowed to know about any of this, presumably to prevent mass hysteria or other cataclysmic psychological reactions, but Richard Dreyfuss, a normal family man that has been "touched by an alien" after a "close encounter," becomes obsessed with finding the extraterrestrials, and despite formidable obstacles thrown in his way, makes it to the final rendezvous and blasts off with the aliens into the great unknown.

The Scene

In the climactic scenes (19 and 20) we are treated to a musical game of table tennis where alien and human attempt to communicate by bandying the same five-note tune back and forth *ad nauseum*. It's a pretty good idea, communicating using a kind of "tonal language" based on sound frequency (however, because a bunch of previously abducted humans exit the "mother ship" after it lands, you would think the contact would have resulted in a more sophisticated repartee). Along with the musical tones, each note is accompanied by a colored light. The humans have set up a large light board and the aliens flash the colors on their taillights. We get the impression that these lights are also a significant element in the attempts at communication. Do the lights reinforce the message?

Let's analyze the sound sequence.

The Physics—Name that Tune. If we record the five-note pattern, we get sound amplitude versus time graphs for each note in the sequence in order, as follows:

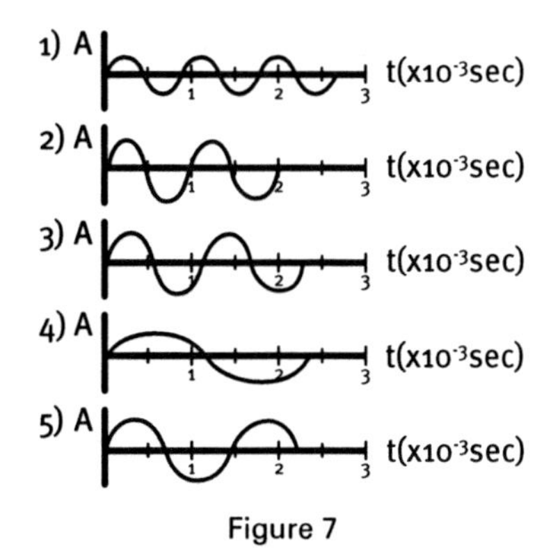

Figure 7

What is the frequency of each note? To determine this all we need to do is look at each graph. The time between each wave peak gives us the period T, and the reciprocal of T is the frequency f. Therefore, the five pitches correspond to frequencies of approximately (1) 985 Hz, (2) 1,110 Hz, (3) 880 Hz, (4) 440 Hz, and (5) 660 Hz. Figure 8 tells us what notes these frequencies correspond to. (These are the actual frequencies from the movie in case you were wondering.)

In musical terms we have the following sequence of notes: B, C-sharp, A, A, and E, where the second A is exactly one octave below the first. Does this pattern have any significance? For example, do the lights rise and fall in frequency along with the notes? Well, the humanoids flash the following sequence of colors: red, orange, pink, yellow, and white. If the pattern is to correspond

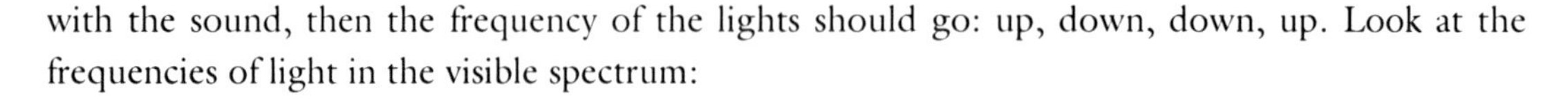

with the sound, then the frequency of the lights should go: up, down, down, up. Look at the frequencies of light in the visible spectrum:

Notes	Frequency (octaves)				
A	55.00	110.00	220.00	440.00	880.00
A#	58.27	116.54	233.08	466.16	932.32
B	61.74	123.48	246.96	493.92	987.84
C	65.41	130.82	261.64	523.28	1046.66
C#	69.30	138.60	277.20	554.40	1108.80
D	73.42	146.84	293.68	587.36	1174.71
D#	77.78	155.56	311.12	622.24	1244.48
E	82.41	164.82	329.64	659.28	1318.56
F	87.31	174.62	349.24	698.48	1396.96
F#	92.50	185.00	370.00	740.00	1480.00
G	98.00	196.00	392.00	784.00	1568.00
A♭	103.83	207.66	415.32	830.64	1661.28

Figure 8

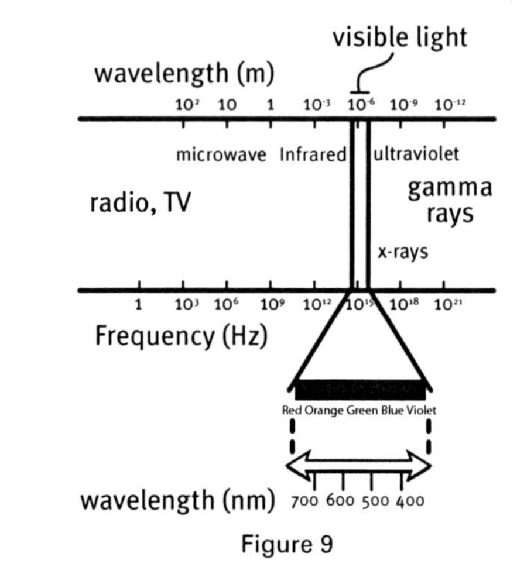

Figure 9

They don't follow the same pattern. In addition, because pink is a combination of spectral colors and white is a mixture of all of the colors together, we see no systematic pattern at all. To make matters worse, the extraterrestrials go with a pattern of orange, pink, yellow, blue, indeterminate color; followed by pink, yellow, pink, yellow, blue.

> **PROBLEM**
>
> Find the wavelengths of the B and E notes, assuming a speed of sound in air of 340 m/s. What are the wavelengths of these frequencies in water where the speed of sound is 1,500m/s?

We won't bother with the issues and difficulties inherent in interstellar space flight until Chapter 8, but assuming these extraterrestrials really have been hanging around Earth all of these years (we know that they have because some pilots from the WWII era come stumbling out of the ship), and assuming during that time they somehow haven't been able to make sense of human language, then the concept of communicating using musical notes and the mathematical patterns contained therein seems like an interesting idea. Nevertheless, unless we're missing something, the colors are merely a directorial touch and strictly for show.

The Doppler Effect. When a fire truck drives past with its siren blaring, the pitch seems to be higher when it approaches, and lower when it recedes. This is due to a frequency shift that occurs due to the relative motion of an observer with the source of the sound. When either the source and/or the observer are moving toward each other, the frequency is shifted upward, and when moving apart, the frequency is shifted downward. For a moving source, because it moves a distance between emitting each successive compression, it will shorten or lengthen the distance between wave peaks, thereby changing the frequency that arrives at the observer. For a moving observer, although the frequency of the sound is not actually changed relative to the air, because the observer is moving toward or away from the source of sound, he or she will encounter the compressions more or less frequently then if standing still, and thus hear a different frequency.

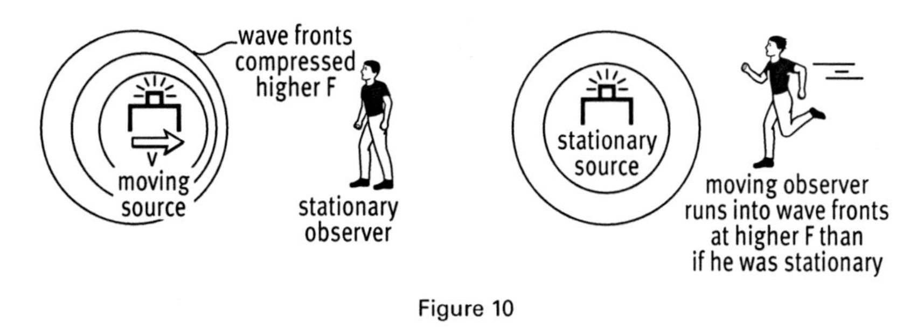

Figure 10

Example 5: Two identical cars drive away from you in the same direction. Each is honking their horn in an irritating way. If car A is moving faster, from which car do you hear the higher pitched horn?

Solution: Because they are moving, you will perceive a lower frequency for both cars. This is due to the longer wavelengths, and correspondingly lower frequencies, of sound arriving at your ear because the cars are moving away from you. However, because car B is not moving as fast as car A, the frequency of the sound arriving from car B will be higher than that arriving from car A.

For a moving *source* of sound, the observed frequency f' depends on the source frequency f, the velocity of the source u, and the speed of sound in the medium v:

$$f' = f(\frac{v}{v \pm u})$$

where the + and – signs refer to receding and approaching sources, respectively. For a moving *observer* the equation becomes:

$$f' = f(\frac{v \pm u}{v})$$

Where u is now the velocity of the observer where + and – refer to moving toward the source and moving away from the source, respectively.

How Fast Are Those Little Ships Moving? Now that we know a little bit about the Doppler Effect, we can go back and reexamine the phenomenon of ships making sounds in space in sci-fi movies in more detail. If "The Force" is strong within us we can estimate the speed of the other ships based on their Doppler shifts.

Recalling our previous "noisy space scene hypothesis," assume the spaceship sound effects were recorded in air where the speed of sound is 340 m/s. We detect a sound frequency of

EARTHQUAKE WAVES

When an earthquake occurs, shockwaves are generated that travel around and through the Earth. These waves can be detected by seismographs, devices that are designed to measure the vibrations due to these waves. The waves that travel through the Earth can be interpreted not only to locate the magnitude and epicenter of an earthquake, but we can also use them to interpret the internal structure of the Earth. Without earthquakes we would know very little about the interior of our planet. There are two types of earthquake waves that travel through the Earth's interior. *P-waves* are longitudinal waves that move with very high velocities. They are able to travel through both solids and liquids. *S-waves* are transverse waves. While also very fast, they are significantly slower than P-waves. S-waves can travel through solids, but they *cannot* travel through liquids.

Because there is an S-wave "shadow zone" on the far side of the Earth from an earthquake epicenter, we conclude that part of the Earth's interior is liquid (specifically the outer core). One of the few things they get right in *The Core* is their explanation of the Earth's interior structure. The Earth's structure is deduced from the information contained in earthquake or *seismic* waves and includes several distinct layers: (1) A relatively thin outer rocky layer called *the crust.* (2) A much thicker layer of rocky material called *the mantle.* The mantle has properties of a solid, but due to the fact that it is at a very high temperature it can flow very slowly. (3) A layer of mostly liquid iron with smaller percentages of other metallic elements called the *outer core.* (4) A solid mostly iron *inner core.* (See Chapter 3, figure 8.)

As the speeds of P- and S-waves increase when they travel deeper into the Earth, they refract and curve. Because S-waves cannot travel through liquid, they don't pass through the outer core. Finally, some of the wave energy is reflected back off of the boundaries between layers. By looking at seismographs in detail it is possible to determine the internal structure of the Earth. (See Chapter 3, figure 8.)

500 Hz on the approaching ships, and 300 Hz on the receding ships. We can use the (previously recorded) Doppler shifts to determine both the source frequency and how fast these space vehicles are moving.

For an approaching ship:

$$f_1' = f\left(\frac{v}{v-u}\right) \text{ so } v = \frac{f_1'}{f}(v-u),$$

and for a receding ship:

$$f_2' = f\left(\frac{v}{v+u}\right) \text{ so } v = \frac{f_2'}{f}(v+u),$$

which gives us $\frac{f_1'}{f}(v-u) = \frac{f_2'}{f}(v+u)$.

We can then solve for u, which gives us a velocity of 97 m/s. Plugging this value back into one of the Doppler shift equations and solving for f we get the frequency emitted by each ship: 357 Hz.

One final comment on this sound in space nonsense: In my opinion, a space battle where all of the action outside the ships occurred in complete silence, in contrast to the explosive sounds heard inside the ships, would be far more exciting (and physically accurate) than the usual fare. Remember that even though a weapon firing would not emit any sound in space, once a spaceship was hit there would be sound generated inside the ship due to the vibrations. Imagine the dramatic tension between the absolute silence of space juxtaposed with the deafening chaos inside an embattled space cruiser. That's a scene I would love to see.

Electromagnetic waves such as light, which we will discuss in more detail next, also exhibit a Doppler shift. Objects that are moving away from us in space have the frequency of their light shifted toward longer wavelengths, or lower frequencies. These objects are said to display a "red shift" because red is at the lower frequency end of the light spectrum. Objects moving toward us are said to display a "blue shift" because their light is shifted toward shorter wavelengths and higher frequencies. The amount of shift can be used to determine the speed at which these objects are moving relative to the Earth. The observation that other galaxies have "red-shifted" e-m frequencies, and that the galaxies further away exhibit the greater shifts, suggests that the universe is expanding. This is also one of the lines of evidence supporting the big bang theory of the origin of the universe.

Light and the Electromagnetic Spectrum

An *electromagnetic wave* is an oscillating electric and magnetic field that propagates through space at an extremely high speed. Because the oscillations are perpendicular to the direction of the wave velocity, electromagnetic waves are transverse waves. Light is merely electromagnetic (e-m) radiation within a certain frequency range. The *speed of light (c)*, and all e-m radiation, is 3×10^8 m/s in a vacuum. The speed is very nearly the same in air, but will be slower in more dense materials such as water or glass. The atmosphere is opaque to certain frequencies of e-m radiation, and transparent to others. For example, visible light, microwaves, some infrared radiation, and radio waves are not affected by the atmosphere, while the other e-m frequencies are absorbed. The entire spectrum of e-m radiation is shown in figure 9 on page 188.

The higher the frequency of e-m radiation the higher the energy the radiation has. Therefore, gamma rays have the highest energy and radio waves the lowest.

Gamma Rays and X-rays. *Gamma rays* are primarily produced as a product of the radioactive decay of unstable elements, while x-rays can be produced when electrons collide with atoms. These frequencies are sometimes famous for being potentially dangerous to biological organisms. Neither gamma nor x-rays can be seen or felt. Gamma rays usually pass through tissue but some can be absorbed by cells, damaging them. Significant exposure can cause cancer, and extremely high doses can be fatal. (So in real life, if Bruce Banner—the Incredible Hulk—was to be exposed to a lethal dose of gamma rays, he probably wouldn't turn into a giant, green, slow-witted, super-powered fighting machine, he would simply expire.)

X-rays have less energy than gamma rays. They pass through most tissue but generally not through bone or metal, and are therefore useful as a diagnostic tool in checking for broken bones. However, overexposure to x-rays can cause cells to become cancerous.

Ultraviolet Radiation. *Ultraviolet* radiation is just higher than the frequency range of visible light. It has gained some notoriety in popular culture in part for the damage it can do to skin cells. Because darker skin cells absorb more ultraviolet radiation, skin often responds to exposure by turning darker so that less radiation penetrates to deeper skin layers. (This is called getting a tan.)

The ozone layer (O_3) in the upper atmosphere absorbs a significant amount of UV radiation, and therefore the depletion in the ozone layer due to reactions with chemicals found in aerosols (chlorofluorocarbons) is increasing our exposure to ultraviolet light.

The Visible Spectrum. Between wavelengths of about 400 to 700 nanometers ($1 \text{ nm} = 1 \times 10^{-9}\text{m}$), e-m radiation is visible to the human eye. We perceive the shortest wavelengths as violet or purple and the longest wavelengths as red.

White light, such as that produced by the sun or by an incandescent lightbulb, is a mixture of all the frequencies of visible light. We can see objects due to the fact that they reflect light back into our eyes. An object that is white reflects all of the frequencies of light while absorbing none. A black object absorbs all frequencies and reflects none. If an object is a specific color such as green, it means that the object absorbs all frequencies *except* green, which is reflected back off of the object and perceived by the eye.

Infrared Radiation. *Infrared* radiation has wavelengths just longer, and frequencies just lower, than visible light. When infrared radiation is absorbed by matter it increases the thermal energy of the matter. Therefore, when standing in sunlight or next to a fire, it is the infrared radiation that produces the sensation of heat in our skin. Infrared radiation is used in heaters and toasters, as well is in fiber optics systems.

Microwaves. Extraterrestrial sources of *microwaves* come mostly from the sun, with a tiny amount from the "cosmic microwave background radiation," which along with the "red shift" that we discussed comprises one of the major lines of evidence supporting the big bang theory of the origin of the universe. While intense microwave radiation can be dangerous, the amount of extraterrestrial microwave radiation is harmless. Microwaves contain wavelengths that are absorbed easily by water molecules. When the water absorbs the microwave radiation it causes the water to heat up, which is how microwave ovens work. Microwaves are also used in radar systems, to transmit signals between mobile phones, and to transmit information from Earth to satellites in orbit.

Radio Waves. The longest wavelength and lowest energy e-m radiation are radio waves. Radio and television programs are transmitted using e-m radiation in this lowest frequency range. Because radio waves are diffracted as they pass by obstructions such as hills or buildings, televisions and radios do not need to be in a direct line of sight of transmitters to receive signals. In addition, radio waves can be transmitted long distances by bouncing them off of the ionosphere, the electrically charged layer in the uppermost part of the atmosphere.

Refraction and Dispersion of Light

When a wave travels from one medium into another, some of the energy is reflected back, and some is transmitted. If a wave-front approaches at an angle to the interface between the media then the transmitted part of the wave will refract. For example, if a narrow beam of light shines on a piece of glass at an angle to the (normal line of the) glass surface we will see some of the light reflected back and some of the light change direction as it travels from air into glass.

PROBLEM

A hollow triangular area of air is cut out of a large piece of glass in the same shape as in the example below. Trace the path a beam of light travels through the glass across the hollow area, and back out again.

The amount of bend depends on the difference in velocities of the speed of light in each of the media. The greater the difference in velocity, the greater the angle through which the light will bend. If the light travels from a "fast" medium, such as air or a vacuum, into a relatively "slow" medium, like glass, the wave will bend towards the normal line. If the light travels from a slow to fast medium it will bend away from the normal line.

Example 6: Light is traveling through air when it enters the glass prism shown here.

What is the path of the light that travels through the prism and out the other side?

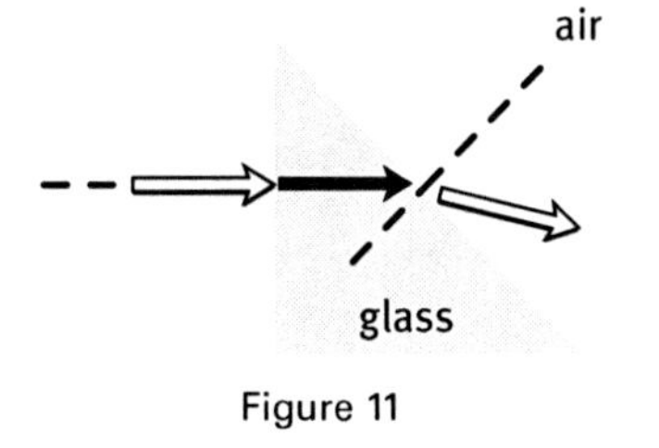

Figure 11

Solution: Since the light enters parallel to the normal line, there will be *no* refraction as it enters the prism. It will continue to move in the same direction. However it does strike the other side of the prism at an angle. Since the light is traveling from a slower to a faster medium it will bend away from the normal line and therefore downward, as shown.

Snell's Law. *Snell's Law* quantifies the relationship of the incident angle of the wave or beam, to the refracted angle depending on the velocity of light in each of the media:

$$\frac{\sin\theta_i}{v_i} = \frac{\sin\theta_r}{v_r}.$$

Sometimes the velocity of light in the medium is expressed in terms of the *index of refraction* (n) of the medium.

$$n = c/v$$

where c is the speed of light in a vacuum, and v is the speed of light in the medium in question. For example, the speed of light in most glass is about 2×10^8 m/s. Therefore n for glass = $3 \times 10^8 / 2 \times 10^8 = 1.5$. *In a vacuum n would be 1 and n for air is extremely close to 1 because the velocity of light in air is very nearly the same as in a vacuum.*

We can rewrite Snell's law in terms of n as follows:

$$n_i \sin\theta_i = n_r \sin\theta_r.$$

Example 7: Shine a beam of laser light at a prism in the direction shown in figure 12 below.

If $n = 1.5$ for this frequency of laser light in glass, in what direction will the light travel through the glass and back out again?

Solution: We'll call the air medium 1, and the glass medium 2. For the first refraction,

$$\sin\theta_2 = \frac{\sin\theta_1 n_1}{n_2} = \frac{(\sin 30)1}{1.5} = 0.33 \text{ and } \sin^{-1}(.33) = 19.5° = \theta_2.$$

For the second refraction we need to use geometry to find the incident angle. Notice that because the normal line is always perpendicular to the surface of the material, the normal line for the second refraction is oriented differently than the first. Playing with the angles we find the incident angle (θ_3) is 40.5°.

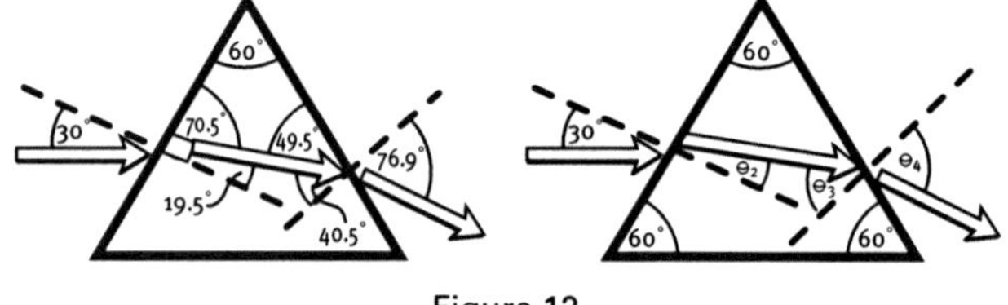

Figure 12

Therefore $\sin\theta_4 = \frac{\sin\theta_3 n_2}{n_1} = \frac{(\sin 40.5)1.5}{1} = 0.53 \text{ and } \sin^{-1}(.53) = 76.9° = \theta_4.$

Dispersion. If white light passes through a piece of glass, such as a prism, the various frequencies can be separated into a spectrum of colors.

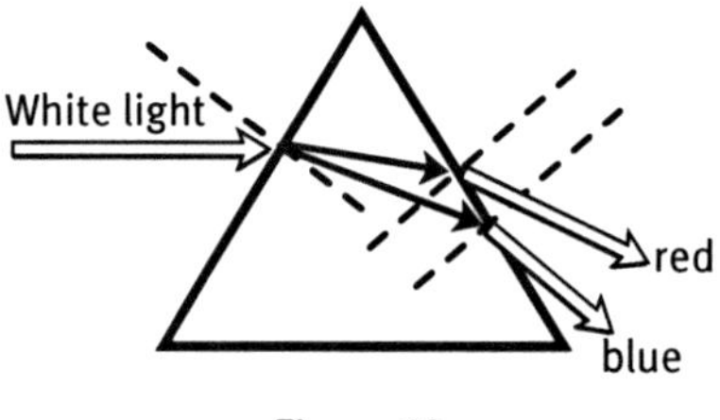

Figure 13

This is because each frequency has a slightly different velocity as it travels through the glass, and will therefore refract at a slightly different angle. (When we say the index of refraction for a material is a certain value it actually depends on the frequency of light. For example, *n* for violet light in a certain glass may be 1.66 while *n* for red light in the same material is 1.62. In some cases we will specify that *n* refers to a specific frequency as in example 7. If we are dealing with refraction of white light then often we will use an approximate value not accounting for dispersion.) As the light passes into the glass, the different frequencies separate out, and as they exit the glass the second refraction separates them even more, so that a rainbow of color can be seen. The blue end of the spectrum refracts the most, and the red end the least.

Rainbows. Perhaps one of the more visually impressive and pleasant natural phenomena is the rainbow, which occurs when sunlight is refracted, dispersed, and reflected as it passes through raindrops (which have a higher index of refraction than air). To be able to see a rainbow, the sun must be behind you, and you need to be looking into an area where there are raindrops in the air. The light from the sun will enter the raindrops and refract inside them as in a prism. While some of the light will then be transmitted through the raindrop, much of the light will be internally reflected off of the back surface of the raindrop. This reflected light will then refract again upon exiting the raindrop, causing the light to be dispersed into a spectrum of colors. The characteristic rainbow shape occurs because light exits the raindrops within a narrow range of angles. This limits the regions in the sky from where the dispersed light can get to your eye.

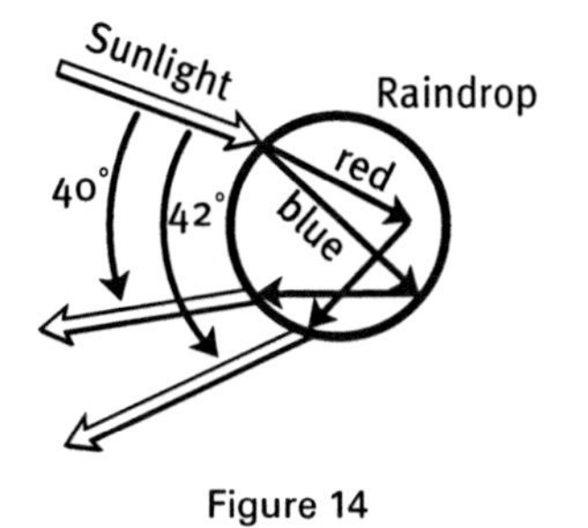

Figure 14

Total Internal Reflection. When light travels from a medium where it moves relatively slowly into one where it has a higher velocity, it will bend away from the normal line, and the refracted angle will be greater than the incident angle. Some of the incident light will be transmitted into and refracted by the second medium, and some of the light will be reflected back.

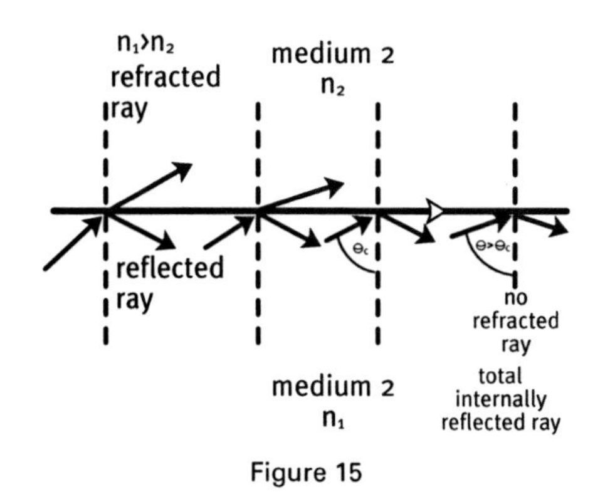

Figure 15

As the incident angle is increased, so is the angle of the refracted light. Eventually, there will be a certain incident angle called *the critical angle* when the resulting refracted angle will be 90 degrees, and move along the interface between the two media. If the incident angle is increased anything beyond the critical angle then none of the light will be refracted into the

ACTIVITY

Find a pond, swimming pool, or ocean. Stand next to the body of water and observe where you can see objects underneath the surface, and where you can only see light reflected off of the top. How could you estimate the index of refraction of water from this observation?

second medium, and all of it will be reflected back. This situation is called ***total internal reflection.*** In the case of total internal reflection, all of the light's energy remains in the first medium, because none can escape.

Example 8: What is the critical angle for light passing from glass of $n = 1.5$ into air?

Solution: $n_{glass} \sin\theta_{cr} = n_{air} \sin 90$ so $\sin\theta_{cr} = \frac{n_{air}}{n_{glass}} = \frac{1}{1.5}$

Therefore, $\theta_{cr} = 42^\circ$.

Fiber Optics. *Fiber optics cables* are thin glass rods that use the principle of total internal reflection to transmit information. While electrical signals and radio waves are methods by which information can be transmitted, sending light (visible or infrared) through a fiber optics cable loses less signal and therefore loses less information than electrical or radio wave transmission. Light going in at one end of a fiber optics cable undergoes multiple internal reflections, even in a bent cable, so very little of the light signal is lost or attenuated by the time it emerges at the other end.

Mirages. Imagine that you are staggering through the desert desperate for a drink of water. Vultures hover overhead. It hasn't rained in weeks. Foolishly, in a fit of hunger, you have just eaten the last bit of food remaining in your pack—a bag of extremely salty crackers. Just when you think you will surely collapse from dehydration, through sweat-encrusted eyes you glimpse a cluster of palm trees up ahead. Not only that, but you can see the trees' reflections shimmering in a delicious-looking pool of blue water. You are saved! Well . . . maybe. You could have fallen for the oldest trick in the book—the desert mirage—where you think you see water when there isn't any. A more common example can occur when you are driving on a road on a hot sunny day. It often appears that there are patches of water in the roadway up ahead. In either case this phenomenon is due to refraction of light. How does it work? On a hot day the temperature near the ground is much warmer than the air further up. Because light travels slightly faster in warmer air, as light passes through layers of warmer air it will be refracted away from the normal line. Light coming from a source at some elevation above the ground will spread out in every direction, but light rays angled toward the ground will refract away from the normal line, at shallower and shallower angles to the ground. Eventually the light will reach the critical angle to the layer of air below and will undergo total internal reflection. These light rays will then refract back upward, fooling your eye into seeing an image below where the light actually originates.

If the light was coming from a treetop off in the distance, you would see light rays arriving from a more or less direct path, but you would also see light arriving from the refracted path shown in the following figure. You would see the actual tree, but you would also see an upside-down image of the tree underneath. This would look very much like the reflection of the tree in a pool of water. So much for that refreshing drink!

Figure 16

Lenses and Mirrors

We can use reflection and refraction of light to construct optical devices with lenses and mirrors. Lenses and mirrors redirect and refocus incoming light such that they can form images. Light emanating from an object will travel outward in all directions; however, if you can "reconverge" some of this light you can form an image of the object. To show how these form we can approximate a continuous wave-front of light to consist of an infinite number of discrete "rays." If you shine a flashlight through a pinhole in a dark room, you could think of the transmitted beam as one of the "rays" of light emanating from the flashlight.

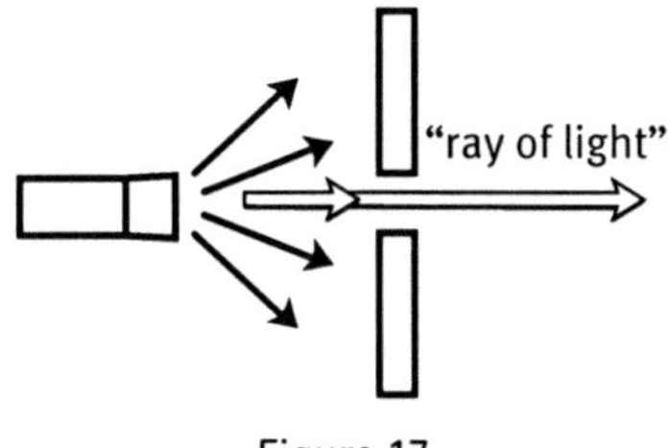

Figure 17

We can then determine the exact paths of a few (at least two) specific rays passing through or reflecting off of a lens or mirror, which can give us enough information to determine the position and size of an image formed.

The distance from the center of the lens or mirror to an object is called the *object distance* (s_0), and the distance to the image is called the *image distance* (s_i). The *focal point* is the position where light rays incident *parallel* to the principal axis (shown in the figure) intersect the axis.

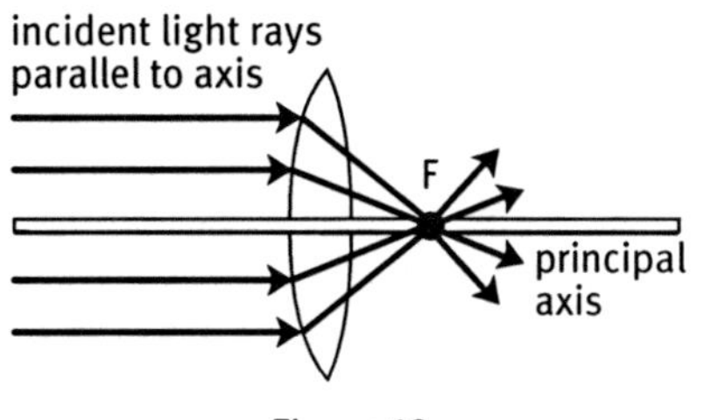

Figure 18

The *focal length* (f) is the distance from the center of the lens or mirror to the focal point.

Images Formed by Thin Lenses. Let's examine in more detail how lenses can refract light emanating from an object and form images. Referring to the next figure we will follow the path of two particular rays coming off of an object (the tip of the arrow) as they pass through and are refracted by a lens (in this case a converging, or convex, lens.)

The first ray we will examine is one that is initially parallel to the principal axis. When this ray of light passes through the lens, it will refract, as shown in the figure. The next ray that will help us to determine where the image will form is one that goes right through the center of the lens. This ray will pass through the lens without any change in direction. Now notice in the figure where these two rays converge. An image of the tip of the arrow exists at this location. If we placed a screen at the location where these rays converge you would see the image formed on the screen. This type of image—where light rays converge at a point in space to form the image—is called a ***real image***. By further examining the diagram, we can see that the image will also be inverted. (It's a common misconception that an image will form at the focal point. This will only be true for an object an infinite distance away such that *all* of the incoming rays are effectively parallel. In any other situation the light rays will not converge at the focal point.)

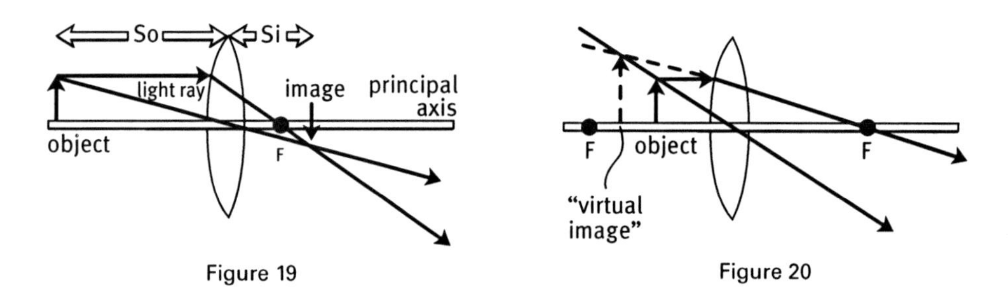

Figure 19

Figure 20

What happens if we place an object inside *f* for a converging lens? If we trace the same two rays we can see that they will never converge. In fact, they diverge.

However, due to the way that rays diverge, your eye will extrapolate the rays backward to an apparent convergent point, where you will perceive an image. In this case if you were to place a screen at the image location you would not get a focused image on the screen because light is not actually converging at that location. This type of image, in which the light only appears to converge, is called a ***virtual image***.

Notice in the previous figure that the image is upright and magnified. If you look through a converging lens at an object placed inside the focal point, then you are looking through a magnifying glass. What happens if you place an object exactly at *f*? Incoming rays exit the lens parallel to each other. They neither converge nor diverge, so no image, either real or virtual, will result and you will see only a vague blurry form of the object.

A diverging lens, which is a concave lens, always produces a virtual image, as we can see in the ray diagrams. For a diverging lens incident parallel rays will not be directed through *f*. Instead, they diverge such that if you were to extrapolate the ray backward it would seem to come from f on the opposite side of the lens.

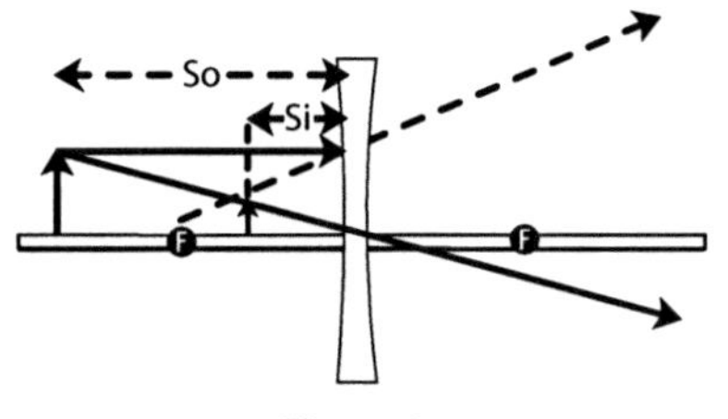

Figure 21

As you can see, by carefully tracing a couple of principal light rays you can geometrically determine the location of an image formed by a lens. We can also do this mathematically using the *thin lens equation:*

$$\frac{1}{f} = \frac{1}{s_0} + \frac{1}{s_i}.$$

For a converging lens f is a positive number, and for a diverging lens f is a negative number. The magnification of the image is equal to the ratio of image to object distance:

$$m = -\frac{s_i}{s_0}.$$

Example 9: You place an object 20 centimeters in front of a converging lens with $f = 8$ cm. Where will the image form and what will its magnification be?

Solution: $\frac{1}{s_i} = \frac{1}{f} - \frac{1}{s_0} = \frac{1}{8\text{ cm}} - \frac{1}{20\text{ cm}} = 0.075\text{ cm}$ so $s_0 = \frac{1}{0.075} = 13.3\text{ cm}$

$m = -\frac{s_i}{s_0} = -\frac{13.3\text{ cm}}{20\text{ cm}} = -0.67$

The image is real, inverted, and 0.67 times the size of the object.

Images Formed by Mirrors. Forming images with mirrors is quite similar to lenses, but reflection, instead of refraction, is used to redirect the light. First let's look at a plane mirror, which is a flat mirror, like the one in your bathroom.

We can also use converging and diverging mirrors to form images by reflection of light. (Diverging mirrors are those big convex mirrors used for security at convenience stores—the ones where you can see everything in the room.) A converging mirror must be concave, and a diverging mirror

ACTIVITY

Look at yourself in the bathroom mirror. Where does your image seem to be located?

Trace some of the rays of light coming from off of a part of your body (your nose for example), as they reflect off of the bathroom mirror and back. Can you use your "ray diagram" to demonstrate the location of the image of your nose? Is this a real or a virtual image?

> **PROBLEM**
>
> Sven has a diverging mirror with $f = -15$cm. He places a 1-inch tall porcelain replica of Ingrid 10 centimeters in front of the mirror. Where is the image formed, and how tall is it?

convex, the opposite as it is with lenses. The procedure for determining where an image will be formed by a mirror is similar to that for lenses. A light ray incident parallel to the principal axis will be reflected through the focal point, and a ray that hits the center of the mirror will have equal incident and reflected angles relative to the mirror. We can use these rays then to find where an image forms as shown in the following figure.

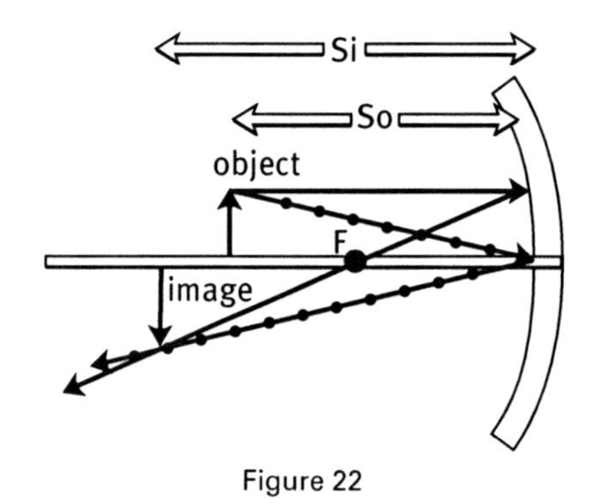

Figure 22

The mathematical equations used to determine image position and magnification are the same as those we used for lenses.

Images Formed by Combinations of Lenses or Combinations of Lenses and Mirrors. To make a more sophisticated optical instrument like a telescope or a microscope we would use multiple lenses and mirrors. To find where an image will form mathematically, (1) calculate the location of the image that would theoretically form due to the first device as if it was by itself, (2) find that position relative to the second device (this distance now becomes the object distance for the second device), and (3) calculate the final image position using the new object distance. The magnification is the product of the magnification due to each lens individually.

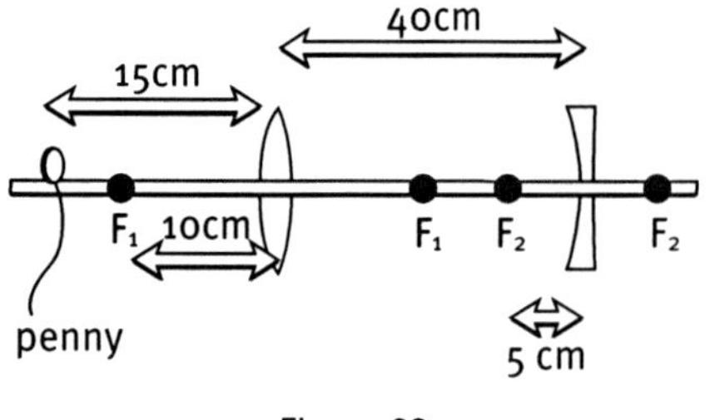

Figure 23

Example 10: You place a converging lens with $f =$ 10 cm a distance of 40 cm from a diverging lens with $f = -5$ cm. If you stand a penny up on its edge 15 cm in front of the converging lens, where does the image form, and what is its magnification?

Solution: First we'll find s_i for the first lens.

$$\frac{1}{s_{i1}} = \frac{1}{f_1} - \frac{1}{s_{01}} = \frac{1}{10\text{ cm}} - \frac{1}{15\text{ cm}} \text{ which gives } s_{i1} = 30\text{ cm}.$$

This is 30 cm to the right of the converging lens, which is 10 cm to the left of the diverging lens. Therefore s_{02} = 10 cm, and

$$\frac{1}{s_{i2}} = \frac{1}{f_2} - \frac{1}{s_{02}} = \frac{1}{-5\text{ cm}} - \frac{1}{10\text{ cm}} \text{ which gives } s_{i2} = -3.3\text{ cm}.$$

The negative sign means that the image is to the left of the diverging lens and it is virtual. Finally,

$$m = (-\frac{s_{i1}}{s_{o1}})(-\frac{s_{i2}}{s_{o2}}) = (-\frac{30\text{ cm}}{15\text{ cm}})(-\frac{-3.3}{10}) = -0.66$$

The image of the penny is inverted and two thirds its actual size.

Star Trek IV and the Search for Intelligent Cetaceans

In *Star Trek IV* we find the now-renegade crew of the defunct Starship Enterprise (it exploded at the end of *S.T. III*) returning to 20th-century Earth in a commandeered Klingon warship in order to find some humpback whales. The problem is that there's a high-energy probe hanging around outside 23rd-century Earth's atmosphere sending messages to the now extinct humpbacks. With no one to respond, the high-energy signals are causing Earth's oceans to evaporate. The only way to save the planet is for Captain Kirk, Spock, Mr. Scott, and the rest of the beloved crew to go back in time and bring some of the whales "back to the future." The most amusing scenes are the "fish out of water" sequences where the 23rd-century crew members clumsily interact with late 20th-century popular culture. One of the best of these scenes finds Checkov asking passersby how to find nuclear submarines in a thick Russian accent because the crew needs to siphon some energy off of one of the nuclear reactors to refuel the badly damaged engines in the Klingon cruiser. (Because the Klingon battle cruiser is designed to travel thousands of light years through deep space at "warp" speeds it doesn't seem like drawing off a small percentage of the energy produced by a 20th-century nuclear reactor would be quite sufficient. In addition, if you watch the scene it's pretty amusing the way in which they do this: by "collecting photons." Photons are particle-like bundles of electromagnetic energy as we'll discuss in Chapter 8. Individual photons don't have that much energy, and how the photons that they "collect" in their little photon box—about the size of a portable telephone—will be converted back into the phenomenal quantities of energy necessary to power the ship is a little hard to imagine. Amazing technology!)

The Scene. Due to unforeseen circumstances, the two whales slated by Kirk for transport to a better future where they don't hunt whales anymore have been released into the Pacific a day too early. They are now in the Bering Sea and imminently threatened by Norwegian whalers. Fortunately the Klingon battle cruiser has a "cloaking device," a gizmo that renders the entire ship invisible when activated, and the crew is able to sneak up on the whalers and scare them away. (Finally after a few more life and death emergency situations, the Earth is saved.)

There are a lot of scientific and technological issues in the Star Trek movies that skirt the extreme theoretical borders of what might someday be possible. (For an entertaining look into the science and technology of Star Trek check out *The Physics of Star Trek* by Lawrence Krauss.) Nevertheless, what about that cloaking device; could it work? Is it possible that we could create technology to perform this amazing optical trick? Believe it or not, this one is actually something that we can make a pretty reasonable case for. Let's start with the basic idea.

The Physics—Our Homemade Cloaking Device. If we were to make ourselves invisible, an observer would not see us, but would see whatever is on the other side of us. In order to do this, we would have to prevent light from reflecting off of us, and also somehow bend the light from objects behind us into the prying eyes of inquisitive observers. We would need to create a mirage.

For our purposes we will play around with a couple of optical concepts. We will redirect light in order to create a crude "cloaking device"/mirage. Maybe we won't be invisible, but at least we might confuse and irritate our adversary for a second.

ACTIVITY

Using the principles that we have articulated in this chapter, come up with two or three ideas or principles that we could use to make our spaceship harder to spot.

Let's first make our imaginary spaceship completely black so that no light will be reflected off of its surface. Then we will try to create the mirage of the background stars and planets where an observer would be looking at the ship. We will consider applying the following methods to make our homemade optical deception: (1) surround the ship with fiber optics cables, and (2) find a way to refract light around the ship.

The fiber optics idea is pretty simple. All we need to do is use total internal reflection inside the cables to bend light around the ship (see the illustration on the next page).

Will this work? Well it depends what we mean by "work." We can divert light such that we create a mirage in front of the ship, but we are going to have problems with reflections off of the exterior of the cables, and not all of the light will necessarily be totally internally reflected because it depends on the angle of the rays relative to the cables. This means *we will be able to see the cables.* In addition, light rays can diffuse as they pass through the cables so that the image/mirage won't be as sharp as the original. Will we be invisible? Not really. Will it be a pretty cool effect? I think so!

To refract light around the ship, we could try to use refraction just like an oasis mirage. Let's surround our spaceship with streams of air. (Obviously these will have to be emitted from inside the ship.) If we can somehow create a temperature gradient such that the air nearer the ship is cooler than the air further away, light will refract as it passes through the air layers. (We could surround the exterior of the ship with space heaters protruding from the hull.)

PROBLEM

Let's say our fiber optics cables are made out of glass with an $n = 1.5$ for visible light. What is the maximum angle of light relative to the cable walls that will be totally internally reflected inside the cables?

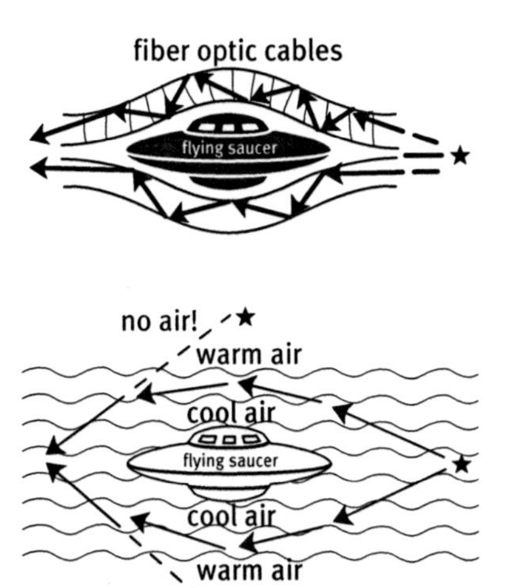

Figure 24

Unfortunately, we might have a few problems. How do we contain the air and keep it from diffusing into space? How do we keep light from reflecting off of the air so that an observer doesn't see that? If you study the diagram you can see that light from a single source will appear to be coming from different directions. Therefore, for example, instead of seeing a single star directly behind the ship an observer would see two stars on either side. This idea probably wouldn't work as well as the fiber optics cables. Maybe we should project images of the background onto the front of the spaceship with a large overhead projector. If worse comes to worse we can just paint a space scene on the outside of the ship, and hope no one pays too much attention.

Conclusion. Our homemade systems are just a way to explore some of the basic issues. It turns out that currently research is being done at Duke University using something called *meta-materials.* These are materials with some really strange properties, such as negative refractive indices, and a reverse Doppler shift (incident waves shift to *lower* frequencies in these materials). In addition, higher frequencies of waves in these materials actually correspond to longer wavelengths. Researchers think it reasonable that it will eventually be possible to surround an object with these meta-materials, and by exploiting their unusual properties, refract e-m radiation around the object rendering it completely invisible to outside observers. *Voila!* A cloaking device. (Unfortunately that also means that outside observers would be rendered invisible to anyone inside. Maybe we could come up with a very tiny periscope.) Interestingly, in the movie *Batman Begins* there is a scene where the Batmobile completely disappears during a pursuit by the police. It sure seems like the Batmobile has some kind of cool high-tech cloaking device. Perhaps Wayne Enterprises (and Morgan Freeman) had some mothballed meta-materials down in the tech department just waiting to be utilized when Bruce Wayne arrived back on the scene.

Starship Troopers and Big Scary Bugs

Starship Troopers is a campy movie, full of exceptionally good-looking actors, which takes place sometime in the Earth's future. It's a future in which humanity is engaged in a deadly conflict with giant insects from the planet Klendathu. Although these bugs don't have any technology to speak of, they are somehow able to fling asteroids all the way across the galaxy where one of them ends up obliterating Buenos Aires. They actually show us a "map" of where Klendathu resides in the Milky Way (our beloved spiral galaxy that we call home) and it's located almost completely on the opposite side. The Milky Way is about 100,000 light years across, so even if they could toss these rocks at velocities close to the speed of light, it would still take about a hundred-thousand years to arrive on Earth. This would result in a very protracted conflict, but we suspend our disbelief. Let's focus instead on how these bugs got so big.

Let's say we want to make our own low-budget giant-insect sci-fi thriller. We don't have fancy computer graphics at our disposable. We don't have a budget of 60 million dollars. We don't even have any actors. What we do have are some real insects and some Starship Trooper action figures. What do we do? We'll use converging mirrors to magnify the bugs.

We are going to start with a simple single mirror system. (With a single device we can only effectively magnify something four or five times, but with combinations of lenses and mirrors we can get much larger magnifications.) The bugs are pretty big relative to our Starship Trooper action figures, but they're not big enough for a horror sci-fi classic. We have to beef up the bugs. Where should we place the insects relative to our converging mirrors to make them look as big and scary as possible? Do we put them inside or outside the focal point? Do we put them close to or far away from *f*?

Well, we know that if we want to magnify an object, just like with a converging lens, we can put it inside *f* to get a magnified virtual image several times the size of the object. We can also get an enlarged real image if we place the object between *f* and 2*f*. In both cases you get the largest magnification if you place the object close to *f*.

So let's say we use converging mirrors with $f = 5$ cm. We set up the shot with a beetle 4.5 centimeters from the center of the mirror. If the beetle rears up three-quarters of an inch off of the ground, how tall is its image in the mirror? Because

$$m = -\frac{s_i}{s_0},$$

we need to find s_0 using the mirror equation

$$\frac{1}{f} = \frac{1}{s_0} + \frac{1}{s_i}.$$

We get

$$\frac{1}{s_i} = \frac{1}{5 \text{ cm}} - \frac{1}{4.5 \text{ cm}} \text{ and si} = -45 \text{ cm, so } m = -\frac{-45 \text{ cm}}{4.5 \text{ cm}} = 10,$$

which means our beetle is now 7.5 inches tall in the mirror and he towers over our Starship Troopers action figure! Much better.

Do you want to see some really big and scary insects? Let's make a microscope. All we need is two converging lenses. Basically we can use the first lens (the objective lens) to form an image inside the focal point of the second lens (the eyepiece). We can get two magnifications for the price of one if we place the object just outside the focal point of the objective lens.

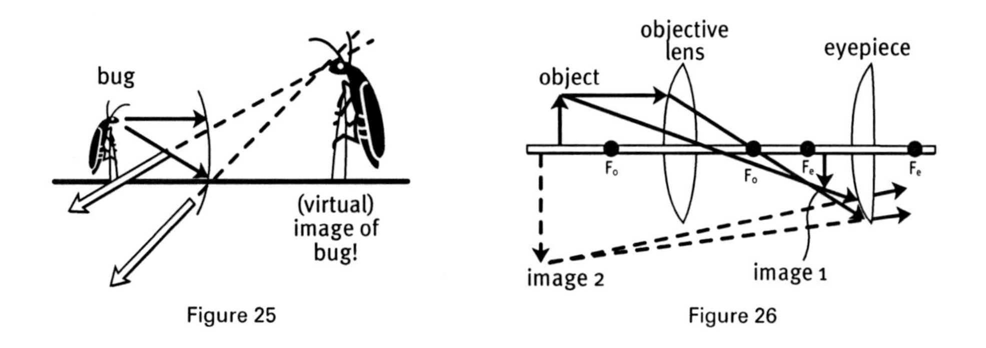

Figure 25

Figure 26

Conclusion

While you have to look around a little bit, the physics of waves, sound, and light is alive (and sometimes well) in the cinema—most notably in the science fiction genre. With a little delving and some imagination you can come up with some interesting scenes to play with that bring up the relevant principles.

ADDITIONAL QUESTIONS AND PROBLEMS

1) A simple harmonic wave is one in which different points in the medium oscillate sinusoidally (or cosinusoidally) as a function of time. These points in the medium are in simple harmonic motion (SHM). A mass oscillating up and down suspended on the end of a spring is a classic example of an object in SHM. Objects will oscillate in SHM if the restoring force—the force that tends to pull an object back to an equilibrium position—is directly proportional to the distance the object is extended from that position. The force a spring exerts on an object fits this requirement because it increases linearly with the amount of stretching or compression ($F = -kx$) (see Chapter 3). The period of a mass oscillating on the end of a spring is given by $T = 2\pi\sqrt{\frac{m}{k}}$, where m is the mass and k is the spring constant—a measure of the stiffness of the spring. Let's say you have a 20-g mass connected to a spring with $k = 400$ N/m, and you extend the spring 10 centimeters and release it. How fast will the mass be traveling when it goes through the equilibrium position? What is the amplitude of the oscillation? Write an equation for the position of the mass as a function of time.

2) Just as you can generate standing waves by shaking a rope, you can generate invisible standing waves of sound in tubes or pipes. You can hear the standing wave when a certain frequency of sound applied to the end of the tube resonates inside the tube. This is the concept behind wind instruments. If the pipe is open at both ends, the condition for standing waves is the same as for the rope fixed at both ends as described previously. For a pipe with one open and one closed end, the condition is $f_n = n\frac{v}{4L}$ where n = 1,3,5,7..., etc. (v is the speed of sound and L is the length of the pipe). Suppose you have a pipe and find that it will resonate at frequencies of 100 Hz, and 120 Hz. At what other frequencies will it resonate?

What is the fundamental frequency at which it will resonate? Is this pipe open at both ends, or closed at one end?

3) In the movie *A Perfect Storm,* a fishing boat is tragically capsized by a huge "rogue wave." The boat probably would have survived the storm if it weren't for this single giant swell. Explain what wave property might have contributed to creating the deadly wave.

4) If an airplane is moving exactly at the speed of sound, it will generate a shock wave, which is an extremely large amplitude wave. Explain how this occurs. Recall the sections on wave interference, and the Doppler shift.

5) The Jell-O aboard the starship *Enterprise* has an index of refraction of about 1.4. The crew is playing with a cylindrical tube of Jell-O by shining laser light into the end of the tube.

   At what maximum angle relative to the end of the tube can they shine the light so that it will stay completely inside the Jell-O?

6) A little project: In the *Star Wars* movies the characters communicate with each other across great distances within the galaxy by broadcasting what appear to be holograms. How messages can be transmitted rapidly and effectively throughout the galaxy remains a thorny sci-fi movie obstacle, but making holograms, well, that is quite doable (though maybe not a "holo-deck" like on *Star Trek the Next Generation*). Holograms are a kind of three-dimensional photograph in which principles of wave interference are used to create 3-D images of objects. Do a little research and look up how these are made. It's quite interesting.

7) A refracting telescope is similar to a microscope in that it uses two lenses to magnify an image. In the case of the telescope the objects are very far away so all of the light rays come in essentially parallel. This means the image formed by the objective lens will be very near its focal point. If you have an eyepiece with a given focal distance how would you arrange it relative to the objective lens such that you could get a magnified image?

Modern Physics in Modern Films

IS INTERSTELLAR SPACE FLIGHT REALLY feasible? What's "warp drive"? Why do nuclear reactions produce so much energy, and why would a matter-antimatter reactor perhaps be the most efficient source of energy production of all? What is antimatter anyway? Does it really exist? While we're at it: Are there other physical dimensions? What exactly is a black hole, or a wormhole for that matter? How big is the solar system, the galaxy, the universe? Is there intelligent life out there?

These are the types of questions that whet the imagination, that allow one to step back from the mundane concerns of daily life and gasp a little with admiration at the vastness, and complexity (absurdity?) of the cosmos in which we somehow live. These are the types of questions that motivate writers of science fiction and makers of science fiction films and that fascinate the general public enough to have created such blockbuster movie and television franchises as *Star Wars* and *Star Trek*.

Sci-fi films have become increasingly more sophisticated, not just in their use of stunning effects but also in the technical sophistication of the jargon. Who doesn't love the idea of traveling to distant and exotic planets, "to seek out new life and new civilizations, to boldly go where no man has gone before!"? Nevertheless, what is the physics behind the cinematic façade? In the last hundred years or so physics has undergone a revolution of sorts, where our everyday notions of space, time, matter, and energy have been challenged—most notably by Einstein's theories of special and general relativity, and the theory of quantum mechanics. New discoveries in cosmology have also shown us that there may be fundamental forces out there that heretofore have gone undetected and unimagined. (Remember the accelerating expansion of the universe that we mentioned in Chapter 2?) These new discoveries not only fuel imagination, but also give creators of science fiction an opportunity to try to find threads of reality within their fictional worlds.

In this chapter we are going to look at some of the classic modern physics issues confronted in science fiction. We will see what the theory of relativity says about interstellar space flight portrayed in sci-fi blockbuster franchises like *Star Wars* and *Star Trek,* in the Disney movie *Flight of the Navigator,* and the sci-fi horror classic *Event Horizon.* We'll end our discussion of modern physics

Photo: Hanging out in space.

with an analysis of quantum mechanics as represented in the low-budget new age film *What the βleep Do We (k)now!?*, examining the borders between the philosophical and the scientific in an attempt to articulate what science is, and what it is not. That said, let's start with the thorny core of the sci-fi universe: interstellar space travel.

THE THEORY OF RELATIVITY: IMPULSE SPEED AND WARP SPEED

The star (Alpha Centauri) closest to Earth (which is actually a system of three stars gravitationally bound to each other) is over four light years away. That means it will take a beam of light traveling at 300,000 km/sec over four years to travel between Alpha Centauri and Earth. (Our galaxy is about 100,000 light years in diameter, so light coming from the other side of the Milky Way gives us images of stars as they were 100,000 years ago. Some of the stars we see might not even exist anymore.) However, in *Star Trek, Star Wars,* and many others these great distances do not present a problem. Using "warp drive" or "hyper drive" you can go hundreds of light years in a matter of days (or even hours). Is this in any way remotely possible in reality or is the entire concept of interstellar travel dead in the water? Let's investigate some of the issues involved.

Special Relativity

Imagine you're standing on the side of a train track watching a train cruise past. Your friend is on the train with a tennis ball, and he throws it forward inside the train. Let's say you measure the train to be moving at a speed of 60 miles per hour, and your friend measures a speed of 20 miles per hour for the tennis ball. How fast is the tennis ball moving relative to you? It probably didn't take you very long to figure that the ball is moving at 80 miles per hour. It's common sense to add the speed of the ball relative to the train to the speed of the train to get the 80 miles per hour that you observe. In any practical situation that you are ever likely to encounter, adding or subtracting velocities this way works perfectly. However, it turns out that if the train and ball are moving at speeds that start to approach the speed of light, this logical and intuitive velocity addition doesn't work anymore. In addition, if you were to do the same experiment but with a beam of light instead of a tennis ball, the speed of light relative to you would not be c + 60 mi/hr, it would be exactly c. If the train was traveling 0.9 c and your friend turned on the light, you would still measure the speed of the light to be c.

Einstein asserted that *the laws of physics must be identical in any inertial frame of reference* (postulate one). He was aware of the results of Maxwell's equations of electromagnetism, which predict that accelerating electric charges give off electromagnetic radiation traveling at a speed of c *without regard to any particular frame of reference.* Einstein realized that if the laws of physics truly are identical in any inertial frame then c must be a universal constant independent of reference frame (postulate two).

Why does simple addition of velocities break down at high speeds? Because no matter what the relative velocity of the source of light, any observer will *always* measure the speed of light traveling through a vacuum to be c. (It has essentially the same speed in air.) The constancy of the speed of light, the fact that it is independent of the frame of reference from which you measure it, is one of the fundamental postulates of the theory of special relativity. (The theory is "special" in that it deals only with inertial or nonaccelerating frames of reference as opposed to the more encompassing and complicated general theory, which takes acceleration into account.)

Time Dilation. Because of the counterintuitive but seemingly innocuous observation that the speed of light is not affected by relative motion, a whole basketful of astonishing and counterintuitive conclusions about the nature of space and time logically follow. For example: The amount of time elapsed between any two events depends on the frame of reference from which you observe those events.

Einstein was famous for constructing "thought experiments" to illustrate physical concepts. A thought experiment is an imaginary experiment that you could probably never actually do due to technological limitations, but the results of the imaginary experiment can be logically deduced based on a basic principle or idea. For example: Suppose you had a device that measures time by emitting a pulse of light, which reflects off of a mirror and back to a detector. The timer starts when the light is emitted and stops when it hits the detector.

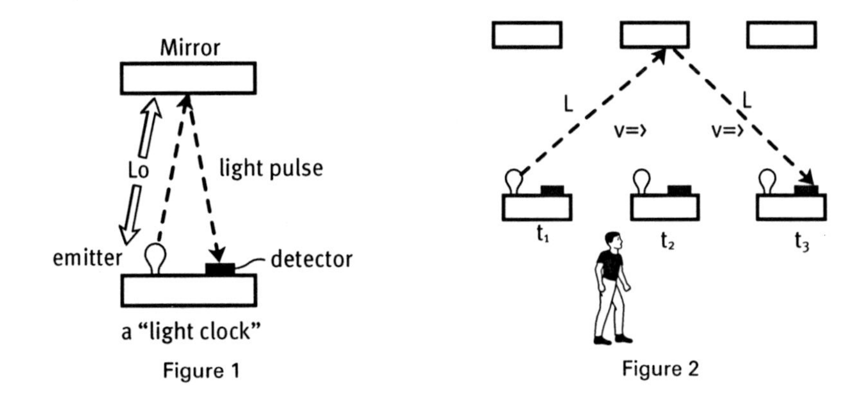

Figure 1

Figure 2

Now suppose that you were watching someone move past you on a train at a very high speed, when they emit a pulse from the "light clock." You will measure a longer travel distance for the light than the person on the train will.

Intuitively, we expect that time measurements should be absolute, that is, we expect that the time between the emission of light and collection of light will be agreed upon by the two observers. That is how we experience the physical world. For example, when we studied kinematics and projectile motion we observed that the time it takes to toss a ball straight upward to a certain height is identical to the time for a ball to get to the same maximum height when it is thrown at an angle. This is because the vertical components of the motion are identical. The ball projected at an angle travels a further total distance by the time it gets to its apex. However, because the horizontal component of its velocity gives the projected ball a greater average velocity both balls arrive at the same height at the same time. Therefore, in the case of the light clock, if the light travels a further distance in our frame of reference then we expect that the horizontal motion of the train must add to the velocity of the light, resulting in the same elapsed time. However, if c is constant then it can't be the speed of the light that is relative to your frame of reference—it must be time. We must measure a longer time between the emission and collection of light than the person on the train does. It turns out that this phenomenon applies not just to light clocks but to the time interval between any two events measured from different frames of reference. Not only will you measure more time for the clock, but also you will measure everything on the train moving at a slower rate than the passengers' experience.

Why don't we see this happen in everyday life? By analyzing the thought experiment mathematically it can be shown that the difference in elapsed time between the two frames of reference depends on their relative velocity v, and the speed of light c.

$$\Delta t = \frac{\Delta t_0}{\sqrt{1 - \frac{v^2}{c^2}}}$$

Δt_0 is the time measured in the frame of reference *where the two events occur in the same location*. In this example the two events refer to the release and the detection of the light pulse. So for the light clock in the above example, the events occurring at the same location would be on the train. Δt is the time measured in the frame of reference *where the two events occur in different locations,* in this case, on the ground outside the train.

If you look at the "time dilation" equation, you can see that for frames of reference with relative velocities in the range of anything you have ever experienced, the term $\frac{v^2}{c^2}$ is so close to zero that the difference between Δt_0 and Δt would be nearly impossible to detect. The fastest (space) vehicles ever built have reached speeds of around 11,000 m/s relative to Earth. That might sound like a lot, but $\frac{11{,}000 \text{ m/s}}{c} = 0.000037$. Therefore, for a space journey on one of these ships the difference in time measured for the trip between an astronaut and an observer on Earth would be:

$$\Delta t = \frac{\Delta t_0}{\sqrt{1 - \frac{v^2}{c^2}}} \qquad \Delta t = \frac{\Delta t_0}{\sqrt{1 - 0.000057}} \text{ which gives } \Delta t = 1.00003\ \Delta t_0 .$$

That's a pretty subtle difference. Nevertheless, what happens when the relative velocities start getting close to c? What if we could build a spaceship that we could accelerate to an appreciable fraction of the speed of light?

Example 1: Luke Skywalker makes a short journey to the planet MacLeod to watch the Highland Games, while Princess Leia waits for him back at Star Base Gamma Hydra Gamma. If the distance to MacLeod is 0.5000 light years (Lyr), and Luke's ship travels at 0.9990 c relative to the Star Base/

MacLeod frame of reference, how long until Luke arrives on MacLeod according to Princess Leia? How long until he gets there in his frame of reference?

Solution: In the Princess's frame of reference Luke travels a distance of 0.5000 light years at a speed of 0.9990 *c*. This gives him a total travel time of

$$\Delta t = \frac{0.5000 \text{ Lyr}}{0.9990\ c} = 0.5005 \text{ yr},$$

or a little over six months. Notice that the two events that define the time (Luke's departure from Star Base and his arrival on MacLeod) occur at different locations according to Leia so therefore we use Δt.

However, according to Luke, his travel time is:

$$\Delta t_0 = \Delta t\sqrt{1 - \frac{v^2}{c^2}} = 0.5005 \text{ yr}\sqrt{1 - \frac{(0.999\ c)^2}{c^2}} = 0.0224 \text{ yr},$$

which is about a week.

If Luke were to make the round trip then he will have been gone over a year on the Star Base, but he will only have spent two weeks on the ship! About 45 times the amount of time will have passed on the Star Base than elapsed in Luke's frame of reference when Luke returns. This is not some mathematical trick; it is actually true. Time is not absolute. This difference in time experienced between different frames of reference is sometimes called *time dilation.* You will notice if you play around with the time dilation equation a little that the effect is only noticeable when v gets very close to the speed of light. Even with $v = 0.5\ c$ the effect isn't that severe, but as you get very close to c, the effect becomes dramatic.

All Inertial Frames Are Equivalent. The following paradox may or may not have occurred to you at this point: All inertial frames of reference are equivalent. Therefore if I observe your clock to be running slow as you move past me at a high enough velocity, then by symmetry you must observe my clock to be running slow, because I am the one moving relative to you. If this crossed your mind, guess what? You're absolutely right! As long as the two observers remain in separate reference frames there's no way to make a direct comparison of their clocks. If, however, you join me in my reference frame we can compare our elapsed times—but then the situation is no longer symmetrical. You would have had to change your frame of reference. You would have had to undergo a period of acceleration to change frames, and because of this when we compare our clocks they would not agree by an amount given by the time dilation equation.

Let's look at Luke's vacation trip one more time. If all inertial frames of reference are equivalent then both Luke and Leia agree that the other's frame of reference is moving at a speed of 0.9990 *c*. However, if in Luke's frame it only took him a week to get to MacLeod, then there's no way it could be half a light year away. In Luke's frame of reference the distance must be much shorter. Not only will two observers in different frames of reference not agree on the time elapsed between two events, but they also won't measure the same distances between two points. Quantitatively:

$$L = L_0 \sqrt{1 - \frac{v^2}{c^2}}$$

where, in this case, L_0 refers to the length of the measured object or distance in the frame of reference where the object or distance is stationary.

Example 2: Once Luke gets underway on his trip how far away is Star Base Gamma Hydra Gamma from the planet MacLeod?

Solution: The Star Base and MacLeod are not in motion relative to each other so the distance between them measured in their frame is $L_0 = 0.5000$ Lyr. Luke measures the distance to be

$$L = L_0 \sqrt{1 - \frac{v^2}{c^2}} = 0.5000 \text{ Lyr} \sqrt{1 - \frac{(0.999\ c)^2}{c^2}} = 0.0224 \text{ Lyr}$$

All of these effects are well-documented experimentally. Although we can't accelerate large masses anywhere near the speed of light, we can look at the decay rate of unstable subatomic particles that move at speeds very close to *c*. The decay rates depend on the frame of reference from which you observe them exactly in accordance with the time dilation equation. On a macroscopic scale, airplanes flown around the Earth carrying extremely precise atomic clocks show different elapsed times after they return (small though they be) compared to identical clocks left on the ground—in an amount exactly as predicted by the theory.

The closer you get to *c*, the distances to places previously hundreds or thousands of light years away become shorter and shorter. In this way (if you could build a ship capable of accelerating to these speeds) you would be able to arrive at distant parts of the galaxy within a lifetime. Referring to the length contraction equation, because as $v \rightarrow c$, $L \rightarrow 0$, in the frame of reference of a light beam, everything is no distance and zero time away.

Flight of the Navigator Goes Back to the Future

There are some potentially very serious problems with interstellar space flight. First of all, according to special relativity, as we will see, no mass can ever travel as fast as light—although you can come close. Therefore, while it might be theoretically possible in months or years to travel great distances from one star system to another across the galaxy in the frame of reference of a very fast spaceship (due to length contraction and time dilation), the same trip might take hundreds of years in the star system frame of reference. When you get to wherever you're going no one will agree on what time it is, or even what year it is.

Remember that when Luke got back to the Star Base only two weeks had passed for him, but over a year had elapsed according to Princess Leia. In effect, Luke has traveled into the future. If Luke were to go back out on his ship at 0.9990 *c* for several months or even years and then return, imagine how chaotic that could be. One year for Luke would correspond to 45 years for the princess. If he took a year long trip at this average speed, upon his return Leia would be old enough to be his grandmother. This could prove to be inconvenient.

The writers of *Star Trek* circumvent these severe complications by stipulating maximum "impulse speed" to be one-quarter *c*, thereby effectively eliminating time dilation. However, this also means that at impulse speeds interstellar travel would not be feasible in any "reasonable" period of time.

Amazingly, the 1986 Disney film *Flight of the Navigator* at first glance seems to deal with the issue of time dilation described in our hypothetical examples. David (Joey Cramer), a 12-year-old boy living a normal suburban life (complete with dog, loving parents, and annoying younger brother) wakes up from a four-hour nap in the woods to find that eight years have gone by for everybody else. This makes for an awkward reunion with his parents and his now older younger brother who has evolved from a nasty little kid into a pretty nice guy. Everyone is mystified and upset until government agents whisk David away for tests in their secret high-tech government testing place. Although David can't remember any of it we discover that David spent those four hours on an alien space ship on a journey out to the planet "Phaelon" and back—a distance of 560 light years each way. It turns out that these government scientists had discovered a space vehicle near the same location where David woke up from his siesta and the fact that he hadn't aged in eight years made them suspicious. Not only that, David's brain is stuffed full of sophisticated information only an extraterrestrial could possibly know.

How Long Does It Take to Get to Phaelon? The good news is that this movie appears to derive its inspiration from the theory of special relativity. Appropriate themes from modern physics are fine examples to represent in children's films, and perhaps the writers heard, obliquely as it turns out, about time dilation.

The bad news is that their promising premise quickly degenerates into the usual incomprehensible pseudoscientific hodge-podge. The scientists in the *Navigator* label the theory that explains time dilation "light speed theory" which we can forgive them for, but this lack of attention to detail is the first sign of a weak link in their whole theoretical edifice, which then comes crashing down.

First and foremost, although it is theoretically possible for David to have only been gone 4.4 hours in his frame if he goes fast enough (remember length contraction), because Phaelon is 560 light years away there is no way that only eight years can have elapsed on Earth. The minimum elapsed time in the Earth frame of reference would be somewhere a little more than 1,120 years, because 1,120 years is the time it would take light to get there and back. If David were to return to an Earth 1,120 years in the future, any reunions would be impossible. In this case the lighthearted but gently serious comedy as we know it would be transformed into either the psychological drama of a 12-year-old boy lost in a terrifying futuristic dreamscape, or a hilarious slapstick "fish out or water" farce about a young boy befriended by a an inept miniature android named Cosmo.

Let's overlook the obvious error of the 560 light year distance to Phaelon, and simply calculate how fast David would have been traveling relative to Earth to account for the time difference experienced by each frame of reference. Because

$$\Delta t = \frac{\Delta t_0}{\sqrt{1 - \frac{v^2}{c^2}}}$$, we can solve for v in terms of c.

Converting 4.4 hours into years we get $v = \sqrt{1 - \frac{\Delta t_0^2}{\Delta t^2}}\, c^2 = \sqrt{1 - \frac{(5 \times 10^{-4} \text{ yrs})^2}{(8 \text{ yrs})^2}}\, c^2$,

which (ignoring significant figures for the purpose of illustration) gives us $v = 0.999999998$ c. That is really fast!

How Far Away Is Phaelon? From the eight years elapsed on Earth for David's round trip journey, we can deduce that Phaelon must be no more than just slightly over four light years away (which happens to be the distance to Alpha Centauri, our *nearest* star system) in the Earth/Phaelon frame of reference. How far apart, though, are Earth and Phaelon in David's reference frame during his flight?

$$L = L_0\sqrt{1 - \frac{v^2}{c^2}} = 8 \text{ Lyr}\sqrt{1 - \frac{(0.999999998\ c)^2}{c^2}} \approx 5 \times 10^{-4} \text{ Lyr}$$

This is one sixteen-thousandth ($\frac{1}{16{,}000}$) of the distance as measured from Earth. Not bad!

What Interstellar Travel Might Feel Like

While we're at it, let's calculate the magnitude of the acceleration David would experience on his interstellar sojourn. Because David needs to achieve an average speed of 0.999999998 c, and because he has to accelerate from rest to a maximum speed and then come to a stop for the first half of his trip, his top speed must be greater than the average—even closer to c. Therefore, for our purposes let's just assume a maximum velocity of ***about*** c. He has to accelerate from rest in a maximum time of 1.1 hours. Therefore his minimum acceleration would need to be

$$a = \frac{\Delta v}{\Delta t} \approx \frac{3 \times 10^8 \text{ m/s}}{3{,}960 \text{ s}} \approx 76{,}000 \text{ m/s}^2.$$

This is around 7,600 g's and might be uncomfortable if you could survive it long enough to feel anything, which is doubtful.

The Ultimate Speed Limit? One of the many major problems that we encounter when contemplating the idea of interstellar space flight is the limitation that no mass can ever attain the speed of light. Why is this?

In formulating his theory of special relativity, Einstein showed that although positions, lengths, and times are relative to reference frame, fundamental conservation laws still hold. Particularly relevant is the fact that conservation of momentum is maintained as long as the mass term is replaced by:

$$m = \frac{m_0}{\sqrt{1 - \frac{v^2}{c^2}}},$$

where m_0 is called the rest mass and is the mass measured when the object is stationary relative to the observer/measurer. This is mass as we normally think of it. However, according to the equation, the "relativistic mass" m depends on the relative velocity of that mass. As you can see an object gets more massive as it gets faster, and m approaches infinity as $v \rightarrow c$. This means that it would take infinite energy to get any mass all the way up to c. We don't have that much energy, so we're stuck with c as the ultimate speed limit.

How Massive Is David?

Let's say David whizzes past at a speed of 0.999999998 c, and you want to determine his mass in your frame of reference. Because you know he has a mass of about 40 kg when he is lounging around in his living room, you can determine his relativistic mass as follows:

$$m = \frac{m_0}{\sqrt{1 - \frac{v^2}{c^2}}} = \frac{40\text{ kg}}{\sqrt{1 - \frac{(0.999999998\ c)^2}{c^2}}} = 630{,}000\text{ kg}.$$

In addition to his high velocity relative to you, David has an awful lot of mass, so you probably want to avoid colliding with him.

What Is Relative, and What Is Not

We have come to the counterintuitive conclusion that measurements of time elapsed between events, and the positions and masses of objects are completely dependent on the frame of reference from which you measure them. Does this mean that all physical measurements taken from different frames of reference will yield different results? It does not. In fact, we already know that the speed of light is not relative. We have also implied that the conservation laws hold as long as modifications are made for the momentum and energy equations. Well, it also turns out that there is a four-dimensional vector that you can construct called the *spacetime interval.* It includes components of the three special dimensions plus one time dimension, and represents a kind of four-dimensional "distance" between two events in space and time. Spacetime is independent of the frame of reference from which you measure it.

The equation for the spacetime interval Δs is given by:

$$(\Delta s)^2 = c^2(\Delta t^2) - [(\Delta x)^2 + (\Delta y)^2 + (\Delta z)^2],$$

where Δ refers to the difference in space and time coordinates between two events.

Mass and Energy

Probably the most famous result of Einstein's theory of special relativity is the equivalence of mass and energy encapsulated in the famous formula

$$E = mc^2$$

For an object in motion, its total energy is therefore

$$E = mc^2 = \frac{m_0 c^2}{\sqrt{1 - \frac{v^2}{c^2}}}$$

This includes the kinetic energy of the object. However, if $KE = 0$, then $v = 0$, and the total energy reduces to

$$E = m_0 c^2$$

This is called the rest energy of the object and suggests the astonishing result that a stationary object has an energy proportional to its mass.

You can think of mass as a form of stored or potential energy, but can we actually turn mass into energy or energy into mass? The answer is yes. We turn mass into energy all the time. In any interaction where energy is released or absorbed by a system there must be a corresponding loss or gain of total mass of the system. Now in the case of chemical reactions we would not even notice this. A very small mass converts into an enormous amount of energy (look at the equation) and the magnitude of energies involved in chemical reactions result in essentially undetectable mass differences. Nuclear reactions convert a much higher and noticeable percentage of particle mass into energy (although actually still less than 1 percent of the total), and that is why they are so good at generating energy in nuclear power plants or, more dramatically (thus their popularity in all kinds of contemporary films), in nuclear bombs.

Turning energy into mass is trickier but we do see it happen in a process called *pair-production* where a photon (a massless "particle" of e-m radiation/energy) is converted into an electron and a *positron*. A positron is a particle with the mass of an electron but positively charged. (For all you Star Trek fans and others excited about "matter-antimatter drives," you may already know that a positron is an example of *antimatter*.)

ALIEN AND ENERGY

One of the many branches of the ever-expanding sci-fi tree is the science fiction horror drama. What better example of the genre than that classic of suspense and terror in space: *Alien.* Remember its famous tag line "In space no one can hear you scream"? We appreciate this double entendre not only declaring what kind of a movie is about to shock your system, but also paying homage to the "no sound in space" rule that we tried to hammer home relentlessly in Chapter 7. (Unfortunately in the actual movie the ship did make noise in space, thus destroying that entire idea.) This is one scary movie, and far exceeds the continuously diminishing quality of the sequels.

The Movie

In *Alien,* the crew of a deep-space mining ship receives a distress call from an unknown star system. When they arrive to investigate, they find a shipwrecked spaceship and her deceased crew. When Cain (John Hurt) discovers some living pods we suspect it would be a lot better for Cain if he would just walk away fast. He doesn't, and something about the size of a house cat that looks like a big crab jumps out and attaches itself to his face. The traumatized landing party foolishly brings Cain back onto the ship and puts him in the infirmary. They can't detach the big crab from his face but eventually it falls off and dies, and everyone (except us) is relieved when Cain wakes back up and joins the rest of the crew for dinner.

Unfortunately, in one of the more unforgettable scenes in science fiction, something starts going wrong with Cain. He seems to be having severe pain in his stomach and chest. Something appears to be pushing its way out from the inside, and when an incredibly vicious-looking creature that looks something like a cross between a slug, a crocodile, and a shark comes bursting out of Cain's chest we have a sense that we know where the rest of the movie is going. The alien creature is small, maybe the size of a medium-sized Dachshund at this point, but it's really fast, and it skitters away and out into the giant cavernous ship before anyone can react. The next time the crew encounters this terrifying creature it's about the size of a Yeti. Then the crew has to figure out how to deal with it.

$E = mc^2$ for Aliens, Too!

Even though *Alien* is a really good movie, there is one aspect of the film that is vaguely troubling. How does that alien get so big so fast without eating anything? Let's consider the possibilities.

1) Perhaps in its larval form the creature is incredibly dense. As it grows, it simply maintains its mass but expands to its giant but less-dense state. However, if this were true, when Cain wakes up he should feel at least three times as heavy as normal, but there is no evidence of this.
2) Maybe this incredible creature is somehow able to convert energy directly into mass. How might this work? We know about mass energy equivalence, so let's calculate the amount of energy we'd need to beef up the little alien into something big and scary.

Let's assume the alien has a density twice that of most Earth creatures, which are just a little denser than water (remember, it's armored and spiny, and has acid in its blood). We'll give it a density of 2,000 kg/m^3. The volume of the alien is probably at least twice that of a normal size man. Therefore, if we assume 80 kg for the mass of an average man, then we'll estimate that the grown up alien has mass of around 320 kg. In his larval form it couldn't have been more than 20 kg. This means that the alien must have added 300 kg of pure muscle in a period of a couple of hours (try that in the gym!). How much energy would this take if we could convert it into mass? The calculation is simple.

$$E = m_0c^2 = (300 \text{ kg})(3 \times 10^8 \text{ m/s})^2 = 2.7 \times 10^{19} \text{ J}$$

This equals the energy of over 6,500 1-megaton nuclear bombs that the alien has to convert into mass in a couple of hours. This means that the power rate of whatever process it uses to convert energy is incredibly fast.

$$P = \frac{E}{t} = \frac{2.7 \times 10^{19} \text{ J}}{120 \text{ s}} \approx 2 \times 10^{17} \text{ W}$$

Where does the alien get all of this energy? How does it convert it into mass? These are questions that no one knows the answers to, even us.

According to the equations of quantum mechanics, every matter particle must have an associated *antiparticle*. For example, a positron is the antiparticle of the electron. Antiparticles are sometimes called *antimatter*. We can create antiparticles in laboratory experiments, but if a particle of matter meets its antimatter counterpart they will annihilate each other, converting their mass into energy in the process.

The opposite process to pair production is called *pair-annihilation*. In this case an electron and a positron interact and annihilate each other giving off pure energy in the form of a photon. In this process 100 percent of the original mass is converted into energy. Therefore, you can see why writers of science fiction might consider the imaginary (but based on real principles) matter-antimatter drive their best hope for generating enough energy to fly their imaginary space vehicles.

The principle of conservation of energy still holds as long as we include mass as one possible form of energy.

Example 3: In one of the original *Star Trek* episodes, (The Alternative Factor) the entire universe is threatened when a "hole" is created connecting to a "parallel" antimatter universe. Lazarus, a lone space traveler, is being pursued by "Anti-Lazarus" a madman from the parallel universe. Anti-Lazarus is irritated that he has a twin and he is determined to destroy his anti-self. Anti-Lazarus obviously is made of antimatter and Lazarus tells us that if he and Anti-Lazarus meet, both universes will be obliterated. The fact that Anti-Lazarus interacts with the *Enterprise* crew doesn't seem to be a problem somehow, even though they are made of matter (apparently, people will only explode when contacting their *specific* parallel universe anti-person). Anyway, if Lazarus and Anti-Lazarus meet, how much energy will they produce?

Solution: Lazarus is a pretty hefty guy so let's assume he has a mass of about 90 kg, as does Anti-Lazarus. One hundred percent of their mass will be converted to energy, so

$$E = m_0c^2 = (180\text{ kg})(3\times10^8\text{ m/s})^2 = 1.6\times10^{19}\text{ J}$$

This is a lot of energy relative to the kind of everyday phenomena that we deal with down here on Earth, equal to about 4,000 1-MT nuclear bombs, but still it isn't enough to even blow up a large asteroid, let alone destroy the entire universe.

The Starship *Enterprise* Under Impulse Power. Let's see how we can apply the principles of mass-energy equivalence to spaceship propulsion. (In *The Physics of Star Trek* by Lawrence Krauss the author discusses issues relating to the physical possibility of interstellar space travel based on principles of special and general relativity. Here we present a similar discussion that differs somewhat in the details.)

Example 4: On the Starship *Enterprise* the "impulse" engines, which accelerate the ship to less than light speeds, use nuclear fusion to generate energy. (The matter-antimatter drive is reserved for "warp" speed.) If 0.4 percent of the mass of nuclear fuel in each fusion reaction is converted into energy, how much fuel would be required to accelerate the ship to one half the speed of light using a mass for the *Enterprise* on the order of 1×10^9 kg from the technical manual?

Solution: If we optimistically (unrealistically) assume that 100 percent of the released energy goes into kinetic energy of the ship, then the total energy released would be:

$$E = m_0 c^2 = \frac{1}{2} m_{ship} v^2 \text{ so } m_0 = \frac{m_{ship} v^2}{2\ c^2} = \frac{(1 \times 10^6 \text{ kg})(0.5\ c)^2}{2\ c^2} = 125{,}000 \text{ kg},$$

where $1/2\ mv^2$ is the kinetic energy. (While technically the kinetic energy term is no longer valid at speeds close to c, at $0.5c$, it is still a pretty good approximation.)

This is the amount of mass actually converted to energy in the reaction, so the total amount of fuel required would be:

$$\frac{125{,}000 \text{ kg}}{0.04} = 3.1 \times 10^6 \text{ kg},$$

which is three times the mass of the ship (and this is a gross underestimate). Remember how rocket propulsion works? Remember Newton's third law, and the principle of conservation of momentum? The only way to propel the ship forward is by ejecting fuel at high velocity in the opposite direction.

The ejected fuel is going to take away a significant percentage of the kinetic energy generated in the reaction. We are also assuming that we are able to harness all of the energy produced in the reaction in the form of macroscopic kinetic energy. However, we know that nuclear reactors generate a lot of thermal energy, so a more reasonable estimate for the mass of fuel required might be as much as 40 or 50 starship masses. This could be a problem—and this is just for a single acceleration. If you want to accelerate the same amount more than once, you have to keep multiplying the fuel mass by the same amount. (Krauss uses conservation of momentum to calculate that we actually need 81 times the mass of the ship to accelerate from rest to 0.5 c. Therefore, to speed up and then slow back to a stop once, you would need 81 × 81 = 6,561 times the mass of the ship for fuel.)

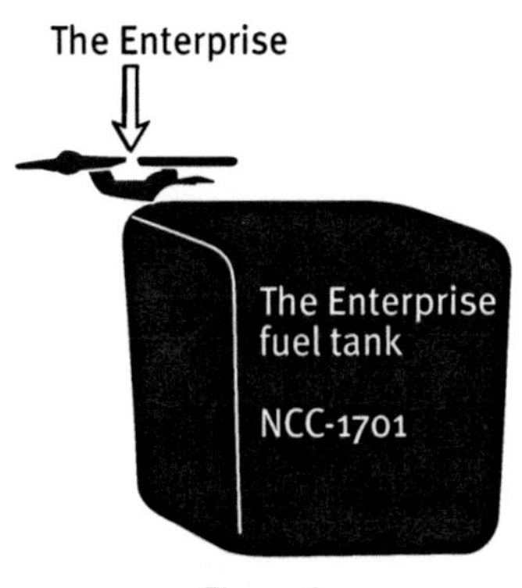

Figure 3

Figure 3 is an illustration of what the *Enterprise* really has to look like if the crew wants to use the impulse engines to go places.

However, what if they use the matter-antimatter engines for the impulse drive? Remember pair-annihilation? Throw some electrons and positrons together and we get 100 percent of their rest masses converted into energy in the form of gamma rays. That's quite a bit better than nuclear fusion, and 125,000 kg of fuel will give you an energy equivalent to the kinetic energy of the

Enterprise traveling at 0.5*c*. This is only one-eighth the mass of the ship, but again we are being overly optimistic, and when we consider the energy going into the expelled photons and thermal losses, we're looking at masses of fuel as about equal to the mass of the ship. This is better than when we used the nuclear reactors, but to do any significant space travel the mass of the fuel is still going to completely dwarf the mass of the ship.

At this point, things look pretty bleak for our interstellar voyage. Our obstacles include

1. The limitation that no mass can travel faster than *c*.
2. The speeds that would make it possible to get a crew to any but the nearest stars in their lifetimes would create such severe time dilation effects that every trip would result in a jump into the far future.
3. The amounts of energy/fuel required to accelerate to the necessary speeds would be prohibitive.
4. The rapid accelerations that would necessarily be involved would be fatal.

The disappointing truth is that interstellar travel under these conditions is just not feasible. At this point, we could just give up and admit that space travel to distant worlds is a pleasant but impossible fantasy, which it most likely is. However, there is still a glimmer of hope—a theoretical loophole, if you will. We can't beat the speed limit, but we are not finished with Einstein and relativity. We haven't yet discussed his theory of *general* relativity, which tells us something more about the nature of space, time, and gravity. Due to the mathematical complexities we will only be able to talk about general relativity qualitatively, but some of the predictions of the general theory are not only as mind-blowing as the special theory but hold out at least a faint possibility of resurrecting our interstellar ambitions.

EVENT HORIZON, GENERAL RELATIVITY, *STAR TREK*, AND WARP SPEED

Event Horizon (1997), starring Sam Neill and Laurence Fishburne, starts off with a heavy dose of sci-fi, providing a perfect lead-in to our discussion of general relativity, before crossing over into the realm of your pretty standard straight-out horror film.

It goes something like this: The *Event Horizon* is a spaceship that disappeared without a trace somewhere outside the orbit of Neptune over seven years before the start of the film. It has suddenly resurfaced in the form of a radio distress signal. A rescue ship under the command of Captain Miller (Fishburne) is sent to Neptune to investigate, accompanied by the designer of the *Event Horizon*, Dr. Weir (Neill). Doctor Weir explains that the *Event Horizon* was designed to travel to distant stars by bending space-time according to the principles of general relativity (although he doesn't actually call it general relativity in the scene), which we will discuss next. However, when the crew arrives and contacts the *Event Horizon* things start to go very wrong. We discover that it has been (in the words of Dr. Weir) to a region of "pure chaos" and "pure evil" outside of the known universe. The ship itself has become a living entity intent on destroying the crew by driving them insane. The rest of the film depicts the desperate struggle of the crew to hold on to their sanity long enough to extricate themselves from their extremely unpleasant circumstances.

Let's focus on Dr. Weir's explanation of the means by which the ship might be capable of interstellar travel. Can we circumvent the obstacles that special relativity has thrown in our path? What does general relativity have to offer?

The Theory of General Relativity

The theory of special relativity is "special" in that it is limited to the case of constant velocity (nonaccelerating) frames of reference (a.k.a inertial frames of reference). Einstein wanted a theory that included accelerated frames of reference. It turned out this was not a trivial task, and it took over a decade after he published his first paper on special relativity to formulate the general theory. We will highlight some of its main points relevant to our discussion.

Acceleration and Gravitational Fields. Consider the following thought experiment. You have a busy day. In the morning you enter a windowless spaceship that is sitting stationary on the surface of the Earth. You walk around inside checking out all of the gadgets for a while.

Soon you grow sleepy and take a nap. When you wake up you don't know how many hours or days have passed, and it turns out that the ship is in fact now somewhere in deep space accelerating at a rate of $a = g$. Is there any way for you to be sure that you are no longer on Earth? Remember the ship has no windows.

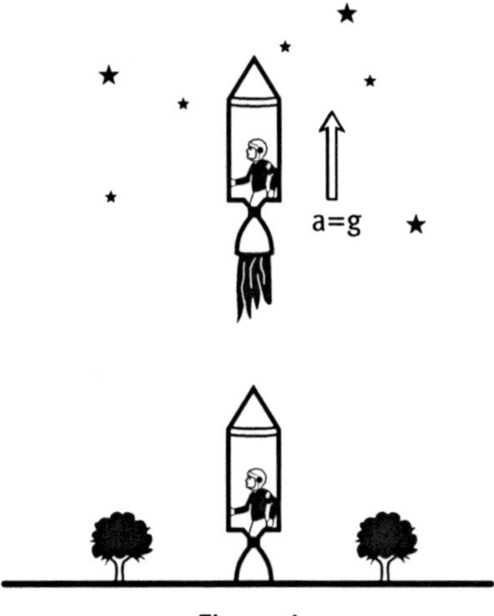

Figure 4

Any experiment you would do would yield the same result for both cases. For example, what if you threw a ball across the ship? What would the trajectory look like on Earth and what would it look like when accelerating at g?

On Earth we might say that gravity is pulling the ball downward with an acceleration of g, and on the ship we might say that the ship is accelerating such that the ball appears to fall/accelerate at rate g relative to the ship. On Earth we might say that gravity is pulling us into the floor giving us the sensation of weight, while on the ship we might say it is the ship's acceleration causing the contact between our feet and the floor giving us the sensation of weight. Without windows, however, how can we tell the difference? The answer is: we can't.

This leads us to the logical conclusion that *the effects of accelerating frames of reference are indistinguishable from the effects of gravitational fields.*

Gravity Curves Spacetime. Let's extend our thought experiment a step further. Now you are back on the accelerating ship and you shine a flashlight. The beam should appear to you to bend toward the floor due to the ship's acceleration, while an observer outside the ship would see the light traveling in a straight line. (Because light is so fast, you would need some very good measurements or a really big ship to notice the bending of the light, so let's make the ship really big.) Now if gravitational fields are equivalent to accelerated reference frames, then the light should be bent in exactly the same way when the ship sits on the Earth's surface.

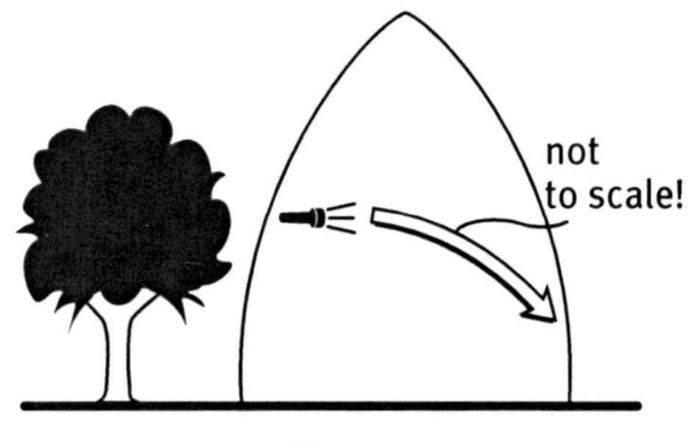

Figure 5

The logical conclusion is that mass causes the spacetime surrounding it to curve. Therefore general relativity is actually a new theory of gravity. In Newtonian mechanics, we think of gravity as a pull that exists between masses, while in general relativity we describe the force of gravity differently—as the result of an object following the curvature of spacetime. Mass causes spacetime to curve, and in turn masses are affected by the curvature of spacetime. For example, we can imagine a two-dimensional analogy where we place a bowling ball on a tightly stretched rubber sheet.

Figure 6

The bowling ball causes the sheet to bend. If you were to place a marble on the sheet it would feel a force directed toward the bowling ball along the direction of the curvature. If instead you were to give the marble a velocity tangentially, it would orbit the ball just as a planet orbits the sun.

Predictions of General Relativity. General relativity completely replaces Newton's law of gravitation as a theory of gravity. It is more accurate—and yes, it has been convincingly tested. As we might expect after our experience with special relativity, general relativity also makes some amazing predictions. For example, there is a time dilation effect due to gravity. Clocks in a strong gravitational field have been shown to run slower than clocks in regions of less gravity.

Perhaps the most popular prediction with fans of science fiction is that of the *black hole*. The theory predicts that if a large enough star were to exhaust its nuclear fuel, the resulting gravitational collapse would result in an astonishingly dense object. In fact, the object would be so dense that the curvature of spacetime surrounding it would be sufficiently severe that all matter *and light* within a certain radius (called the *Schwarzschild radius* or the *Event Horizon*) would never be able to escape. A black hole is black because you would never be able to see it—and at its center there may be something called a *singularity,* a point of infinite density and curvature. Astronomers have seen some pretty convincing evidence for the existence of black holes in binary star systems (systems where two stars revolve around each other). However, in some cases, one "star" is invisible. We know there are two objects because we see one in orbit around something unseen, and we also see material from the visible star being pulled off and spiraling towards the position of the unseen object. Astronomers also believe that centers of galaxies consist of super-massive black holes with masses between millions and billions of stars.

This is some pretty cool, amazing, mind-boggling stuff. Now we're going to see if it can help us travel to distant star systems and make the world of science fiction that we so long for seem less remote and impossible.

Dr. Weir Explains It All. The *Event Horizon* is designed to use the principles of general relativity to circumvent the light speed limit. In scene 4, Dr. Weir explains to the ill-fated crew of the ill-fated rescue ship the *Lewis and Clark* how the ill-fated *Event Horizon* is able to traverse great distances across the galaxy in seemingly impossibly short times. It is all based on the fact that gravity can curve spacetime and theoretically provide us with shortcuts. He gives us a two-dimensional analogy using a piece of paper to illustrate the point. The piece of paper represents spacetime, and if there are no large masses in the vicinity, it is flat. He explains that if we want to travel from point A to point B on the paper, we're going to have to go all the way across.

What if we have a mass sufficiently large to bend the spacetime paper like the one shown here?

If we could somehow find a way to bridge the gap between A and B, then the distance would be much shorter than it was when space was flat. Dr. Weir then explains that if we can create a "wormhole" or a tunnel between the two points then we could do it. Not only would this be theoretically possible, but it may also be possible, according to the relativity equations, actually to go backwards in time as you travel from one end of the wormhole to the other.

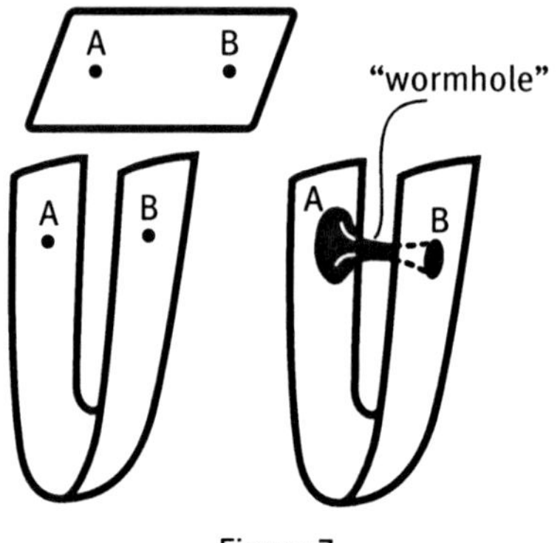

Figure 7

Did we just solve all our space flight problems? Are we ready to travel to Orion with Captain Picard? Not so fast.

What exactly would it take to bend spacetime to this degree? How might we produce a sufficient gravitational field? How much energy would this take? (Hint: A lot.) If we were able to create a field this strong, how might it affect any living things in the vicinity? (Hint: By killing them.)

On the *Event Horizon* they have a giant metallic ball called the "gravity drive" that bends spacetime. However, Dr. Weir offers no explanation for how the gravity drive works, which doesn't alleviate any of our concerns. It seems appropriate that he goes insane well before the end of the movie.

We can mathematically create a wormhole where two singularities meet in spacetime, and momentarily join, but because of the intense gravity, they cannot exist for any length of time. They collapse instantly. Some physicists have proposed the idea that the wormhole could be kept open if there were some kind of negative energy that would have the opposite effect as gravity. There is conjecture that something like this is causing the accelerating expansion rate observed in certain locations in the universe, but there might be some other explanation. The jury is still out on this one.

"Warp Speed," Our Last Best Hope. We still have some problems, not the least of which is that traveling into a wormhole might, to put it gently, violently rip apart and crush into black-hole mush anything or anyone that ventured in. Nevertheless, we are going to make one last attempt to salvage our space adventure dream within the context of general relativity. We are going to go to Warp Speed.

Any Trekkie can tell you that if you want to travel somewhere faster than the speed of light, you use the warp drive. If you look in the Star Trek technical manuals they'll tell you by what factor each warp speed number exceeds the speed of light. However, we know that we can't actually travel faster than *c* relative to our surroundings. So how does this work? The equations of general relativity tell us not only that mass can bend spacetime as we discussed, but we can also stretch and compress it. If somehow the warp drive is able to expand space behind the ship while compressing it in front, then the ship can leap great distances through space without ever exceeding *c* relative to the *local* surroundings. Better yet, if whatever humongous gravitational field required to accomplish this feat is kept far enough from the ship, we might not destroy it in the process.

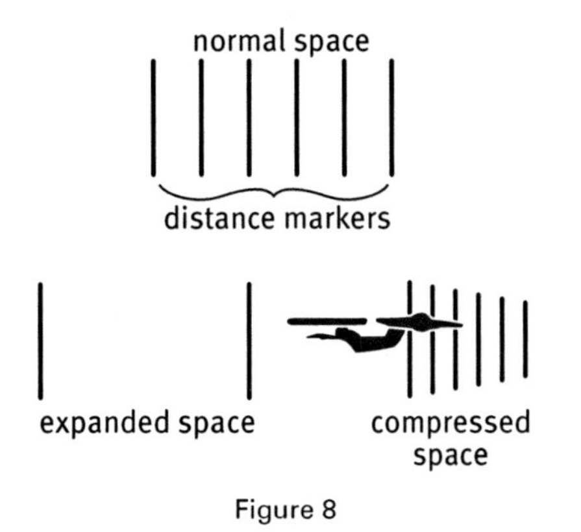

Figure 8

Unfortunately, we still have some formidable technical issues standing in the way of our aspirations towards interstellar travel. We have no idea in practice how to make a device that could stretch spacetime, and if we did we still have to consider the incredible (impossible?) amount of energy this certainly would require.

Conclusion. Interstellar travel like that portrayed in the movies has tremendous theoretical and practical obstacles standing in the way. Is it likely that we will *ever* be able to achieve this? No, but can we prove at this point that it is absolutely impossible? No.

NUCLEAR PHYSICS IN FILM

Nuclear reactors, nuclear bombs, and dangerous radiation levels all play a role in many of our science fiction favorites. Why does Spock die at the end of *Star Trek II: The Wrath of Khan*? What are the specific nuclear reactions that produce such explosive energy in that icon of the modern cinema (so beloved in movies like *The Core* and *Armageddon*), the nuclear bomb? Does it make any sense for the *Enterprise* crew to "collect photons" from the reactor of a naval submarine in *Star Trek IV*? What are fission and fusion? What are alpha, beta, and gamma radiation? Those are the kinds of important questions that we are going to address in this section.

Star Trek II: The Wrath of Khan *and the Death of Spock*

At the end of *The Wrath of Khan* Spock gives his life in order to save the ship when he enters the ships reactor and makes the necessary repairs. However, what is it that kills him? Let's get specific.

Radioactivity. We start with some definitions:

Atomic number (Z) equals the number of protons in the nucleus of an atom.
Neutron number (N) equals the number of neutrons in the nucleus.
Mass number (A) equals the total number of nucleons (protons plus neutrons).

The general format for expressing the atomic number and mass number of a given element X is ${}_{Z}^{A}X$. For example, ${}_{6}^{12}C$ is the symbol for the most common variety of carbon. It has 6 protons and 6 neutrons.

An element is defined by the number of protons in its nucleus. However, a given element may contain varieties with different numbers of neutrons. These different versions of the same element are called *isotopes* of that element. For example, while the isotope ^{12}C is the most common variety of carbon, it can also exist as ^{11}C, ^{13}C, and ^{14}C. In each case the carbons must have 6 protons, but the number of neutrons may be 5, 6, 7, or 8 depending on the isotope.

An *atomic mass unit (u)* is defined as one-twelfth the mass of one atom of the isotope ^{12}C. A neutron and a proton each have a mass of about $1u$, which is equal to 1.66054×10^{-27} kg.

Because nuclei contain closely packed positively charged protons, there will be a repulsive electrostatic force between them, so the only way nuclei can stay together is if there is some other stronger attractive force acting between the nucleons. This force, which acts over very short distances, is sometimes called the *nuclear strong force*. Without it, stable nuclei could not exist.

The mass of a nucleus is always less than the masses of the individual nucleons that make up the nucleus. Remember mass-energy equivalence? The missing mass is in the form of energy called the *binding energy* of the nucleus, and it is equal to the amount of energy that you would have to add to the nucleus to break it apart.

It is usually convenient to calculate binding energy in terms of units of MeV (mega electron volts) where 1 MeV = 1.6×10^{-13} J.

Example 5: A proton has a mass of 1.67262×10^{-27} kg, a neutron has a mass of 1.67493×10^{-27} kg, and an alpha particle (${}_{2}^{4}He$ nucleus) has a mass of 6.64466×10^{-27} kg. What is the binding energy of an alpha particle?

Solution: The combined mass of two individual protons and two individual neutrons is $[2 \times (1.67262 \times 10^{-27}\ \text{kg})] + [2 \times (1.67493 \times 10^{-27}\ \text{kg})] = 6.6591 \times 10^{-27}$ kg. The difference in mass between the sum of the individual particles and the alpha particle is $(6.6591 - 6.64466) \times 10^{-27}\ \text{kg} = 5.044 \times 10^{-29}$ kg. The binding energy is, therefore, $E = m_0c^2$ where m_0 is the mass difference. Therefore,

$$E = (5.044 \times 10^{-29}\ \text{kg})(3 \times 10^{8}\ \text{m/s})^2 = 4.54 \times 10^{-12}\ \text{J}(= 28.4\ MeV)$$

Nuclear Stability. Many elements contain isotopes that are not stable, and some elements contain *only* isotopes that are unstable. We call these unstable isotopes *radioactive.* A radioactive isotope will always eventually break apart and/or decay into a stable isotope. The time it takes for half of the amount of a sample of a radioactive isotope to decay into a "daughter" isotope is called the *half-life* of that isotope. Half-lives can vary from millionths of a second to billions of years depending on the isotope.

Example 6: Originally you have 24 grams of a radioactive isotope. Twelve days later the sample consists of 3 grams of the radioactive isotope and 21 grams of a daughter isotope. What is the half-life of the original isotope?

Solution: The sample must have gone through a period of three half-lives if 3 grams remain. This means the half-life must be 4 days.

Decay Processes. A radioactive isotope can spontaneously decay by either *alpha*, *beta*, or *gamma* decay.

In *alpha decay* the nucleus of the atom spits out an alpha particle $({}^{4}_{2}He)$. In general:

$${}^{A}_{Z}X \rightarrow {}^{A-4}_{Z-2}Y + {}^{4}_{2}He$$

where X is called the parent nucleus, and Y is called the daughter. For example

$${}^{238}_{92}U \rightarrow {}^{234}_{90}Th + {}^{4}_{2}He$$

describes the decay of a uranium 238 into a thorium 234 via alpha decay. Notice that the total number of nucleons is the same before and after the decay.

Beta decay involves the emission of either an electron or a positron. The daughter has the same number of nucleons as the parent but the atomic number changes by 1. This means that in the case of an electron decay a neutron in the nucleus is converted to a proton and electron, while in a positron decay a proton is converted to a neutron and a positron.

$${}^{A}_{Z}X \rightarrow {}_{Z+1}^{A}Y + e^{-} \text{ or}$$

$${}^{A}_{Z}X \rightarrow {}_{Z-1}^{A}Y + e^{+}$$

where e^- is an electron and e^+ is a positron. Interestingly, the ejected electron or positron in a beta decay does not account for all of the energy released in the process. Because of this, Wolfgang Pauli in 1930, proposed the existence of an additional previously undetected particle called a *neutrino* that was being released in these reactions. The neutrino had to have very little to zero mass, and it had to travel completely through most matter without affecting it in any way. Neutrinos have, in fact, been detected in very sensitive experiments. Electron decay produces an *antineutrino*, and positron decay produces a *neutrino*.

Two examples of beta decay:

$$^{14}_{6}C \rightarrow {}^{14}_{7}N + e^{-}\,(+\bar{\upsilon})$$

$$^{12}_{7}\mathrm{N} \rightarrow {}^{12}_{6}C + e^{+}\,(+\upsilon)$$

where $\bar{\upsilon}$ is an antineutrino and υ is a neutrino

Just as we saw with alpha decay, the total number of nucleons and the net charge is the same before and after the decay.

Often after a radioactive decay an atom is left in an "excited" or energized state. The atom then releases the excess energy in the form of a photon. This is called *gamma* decay, and the emitted photons are called *gamma rays*. For example, after the beta decay in which ^{12}N decays into ^{12}C, the carbon may be left in an excited state. The C will then release energy in the form of a gamma ray (γ).

$$^{12}_{6}C^{*} \rightarrow {}^{12}_{6}C + \gamma$$

where the * means that the carbon is in an excited state.

Nuclear Reactions

If you bombard the nuclei of many atoms with high-energy particles, you can induce a *nuclear reaction*. In the following reaction, a nitrogen nucleus is bombarded with an alpha particle giving rise to an isotope of oxygen and a proton.

$$^{4}_{2}He + {}^{14}_{7}N \rightarrow {}^{17}_{8}O + {}^{1}_{1}H$$

Nuclear Fission. If you bombard a massive nucleus with low-energy neutrons it may split into smaller fragments, giving off energy in the process. This is called *nuclear fission*. A given element may fission in a variety of different ways. For example, ^{235}U can fission via many possible reactions, two of which are shown here:

$$^{1}_{0}n + {}^{235}_{92}U \rightarrow {}^{141}_{56}Ba + {}^{92}_{36}Kr + 3\,{}^{1}_{0}n$$

$$^{1}_{0}n + {}^{235}_{92}U \rightarrow {}^{140}_{54}Xe + {}^{94}_{38}Sr + 2\,{}^{1}_{0}n$$

In each of these reactions energy is released in an amount around one hundred-million times greater than that released per molecule in a typical chemical reaction. Where does this energy come from? From good old $E = mc^2$. The total mass of the original uranium plus neutron is greater than the combined masses of the products. The difference is given off mostly as kinetic energy in the fission fragments, but also as gamma rays, beta particles, and neutrinos.

Fission Reactors or Fission Bombs? Each time a nucleus fissions, it emits neutrons, and these neutrons can be absorbed by the surrounding nuclei, which can then fission, causing a chain reaction. If the reaction proceeds too fast it can result in a violent explosion, which is sometimes called an "atomic bomb." If the reaction is controlled at just the right rate, it can be self-sustaining and yet controlled. This is the principle behind nuclear fission reactors.

Nuclear Fusion. For lighter elements energy is released if you fuse them together in a *nuclear fusion* reaction. The energy produced in stars comes from nuclear fusion reactions. For example, if two protons (${}^{1}_{1}H$ hydrogen nuclei) are slammed together they can form a deuterium nucleus (${}^{2}_{1}H$).

$${}^{1}_{1}H + {}^{1}_{1}H \rightarrow {}^{2}_{1}H + e^{+} + \upsilon$$

Because the total mass of the original protons is greater than the mass of the deuterium the difference is released as energy (in this case 0.42 MeV). In a star the deuterium that is created in the above reaction can then react with other hydrogen nuclei forming a ^{3}He nucleus.

$${}^{1}_{1}H + {}^{2}_{1}H \rightarrow {}^{3}_{2}He + \gamma \text{ releasing 5.49 MeV in the process.}$$

The ^{3}He nuclei formed in this reaction can then fuse to form nuclei of ^{4}He

$${}^{3}_{2}He + {}^{3}_{2}He \rightarrow {}^{4}_{2}He + 2\,{}^{1}_{1}H \text{ releasing 12.86 MeV.}$$

Along with these reactions the positron formed in the first step can react with an electron. These matter and antimatter particles will completely annihilate each other releasing two gamma rays with a total energy of 1.022 MeV.

$$e^{+} + e^{-} \rightarrow 2\gamma$$

Several decades ago nuclear fusion seemed to present great promise as a potential energy source. Unfortunately, the construction of terrestrial fusion reactors able to produce energy economically has proven to be a formidable challenge. Meanwhile, it is much easier to create uncontrolled fusion reactions. Thermonuclear bombs in which the majority of the energy is supplied by fusion (and which yield a thousand times more energy than the earlier fission bombs) constitute the majority of nuclear weapons in the world's arsenals.

The Biological Effects of Radiation, and the Demise of Spock

In scene 15 of *Star Trek II—The Wrath of Khan*, Spock enters the reactor, despite McCoy trying to hold him back from certain death. (Little does McCoy know that Spock has placed his "katra" in McCoy's subconscious, so that he can be revived in *Star Trek III*.) If Spock is not able to restore the warp drive by the time that wacky psychotic Khan (Ricardo Montalban) detonates the "genesis device," the *Enterprise* will be destroyed. In the scene Spock has to pull the cap off of what we assume is the warp drive reactor, thus exposing him to a deadly flux of radioactivity. Spock is willing to give his life because "The needs of the many outweigh the needs of the few," and therefore he subjects himself to the lethal dose of radiation. However, what is it that kills him?

Although we do not sense the alpha, beta, and gamma particles that are emitted in radioactive decay, they do have sufficient energy to destroy cells and to cause cancer and other unpleasant cell mutations. In high enough doses death can occur very rapidly. If you were in the presence of an unshielded fission reactor, you would be bombarded by neutrons, beta particles, and gamma rays. If you were loitering by a fusion reactor, alpha and beta particles, gamma rays, and maybe some neutrons would be headed your way. We are told that the *Enterprise's* warp drive is a matter-antimatter drive, however, which means the radiation consists essentially of gamma rays of energy 511 KeV produced by electron-positron annihilation. If the electrons and positrons (all beta particles) in the hypothetical matter-antimatter drive have sufficient energy, they could pose a serious problem as well. Overexposure to the gamma rays alone can be lethal, and therefore we will hypothesize that it was the gamma rays, with perhaps some assistance from beta particles, that did Spock in.

QUANTUM MECHANICS

Around the same time as Einstein's theories were being formulated physicists were troubled by what seemed to be a few minor unexplained phenomena relating to the behavior of matter, light, and energy. The solution to these problems resulted in a revolutionary new theory describing the behavior of matter and energy at the atomic scale: quantum mechanics. As we might expect after our experience with relativity, some of what quantum mechanics tells us about the nature of the universe is unexpected and counterintuitive. In an attempt to explain the nature and behavior of matter and energy at the atomic level, quantum mechanics presents a reality that is hard to discuss in terms of our macroscopic experience of the world. This bizarre quantum reality has provided not only fuel for our beloved science fiction movies and television shows, but has also been the cause of much philosophical speculation. It has become popular in some circles to extrapolate quantum mechanical models to the macroscopic level in order to explain philosophical ideas like "free will," or the nature of consciousness and its effect on physiology. Are these speculations scientifically justified?

The final film up for review is not actually a science fiction film, although it deals with both science and fiction. It (unintentionally) serves as a cautionary tale about what science does and does not tell us about the universe. *What the βleep Do We (k)now!?* is essentially a new age propaganda piece disguised as a scientific investigation of the nature of consciousness. (The fact that the movie was produced by The Ramtha School of Enlightenment, a group whose leader JZ Knight claims to be channeling the spirit of a 35,000-year-old spiritual warrior named Ramtha, should tell us something.) *What the Bleep* consists of a series of interviews with a panel of scientists and some other "experts" including "Ramtha as channeled by Ms. Knight" explaining to us how physiology and quantum mechanics tell us that we can affect the future and the nature of physical reality with the power of our thoughts. Our task will be to evaluate the claims made in this movie, which purport to prove the connection between quantum mechanics human consciousness and macroscopic physical phenomena. To do this we need to have a sense of what the theory of quantum mechanics actually says. Therefore, let's start with a brief discussion of some of the fundamental ideas.

The Particle Nature of Light and Energy

By the turn of the 20th century, Maxwell's theory of electromagnetism had pretty convincingly established the wave nature of light—or so physicists thought.

In 1900, the physicist Max Planck came up with the only mathematical equation that could successfully describe the frequencies of energy emitted by blackbodies (objects that are perfect absorbers and emitters of e-m radiation). However, the equation had a surprising and completely unexpected implication: The energy emitted by vibrating molecules is quantized. This means that energy can only be emitted in discreet amounts. Specifically,

$$E = nhf$$

where n is an integer, f is the vibration frequency, and h is a very small constant (Planck's constant) equal to 6.626×10^{-34} J·s or 4.14×10^{-15}eV·s.

While this result might not seem astonishing at first glance, the implication that energy in this case does not exist over a continuous range of frequencies but only in discreet quantized amounts leads to profound consequences. It is a first hint of a particle property intrinsic to e-m radiation.

The Photoelectric Effect

Another example of the particle nature of light is demonstrated by the *photoelectric effect.* The photoelectric effect occurs when light shines on a metallic surface causing electrons to be emitted from the surface. The circuit on the next page can be set up to demonstrate the effect. When light hits the metal electrode, electrons are ejected. They jump across the gap resulting in a current flowing in the circuit. You can measure the maximum kinetic energy of the electrons by attaching a variable voltage source to the circuit. If you set up the voltage source to counter to the direction of the current at a certain voltage, you will stop the flow. This is called the *stopping potential.* The stopping potential is related to the maximum KE as follows: $KE_{max}=eV_0$ where e is the charge on the electron and V_0 is the stopping potential.

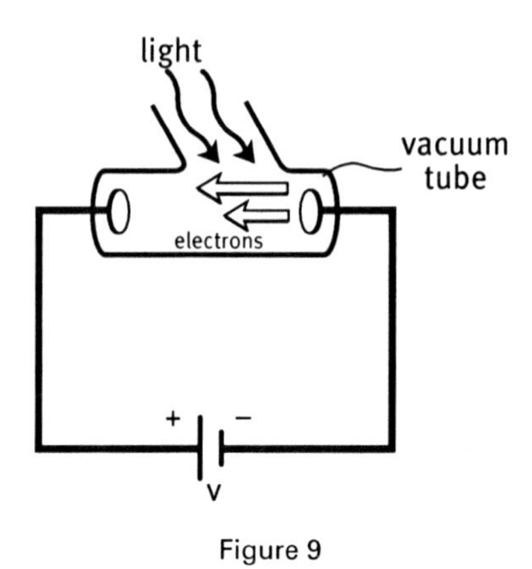

Figure 9

While this effect was well-known to physicists at the turn of the 20th century, the wave explanation of light just didn't fit with the specific behavior of the ejected electrons. The emission of electrons is consistent with the wave theory in so far as the electric fields of e-m waves can exert forces on the electrons knocking them out of the metal. However, the wave theory predicts that (1) as the light intensity is increased the maximum kinetic energy of ejected electrons should increase, and (2) the frequency of the light should have no effect on the kinetic energy of ejected electrons.

What actually happens is that the intensity of light has no effect on maximum KE of the electrons, and below a certain cutoff frequency no electrons will be ejected at all, regardless of the intensity of the light.

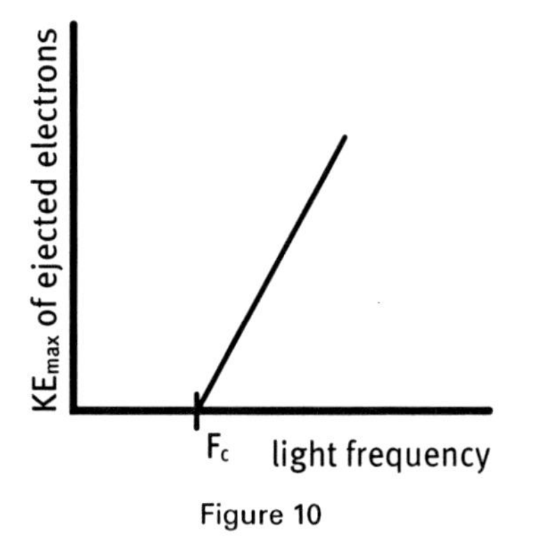

Figure 10

Einstein's Explanation. Einstein explained the photoelectric effect as follows: Light is transmitted in tiny bundles or particles called ***photons*** each with an energy $E = hf$. When an individual photon collides with an electron it disappears, transferring its energy to the electron. If the photon has as much or more energy as that binding the electron to the metal, the electron will be ejected. If not, then it does not matter how many photons strike an electron (more photons means greater intensity, brighter light), because if none have the minimum energy, electrons will never be freed. This explanation exactly matches the experimental results.

The minimum amount of energy necessary to free an electron from a given metal surface is called the *work function* (Φ) of the metal. Therefore, the maximum kinetic energy of ejected electrons will be the difference between the photon energy and Φ.

$$KE_{max} = hf - \Phi$$

Example 7: You shine light of a frequency 1.1×10^{15} Hz onto a metal surface with a work function of 3.3 eV.

a) What is the maximum kinetic energy of ejected electrons released from the metal surface?
b) What minimum voltage applied by a variable power supply will bring the photoelectric current to a halt?

Solution:

a) KE_{max} (in electron volts) = $hf - \Phi = (4.14 \times 10^{-15}\ \text{eV·s})(1.1 \times 10^{15}\ \text{Hz}) - 3.3\ \text{eV} = 1.26\ \text{eV}$
b) $V = 1.26\ V$

What is going on here? We thought that light (e-m radiation) was a wave—multiple experiments seemed to confirm that—but now we find the only explanation for some other experiments is to describe light as consisting of particles. We can only resolve the dilemma by realizing that light can be either, or you might say that it is neither. It is not possible to macroscopically visualize or model the true nature of light, so we make conclusions based on the experiments that we do. Light exhibits properties of both waves and particles, and which property you detect depends on the experiment that you conduct. This phenomenon is called *wave-particle duality.*

The Wave Nature of Matter. Is it possible that if light sometimes behaves like a particle, then something that we think of as a particle like an electron or other piece of matter might behave like a wave? In 1923, Louis de Broglie proposed that the wavelength of a particle is related to its momentum. The momentum of a photon can be expressed as

$$p = \frac{E}{c} = \frac{hf}{c} = \frac{h}{\lambda}.$$

de Broglie proposed that the wavelength of a particle can be expressed by the same relation.

$$p = \frac{h}{\lambda} \text{ or } \lambda = \frac{h}{p} = \frac{h}{mv}$$

The amazing thing is we actually can observe wavelike properties of particles. Say you shine laser light through some opaque material that has two small slits that are around the same width as a wavelength of light (a grating). The light that passes through the slits will show dark and light spots if projected on a screen at some distance from the grating. This is because the light passing through one slit interferes with the light coming through the other, producing a wave interference pattern. It turns out that if you fire a stream of electrons through a double slit of appropriate width and spacing, they too will produce a wave interference pattern on a screen. Matter also exhibits wave-particle duality.

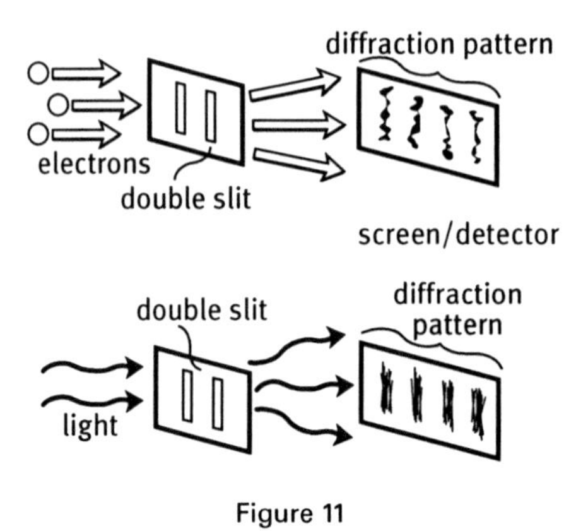

Figure 11

If particles have wave properties, why don't we notice this all of the time? It has to do with the size of Planck's constant.

Example 8: Compare the de Broglie wavelength of an electron moving at a speed of 1.0×10^7 m/s with that of a rock of mass 1 kg thrown with a speed of 20 m/s.

Solution: $\lambda = \frac{h}{p} = \frac{h}{mv}$ so for the electron

$$\lambda = \frac{6.63 \times 10^{-34}\ \text{Js}}{(9.11 \times 10^{-31}\ \text{kg})(1 \times 10^{7}\ \text{m/s})} = 7.3 \times 10^{-11}\ \text{m},$$

which is within an order of magnitude of the size of an atom. This implies that because the electron exists and interacts at the atomic scale its wave properties will be relevant. In contrast for the rock,

$$\lambda = \frac{6.63 \times 10^{-34}\ \text{Js}}{(1.0\ \text{kg})(20\ \text{m/s})} = 3.3 \times 10^{-35}\ \text{m}$$

That's a lot smaller than the size of a rock and other things in the macroscopic world. (Throwing rocks through gaps in a fence will not result in a diffraction pattern.)

In the case of particle waves, the wavelength is describing the *probability* that a particle will be found at a given location. In the case of the double slit experiment, if a relatively small number of electrons are fired through the openings, you won't detect any discernable pattern. However, if enough go through then a pattern starts to form with the bright parts of the screen corresponding to the regions with higher probabilities of electrons being found there. This idea that particle waves are waves of probability forms the core of the theory of quantum mechanics.

The Schrödinger Equation

The first attempts to come up with a more formal mathematical description of the wave nature of matter introduced the idea that a matter wave could be quantified by a *wave function*. The wave function ψ represents the amplitude of the matter wave, and ψ^2 represents the probability of finding the particle at a given position at a given time. What this means is that if we treat objects like electrons as particles, we can only talk about the probability that they are at a certain place at a certain time. We cannot know or predict the path of a single electron as it moves through space—not because we don't have enough information but because it is fundamentally theoretically impossible. The fundamental nature of what a particle is does not correspond to our everyday notions based on our experience of the macroscopic world. The first equation to formulate this was developed by Erwin Schrödinger and is therefore called The *Schrödinger wave equation.*

The Heisenberg Uncertainty Principle

At this point we are starting to see that on scales of electron size, the notion of what matter and energy are becomes a fuzzy thing. Is a photon wave or particle? Is an electron particle or wave? Both have properties of each. We might even be starting to get the possibly disquieting idea that there is a fundamental limit to how well we can know the details of a physical system. In Newtonian mechanics it should be theoretically possible if you have precise enough measuring instruments to make uncertainty in a measurement infinitely small. In addition, if you have enough information on a system then theoretically you should be able to predict precisely the future behavior of the system (a phenomenon called ***determinism***). However, according to quantum mechanics, neither of these proposals is true, and the limitations are not imposed by our ability to build precise instruments but rather are inherent in the nature of the universe as we know it. There are two reasons for this. The first reason is wave-particle duality, which we were just talking about. The second has to do with a thought-provoking idea: It is not possible to make a measurement without disturbing the system that you are measuring. Consider what you must do in order to measure some physical property of an object. In order to "see" the object you must bounce photons off of it such that they are reflected back into a detector (like your eye).

At least one photon must bounce off of the object and into the detector, but photons have energy and momentum, and therefore when they collide with the object they will transfer some of that energy and momentum to the object. Obviously photons will have very little effect on the position and momentum of macroscopic objects, but what about something the size of an electron? Any attempt to determine both the position and momentum of the electron with infinite accuracy will be impossible because any attempt to measure these quantities will change them. According to Heisenberg's uncertainty principle, ***it is physically impossible to measure simultaneously the exact position and the exact momentum of a particle.*** Approximately,

$$\Delta x \Delta p \geq \frac{h}{2\pi}$$

In addition to momentum and position it is also impossible to measure the energy of a system over an arbitrarily small time interval. Approximately,

$$\Delta E \Delta t \geq \frac{h}{2\pi}$$

This implies that atomic and nuclear energy levels are always inexact.

At this point it's possible that you have an objection. It might go something like this: While maybe these probabilities and uncertainties limit what we will ever be able to measure in any practical way, ultimately the electron must have a specific momentum and position, at a specific time even if we can't determine it. However, the general consensus is that is not the case. Furthermore, according to the Copenhagen interpretation of quantum mechanics articulated by Niels Bohr, it actually makes no sense to discuss the physical state of an atomic system before it is measured. It is undefined. The system merely exists as a multitude of probabilities. It is the act of measurement that actually forces the system into a particular state. Whether or not the Copenhagen interpretation is true or not it has certainly become fuel for a lot of pop-culture philosophical speculation on the nature of reality, as we find in *What the Bleep*.

Taking Liberties with Quantum Mechanics

Now, what about the movie; how does it use/misapply principles of quantum mechanics to make its points? First of all there *are* scientific concepts in *What the Bleep* that are presented accurately, like the sections on brain chemistry and neurotransmitters (there is evidence demonstrating that our emotions and thoughts can affect our physiology). Moreover, we don't object to the sincere desire of the filmmakers to try to find some spiritual comfort in a big scary universe. The problem is, while in some sections of the film the science is presented accurately, in many other sections, particularly those incorporating principles of physics, it misleads viewers by claiming "scientific evidence" supports a series of subjective and sometimes absurd conclusions.

What the Bleep presents several completely unverifiable and absurd "examples" of how thought affects the macroscopic world. In one of these we follow a group of Maharishi Mahesh Yogi devotees to Washington, DC, in the summer of 1993. They have a goal: to lower the crime rate by meditating. Lo and behold, violent crime in DC actually *was* lower that summer. Of course, they conveniently ignore the fact that there could have been other factors responsible for that outcome. Perhaps a look at the local crime statistics over a long period of time (which show a consistent decrease in violent crime over several years) might say something about whether or not the meditation had any statistical significance.

Although not discussed in the movie is the fact that the same group claims they were able to alter weather patterns when an impending deep freeze threatened to ruin the recently poured concrete in the "golden dome" at Maharishi International University in Fairfield, Iowa. Apparently, by meditating, they increased the temperature by over 20 °F (by causing a warm front to travel across the western United States?). That's an enormous effect on the physical world. Could they have done it? Well, the atmosphere is a giant thermodynamic system. Let's apply a little physics and estimate how much energy would be required simply to change the temperature of weather-system-size air masses.

Changing the Weather. How much energy would it take to raise the temperature of a modest-size volume of air 10 miles by 10 miles by 1 mile (4.1×10^{11} m^3) in the vicinity of Fairfield, Iowa by 20 °C? The specific heat of air varies depending on temperature and pressure, but we can take the approximate value to be in the neighborhood of 1,000 J/kg °C. The density of air is just over 1 kg/m^3 so the mass of the air will be around 4.1×10^{11} kg. Therefore the total energy to heat up the air will be

$$Q = mc\Delta T = (4.1 \times 10^{11}\ \text{kg})(1{,}000\ \text{J/kg °C})(20\ \text{°C}) \approx 8 \times 10^{15}\ \text{J},$$

which is on the order of the energy released by a 2-MT hydrogen bomb! If meditation could do that then it should be no problem to do much simpler things—and on a much smaller scale—such as levitate a tennis ball, light a candle, or even just roll a marble a few centimeters across the table.

Let's calculate the amount of energy it would take to increase the speed of a marble from rest to, say, 10 cm/s using the work energy theorem.

$$W = \frac{1}{2}mv^2 + \frac{1}{2}I\omega^2 \text{ and } I = \frac{2}{5}mr^2 \text{ for a uniform sphere around its center.}$$

Therefore, for a 10-g marble:

$$W = \frac{1}{2}mv^2 + \frac{1}{2}\left(\frac{2}{5}mr^2\right)\frac{v^2}{r^2} = \frac{7}{10}\,mv^2 = 7 \times 10^{-5}\ \text{J}.$$

That's 100 billion-trillion times less energy than it would take to change the weather. How come no one ever does that? (Hint: It's a lot easier to make claims to phenomena that are completely unverifiable.)

The Power of Positive Thinking. Speaking of completely unverifiable, another "experiment" in the film supposedly demonstrated that ice crystals formed differently depending on what words were written on the container in which they were frozen. Astonishingly, positive words like "love" or "thank you" resulted in beautiful soothing deep blue crystals, while words like "sick" or "kill" created ugly irregular yellowish structures. Supposedly, this is evidence that thought affects the physical world. "If thoughts can do that to water, imagine what our thoughts can do to us," a man sagely exclaims in scene 6, implying that the fact that this "experiment" can't be duplicated shouldn't trouble us and the fact that it is impossible to identify a physical mechanism that could accomplish this is irrelevant.

The Scale of Quantum Mechanics

The previous examples are some of the more obviously outrageous assertions that should undermine any credibility for the casual viewer. However, it is the way they use and abuse quantum mechanics in *What the Bleep* that we must take particular exception to. A lot of "new-age philosophies" seem to be predicated upon unverifiable notions that border on the magical. They incorporate scientific terminology and amalgamated phrases such as "magnetic fields of zero point energy" (scene 6), or "affecting reality at the quantum level" used in completely nonrigorous pseudoscientific contexts. The terminology of quantum mechanics is particularly popular in new age circles when trying to demonstrate "proof" of things like free will, the effect of the mind on the physical universe, and other things like that.

One of the more commonly expounded lines of reasoning (which forms one of the major points the film is trying to convey) applies the Copenhagen interpretation and the Heisenberg uncertainty principle as follows: The state of a physical system isn't specified until a measurement is made, and the measurement affects the system being measured. Therefore, they conclude, that the mind as an observer/measurer of physical reality affects the physical outcome of the observation. Therefore, the mind changes the external physical universe, and these changes can be observed macroscopically. Therefore obviously, if we truly believe, we can accomplish amazing physical miracles with the power of thought—like change the weather, for example.

However, quantum mechanics is very specific in terms of what it can and cannot be meaningfully applied to. It is a theory designed to explain behavior of matter at atomic scales. The idea that it is impossible to make a measurement without affecting the outcome, expressed by Heisenberg, is only relevant in any practical sense when the thing being measured is *really small.* As we stated previously, the minimum theoretical uncertainty in an experiment is on the order of Planck's constant, which is around 10^{-34} J·s. How much meaning does an uncertainty that tiny have when measuring the momentum of something like a baseball? Even biological systems like the ones discussed in the movie are billions of billions of times larger than the Plank scale (the mass of a single cell is about 20 orders of magnitude larger than an electron).

The equations of quantum mechanics when applied to predicting the behavior of atomic systems have proven to be remarkably accurate. However, they are not appropriate for making meaningful predictions of macroscopic systems. Any discussions of how the rules of quantum mechanics manifest themselves at scales much larger than the atom and, in particular, attempts to explain human consciousness in terms of quantum mechanics are philosophical speculations, not scientific investigations.

Conclusion. The uncertainty principle simply cannot be applied to macroscopic systems in any meaningful way (even though they do it on *Star Trek the Next Generation* from time to time). In *What the Bleep,* as in many other works of the genre, the filmmakers love to expound on how the mind affects the universe "at the quantum level." In scene 6, Joe Dispenza says, "I create my own reality. Quantum mechanics is telling us that." It isn't. We really don't know what this means (and, really, neither do they). In the movie, they use the uncertainty principle to conclude (without evidence) that—somehow—the mind affects the universe at the atomic scale. Convinced that this is true, they then surmise that it is clear that these effects will also extend to the macroscopic level—to our everyday life in some obvious way. They use it to prove that free will exists. They use it to show how all matter is part of an interconnected spiritual brotherhood. Whether or not there is a spiritual reality to the universe, quantum mechanics has nothing to say on the matter. It is scientifically disingenuous to claim that it does.

FINAL REMARKS

Well, we've looked at a lot of movies and analyzed a lot of physics. Hopefully, in addition to the time spent experiencing, learning, and applying the elegance and beauty of the laws that govern the universe in which we live, you've come to appreciate how Hollywood movies supply an entertaining and original springboard for the study of physics. Some may say that by exposing the truth behind some of the outrageous and impossible physical stunts so prevalent in the cinema, we take the fun out of the fantasy—not so! We can still have a great time watching *X-Men* while making fun of the incredible amounts of food that Magneto would have to eat to generate such enormous quantities of electromagnetic energy. In addition, when it comes to bad movies with even worse physics (like our friend *The Core*), our investigation falls into the category of public service. Let us defend the concept of a scientifically literate public! How impressive (and refreshing) it is, indeed, when a movie gets it (more or less) right.

ADDITIONAL QUESTIONS AND PROBLEMS

1) What if instead at traveling at speed *c*, light plodded along at the every day speed of 40 miles per hour (but still independent of frame of reference). How would this change our day to day lives?

2) A fast train travels from New York to Los Angeles, a distance of 3,000 miles in the Earth frame of reference at a speed of 0.98 c. If it departs the station at 9:00 AM. At what time does it arrive in L.A.? How long does it take to get there according to the passengers on the train, and how far is it to L.A. once they get underway? If the train is 100 m long in its reference frame how long would people on the side of the tracks measure it to be as it hurtles past?

3) A cubic box is 1 m on a side and has a mass of 20 kg when at rest relative to you. If it happens to be traveling past you at a speed of 0.90 *c*. What are the dimensions of the box? What is its density? What is its total energy?

4) In the movie *Close Encounters of the Third Kind,* when the extraterrestrials land they deposit all of the people they've abducted over the years, including some WW II–era fighter pilots. These guys haven't aged more than a few months! Assuming the pilots have been traveling with the ETs the entire time, estimate how fast they've been moving relative to the Earth frame of reference.

5) An atom of ^{131}I experiences an alpha decay. What is the daughter element? Write the decay equation.

6) A deuterium nucleus consists of a proton and a neutron. It has an atomic mass of 2.014102 *u*. Calculate its binding energy.

7) Why do heavier elements require higher ratios of neutrons to protons to remain stable?

8) Why would a fusion reactor produce less radioactive waste than a fission reactor?

Appendix A: Additional Brief Film Reviews

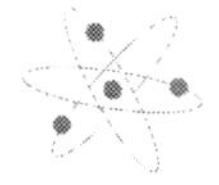

GONE IN 60 SECONDS

Nicolas Cage is Randall "Memphis" Raines, a former professional car thief and really good guy who has gone straight. Unfortunately, his kid brother Kip (Giovanni Ribisi) is in trouble with a local crime boss, and unless Memphis is able to steal 50 hard-to-find cars in a single night, Kip is going to get "whacked." Memphis assembles his old crew in an attempt to pull off the seemingly impossible task. To make matters worse, Detective Roland Castlebeck (Delroy Lindo) is mad at Memphis, because he was never able to catch him in the old days, so he's out to nail him in the act this time around. Can Memphis do it?

Gone in 60 Seconds is a pretty standard but reasonably entertaining action movie in the fast-cars-in-the-underbelly-of-society genre (like *The Fast and the Furious*). Not surprisingly, it also has a few scenes with some problematic physics moments. Let's look at one of these in detail.

Toward the end of the movie, in scene 28, Memphis and his crew have successfully "boosted" 49 of the 50 cars, and he has one last car to steal—a 1967 Mustang Shelby GT. Unfortunately for Memphis, just as he pulls the job, Castlebeck arrives on the scene, and the long-awaited high-speed car chase ensues. Dozens of police cars pursue Memphis. They zoom in and out of alleys, onto busy city streets, through the docks, and finally onto a bridge where Memphis appears to be trapped. There is an accident up ahead and the road is completely blocked—when we see a large tow truck ramp being lowered into place, we have a pretty good idea where this scene is going.

The problem is, Memphis is going to have to jump over the entire accident scene, and it's a big one. In addition to several wrecked cars, there is the tow truck and two or three emergency vehicles. Although it's hard to get an exact length from the camera shots, it seems that the entire blockade spans a distance of at least 50 or 60 meters. Memphis is going to have to jump over all of that and then stick the landing. Can he do it?

Memphis's Big Jump

We can see that just before he hits the ramp, Memphis has the Mustang up to 95 miles per hour (42m/s). If we look carefully at the incline it appears to be around 2 meters high, and the length of the ramp is just a bit longer than the car. Because the 1967 Shelby is 4.7 m in length, we'll say the ramp is 5 m long. From this we can calculate the angle of the takeoff.

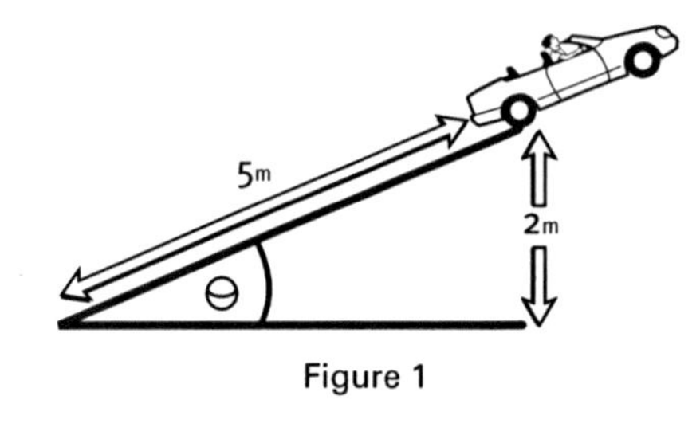

Figure 1

$$\sin^{-1}\theta = \sin^{-1}\frac{2\text{ m}}{5\text{ m}} = 24°$$

From the data we can then calculate the theoretical maximum range of the projectile (the car) assuming *no air resistance.*

From the vertical motion we can determine the time of flight:

$$\Delta y = v_{0y}t + \frac{1}{2}at^2 = 0 = (17\text{ m/s})t - (4.9\text{ m/s}^2)t^2.$$

Solving, we get $t = 3.5$ s.

Therefore, the horizontal displacement will be:

$$\Delta x = v_x t = (38\text{ m/s})(3.5\text{ s}) = 133\text{ m}.$$

That's really far—about twice the distance he needs to cover. Does this mean that Memphis can easily make the jump? If so, why does the shot look computer generated?

Not surprisingly, it turns out that air resistance does have a significant effect on projectiles, with the air resistance force proportional to the square of the velocity. Using a computer model we can find the approximate trajectory for the Mustang with air friction and compare it to the case without.

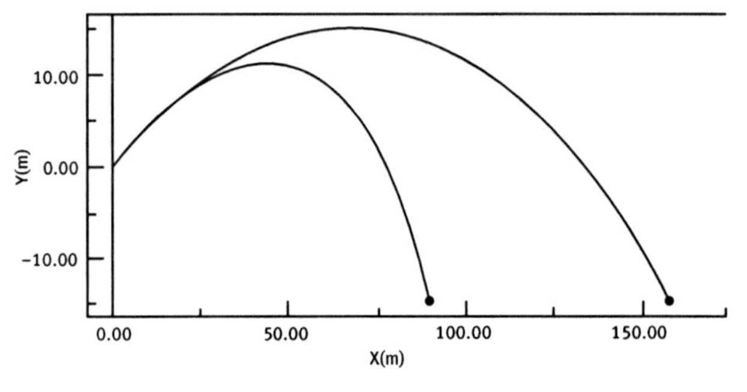

Figure 2

It looks like (in the more realistic case) the car might be able to clear about 70 m. (One of the longest ramp to ramp car jumps on record is 71 m in a Buick Skylark, so at least we know a jump of this magnitude has been recorded.) If Memphis hits the takeoff just right, if he gets really lucky, if the suspension can handle driving onto a ramp angled upward at 24° at a speed close to 100 miles per hour (doubtful), it seems he just might be able to clear the accident scene. Nevertheless, our next—and now bigger and more pressing worry—is the landing.

Have you ever seen an automobile stunt jump either in person, in a video, or in a photograph? If so then you know why they put a ramp up for the landing. The car comes in nose down as the car follows the projectile trajectory.

If the 1967 Mustang follows the projectile trajectory, it's going to land at a steep angle—in the case of no air friction, 24° below the horizontal. With air friction (as you can see from Figure 2), the car will come in at an even steeper angle. There is no way Memphis is going to stick that landing. The tires would probably never even touch the road, and even if they did, the suspension (and Memphis himself) wouldn't be able to handle the force of impact. The Mustang would likely lose an axle and spin out of control, the front of the car would crumple, and the crash site would make the accident he just jumped over look like a fender bender. Unfortunately, Randall will not deliver car 50, and his little brother will have to figure out some other way to avoid getting "whacked."

X-MEN

There are certain movies where you must completely suspend your disbelief in order to enjoy the fantasy. In some of these, the fundamental premise requires that the principles of physics (and biology and chemistry) as we know them are completely dispensed with. *X-Men* is one of these movies. The film takes place in "the not too distant future" where a fearful public persecutes "mutants" with superpowers. The mutants split into two opposing camps. The first is led by the stately, dignified, and compassionate Dr. Charles Xavier (Patrick Stewart), who is hopeful that mutants and humans can learn to live together in mutual respect and peace, while the disaffected and irritable Brotherhood of Mutants, ruled by the distrustful and megalomaniacal Magneto (Ian McKellen), is prepared to destroy humanity to assure the ascendancy of the mutants.

Because the main premise of *X-Men* is completely implausible (after all, it is based on a comic book), we can't be upset by the lack of any adherence to physical reality. The idea that random mutations are somehow creating a race of "super-people" with incredible powers (while a lot of fun) is ridiculous and silly (ask a biologist). As far as the physics goes, much of what happens is impossible to comment on because it's so far outside the realm of possibility that we wouldn't know where to begin. Nevertheless, we can point out some of the more obvious and specifically identifiable violations of major principles that run throughout this film and others like it. In *X-Men* these violations tend to involve Newton's third law and conservation of energy.

Magneto is an extremely powerful mutant who is somehow able to generate huge electromagnetic fields and therefore exert forces on any metallic objects in his vicinity. Considering the fact that magnetic fields are produced by moving electric charge, Magneto must have some phenomenal electric currents running through his body. We often see Magneto using his power to move or crumple large objects like trains and bridges. The amount of force that he applies must be incredible, and yet during one of these amazing feats we never see any physical effect on Magneto himself. However, according to Newton's third law, if Magneto exerts a force on the bridge then the bridge *must* exert a force of equal magnitude back on Magneto. Moreover, because Magneto has a much smaller mass than trains and bridges he should be flung backwards like a rag doll. Of course, he never is. This phenomenon is seen over and over again in *X-Men,* and in superhero movies in general. For example, Dr. Jean Grey (Famke Janssen) is "telekinetic" (she can move objects with her mind). As we understand physics, the only way she would be able to move (accelerate) masses this way is to apply a force to them. We will avoid an argument of what "telekinesis" could possibly be and just point out that when Jean tosses some large object across the room, she should be tossed backwards in the opposite direction. She isn't, and neither are many other characters who fling really heavy objects around the old-fashioned way.

Many of the characters in *X-Men* have powers that are hard to comprehend in terms of energy. Where do they get it? One of the clearer examples involves Storm (Halle Berry), who has the ability to change the weather. She can form clouds, produce incredible wind storms, and rain lightning bolts down on her adversaries. The amount of energy involved in an average thunderstorm is on the order of a 20-kiloton nuclear bomb (around 10^{13} J). This is around 10 billion times the chemical energy required to run a marathon. Perhaps Storm has some organ in her body that acts like a nuclear reactor. It either initiates nuclear fusion reactions, or better yet, creates positrons in order to induce pair-annihilation reactions. According to $E = mc^2$, you would need only a fraction of a kilogram to produce the requisite energy, because a little bit of mass converts to a lot of energy (even in nuclear bombs with yields in the megatons only around a kilogram of mass is actually converted). Nevertheless, how she would able to confine and channel the energy without blowing herself up is another matter! Watch the movie and the sequels for other examples of mutant energy running amok.

INDEPENDENCE DAY

What would humanity do if out of nowhere we were suddenly attacked by a vicious extraterrestrial race bent on destroying the planet? Of course this is a common theme in the science fiction genre. Who can forget the original nightmarish *War of the Worlds* based on the novel by H.G. Wells? Well, in *Independence Day,* we get the 1996 incarnation of this standard plot. In this one, the malevolent alien spacecrafts appear to be invincible and the extraterrestrials are ruthless, soulless killers. Things look hopeless, but, even though the aliens' technology is incomprehensibly advanced, with the usual grit, ingenuity, and good old never-say-die attitude, the scrappy humans are somehow able to hack into the alien computer system and blow up the gigantic mother ship. The setup isn't bad but there's not much originality here, and we're also affronted by some really awful movie physics.

First of all, we are told in the film that the mother ship has a mass one-quarter that of the moon and it is parked in geosynchronous orbit above the Earth. Geosynchronous orbit requires an orbital period of one day, and from this we can calculate the orbital radius of the mother ship:

$$v = \sqrt{\frac{GM}{r}} \text{ and } T = \frac{2\pi r}{v} \text{ and therefore } r = \sqrt[3]{\frac{T^2 GM}{4\pi^2}}.$$

Plugging in the mass of the Earth and the orbital period in seconds, we get $r = 4.2 \times 10^7$ m. Subtracting the radius of the Earth, an object in geosynchronous orbit will be 3.6×10^7 m (36,000 km) above the Earth's surface. This means we've got a mass one-quarter that of the moon in an orbit over ten times closer than the moon. Remember the tidal forces we discussed in our analysis of *Armageddon* when those large asteroid bits went flying by (Chapter 4)? Here we are faced with the same issue. This ship is going to cause huge tides resulting in absolute destruction of coastal areas, and probably even flexing the Earth's crust sufficiently to cause major catastrophic earthquakes. The aliens don't need to bother with their city-destroying death rays. They just have to sit there and let nature do the work.

Using the Third Law to Destroy a City

Nevertheless, in *Independence Day,* the nasty E.T.s accomplish their destruction of the world's major cities by sending out a dozen or so smaller crafts (about 15 miles in diameter we are told) to hover a few thousand feet over a city before unleashing a flaming death ray. The death ray engulfs each city in a giant expanding fireball. It is a common sight in science fiction films to see hovering craft. Somehow, the technology of the future allows for hovering without propellers, jet engines, or other 21st-century methods of applying forces to balance the gravitational force. Perhaps they are using magnetic fields as they do for maglev trains. However they accomplish this levitation we can not ignore Newton's third law. Consider a spacecraft 15 miles in diameter hovering over the city of Los Angeles. Let's do a rough estimate of the gravitational force acting on the ship. For simplicity we'll assume the ship is a flattened cylinder one mile thick with a radius of 7.5 miles, and an average density of 1000 kg/m^3.

Calculating the mass of the ship we get:

$$m = DV = D(\pi r^2 h) \approx 1.4 \times 10^{15} \text{ kg}$$

and because the weight of the ship that near the Earth's surface is mg, we get $W = 1.4 \times 10^{16}$ N. This means there must be an upward force equal to the weight acting on the ship to allow it to hover.

However, according to the third law, whatever is applying this upward force on the ship, the ship must be exerting an equal and opposite downward force back on it. If this is the air then the city lying underneath is going to be crushed simply due to the astronomical amount of air pressure created by the reaction force of the ship acting on the air. Therefore, why waste energy on those death rays?

Have you ever been close to a bonfire? Have you ever felt the heat (radiant energy) emanating from it warming your skin? Did you have to be touching the fire to feel the heat? Imagine how much heat must be radiating off of the giant expanding city-sized fireball engulfing the city of Los Angeles in *Independence Day.* Apparently, however, unless you are in direct contact with the flames, you'll be okay! We know this because when Will Smith's girlfriend is fleeing the approaching inferno in a freeway tunnel, she is able to jump into an alcove at the last second before the flames sweep past. Fortunately it's recessed a couple of feet into the wall, so she survives. Apparently, the fact that flames suck in oxygen creating a vacuum around large fires isn't a problem. We don't see her suffocate either. Apparently, it's one of the advantages of playing a romantic lead in a Hollywood movie.

It seems that in every end of the world disaster science fiction movie there is only one (eccentric and quirky) scientist *in the entire world* who has the genius to figure out how to save the planet. In *Independence Day* this job falls to Jeff Goldblum (always convincing as a scientist if he puts on a pair of glasses), a sort of ne'er-do-well underachiever with the creative insight to realize that the radio transmissions intercepted from the alien ships before the initial attack are actually a countdown—a countdown to destruction! After some major setbacks, after it looks like all hope is lost, it is "Dr. Goldblum" who realizes if he plants a virus in the alien mother ship's computer using his (alien-compatible) laptop then the shields will drop long enough for the Earthlings to "Nuke 'em!" At this point the movie degenerates into a series of feel-good clichés. With dialogue like "Payback's a bitch ain't it?!" and "All right you alien a** holes," we feel we've seen this movie somewhere before, and know that the aliens don't have a chance.

THE PLANET OF THE APES

The *Planet of the Apes* (1968) is a classic. Charlton Heston and his small crew of space explorers are sent on a first-time interstellar voyage to another planet. After waking from 18 months in suspended animation the crew crash-lands on a strange planet. They discover, to their horror, that while humans exist on the planet they are more like beasts. They cannot speak, and they live and act more or less like wild animals (although they do wear loincloths for modesty's sake). These unfortunate hominids are hunted and otherwise persecuted by the only intelligent species on the planet—Apes! (The apes are also hominids.) At the end of the movie, Heston (a hominid) makes a terrifying discovery. He finds the Statue of Liberty half-buried in the sand in the middle of a deserted wasteland. It is a terrific, dramatic, and shocking final moment if we haven't seen the movie before, and when Charlton falls to his knees we feel that we know what he must be going through. It was actually Earth all along! The entire expedition was a waste of money. They never landed on another planet, they simply returned to Earth 2,000 years in the future! Humanity must have destroyed civilization in some final imprudent act of apocalyptic self-indulgence.

Relativity of the Apes

Planet of the Apes must be commended for giving special relativity a shot. (Interestingly, they refer to it as *Dr. Halstein's* theory of space and time instead of Einstein's theory of special relativity.)

The entire premise of the movie hinges upon the concept of time dilation. Let's look at the facts. The ship leaves Earth sometime around 1973. After the crash landing we find out the crew has been in suspended animation for about a year and a half, while the "earth time" clock reads 3978. Therefore, 2,000 years have elapsed on Earth. It's a classic example of the "twin paradox"—if you leave Earth on a round trip traveling at velocities close to the speed of light, when you return, more time will have elapsed on Earth than in your (the spaceship's) frame of reference. Based on the data we can estimate approximately how fast and how far the crew must have traveled on their round trip journey to "jump into the future." Applying the time dilation equation,

$$\Delta t = \frac{\Delta t_0}{\sqrt{1 - \frac{v^2}{c^2}}},$$

with Δt being the time elapsed on Earth and Δt_0 the time elapsed on the ship, we can solve for the ship's velocity.

$$v = 0.9999995c$$

That is really fast! To think that they were able to achieve that back in 1973. It must be one of those technologies (like the electric car back in the 50s) that the automobile industry secretly swept under the rug. Anyway, ignoring the obstacles that we discussed in Chapter 8 regarding the fuel requirements for accelerating a ship to that speed, we appreciate the appropriate use of special relativity in cinema. Let's also calculate the distance traveled by the ship in the Earth's frame of reference.

$$L_0 = vt = (.999995 \text{ Lyr/yr})(2000 \text{ yrs}) = (\text{almost})\ 2000 \text{ Lyr}$$

It's a shame they never got a chance to stop at another planet along the way.

A Few Physics Faux Pas

There are a few moments of problematic physics in *Planet of the Apes,* however. One involves a possible misconception with time dilation and frames of reference. Unfortunately one of the crew members (Stewart) has a malfunction in her suspended animation tube, and when the rest of the crew revives she is dead. We assume this happened because she starved to death, but she looks like a mummy. This seems to imply that she has aged hundreds (or two thousand years). This makes no sense. In the ship, 18 months have gone by. Anyone on the ship will have experienced exactly 18 months of time. The two thousand years of Earth time are a different frame of reference and irrelevant to the time passage in the ship's frame.

The other issue is the crash landing. It looks like the spaceship is coming in really fast, and when it crashes into a lake you would expect everyone, and anything loose in the ship, to go flying according to Newton's first law, due to their inertia. However, the crew members are still comfortably situated on their little cots inside their suspended animation tubes. You would think they would at least have slid forward a couple of feet.

Minor objections aside, *Planet of the Apes* is one of those movies that stands the test of time. Even though the acting isn't so great, it's still a compelling story and a lot better than the remake. The fact that there's a little relativity in there only adds to its mystique.

Appendix B: Fairy-Tale Physics

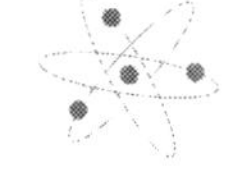

NOT ONLY DOES PHYSICS PERMEATE Hollywood movies, but some powerful physical principles are subtly at work in the beloved fairy tales of yore. Stories from The Brothers Grimm, and even the fables of Aesop, present us with not only entertainment and old-world wisdom, but also some truly groundbreaking physics and technology. Here are a few examples.

RUMPLESTILTSKIN IS MY NAME, AND NUCLEAR FUSION IS MY GAME!

Most of us know the tale of Rumplestiltskin, that little gnome-like creature who helps a poor young woman (we'll call her Zelda) out of a dire predicament by showing her how to spin straw into gold. This causes her predictable meteoric rise in the kingdom culminating in her marriage to "the prince." Unfortunately for Zelda the agreement is that she has to give her first born to Rumplestiltskin in exchange for his teaching her his gold-spinning technique. Rumplestiltskin does give her a way out, though. If she can guess his name, which he has never told to anyone, he'll let her off the hook. It seems hopeless and eventually payment comes due. Zelda is in despair so she decides to spy on Rumplestiltskin to see if she can get some useful intel. Foolishly Rumplestiltskin has a bad habit of prancing around chanting his name out loud when he thinks he's alone in the woods. Of course Zelda secretly catches him in the act, and who can forget the famous final scene where she guesses Rumplestiltskin's name, and he goes into such a rage that he evaporates (disintegrates?)?

What is going on here? How can Rumplestiltskin spin straw into gold? Why does he evaporate at the end of the story? A few seconds of thinking will obviously lead to one of two conclusions: (1) the story is completely made up or (2) Rumplestiltskin has a nuclear fusion reactor in his workshop.

The only way to make gold out of organic fibers (mostly carbon, hydrogen, and oxygen) is to use nuclear fusion. The carbon, hydrogen, and oxygen nuclei can fuse together in a sequence of reactions forming heavier and heavier elements until eventually we get gold. We know this happens in the interiors of stars, and in fact, according to astrophysicists, elements heavier than iron require the stars to explode into supernovas to be formed.

Regarding Rumplestiltskin's explosion at the end of the story, could it be that his workshop is just underneath where he is standing in that final scene? Could it be that when he stamps his foot so hard he initiates a supernova-like fusion reaction in which he is consumed? It seems suspiciously likely doesn't it?

RAPUNZEL, RAPUNZEL, WHAT IS THE TENSILE STRENGTH OF YOUR HAIR?

Everyone knows the tragic tale of Rapunzel who, trapped in a tower by an "enchantress," would let down her long tresses of golden hair in order for her secret love "the prince" to ascend for their secret trysts. Rapunzel had remarkable hair not just because of its physiologically improbable length of over "twenty ells" but also because of its *apparently* formidable tensile strength. Let's do a quick calculation to determine the amount of force exerted on Rapunzel's head as the prince ascends her magic locks, and the approximate force exerted on an individual hair. Following is a force diagram illustrating the appropriate physical situation.

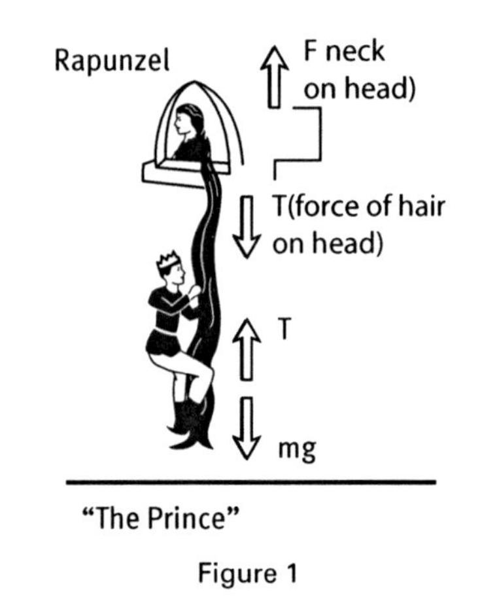

Figure 1

Let's assume the king's son has a mass of 75 kg, and in order to climb he will have periods of upward acceleration of up to 1 m/s². Then the maximum tension acting on Rapunzel's hair will be:

$$F_{net} = T - mg = ma$$

$$T = m(g + a) = 75 \text{ kg}(9.8 \text{ m/s}^2 + 1.0 \text{ m/s}^2) = 810 \text{ N}$$

Because a human head has around 100,000 hairs, then each of Rapunzel's must be able to withstand a force of

$$\frac{810 \text{ N}}{100{,}000 \text{ hairs}} \approx 0.008 \text{ N}$$

Because human hair can have a tensile strength of up to *almost a Newton* this should be no problem for Rapunzel or "the prince." Other than its great length, her hair does not in fact have to be in any way remarkable. According to the story, Rapunzel would even wrap her hair around a hook from the window above to eliminate what could be excessive strain on her neck, and as she got used to the activity I'll bet it didn't even hurt her head that much. (Sometimes you can even see people being swung around by their hair in the circus. It's actually pretty strong stuff.)

HANSEL AND GRETEL—HOW BUOYANT IS A DUCK?

Most people are aware of the gory tale of Hansel and Gretel and how a witch who wants to cook them in her oven, only to be tricked into getting roasted herself, imprisons them. However, not many people are aware that the two resourceful children, when making their escape, are assisted in a dangerous river crossing by a duck. This friendly waterfowl ferries them across one at a time, because, as we're told in the story, two at once would submerge the duck. Nevertheless, even with one child at a time, that is one formidable and buoyant duck. Let's use Archimedes's principle to analyze issues pertaining to floating waterfowl.

Think of your standard duck as he floats in the local pond. He seems to float comfortably with maybe 50 percent of his volume submerged. His density must be about half that of water. How much weight would it take to submerge an average duck? Well, a duck has a weight of about 2 pounds, or 10 N, give or take. This means the buoyant force acting on him must equal 10 N as he floats. According to Archimedes's principle the buoyant force is equal to the weight of the displaced fluid. If a duck displaces 50 percent of his volume and this results in a 10-N buoyant force acting on him, then if you completely submerge him the buoyant force will be twice that much, or 20 N. Therefore, a duck (or any other floating object) that has a density half that of water will be completely submerged if you place a weight equal to one duck on top of the original duck. This means that both Hansel and Gretel must weigh less than the duck or they are going to sink him. Now we know that the witch had been stuffing Hansel with rich food in order to fatten him up, so we must conclude that it was a giant duck weighing at least 40 or 50 pounds that came to the rescue!

Make of that what you will.

THE TORTOISE AND THE HARE

Aesop's fable teaches us something about persistence but it also teaches us about kinematics. Consider the details of the race. A tortoise and a hare challenge each other to a match race. The hare is the heavy favorite, but all of the pre-race hype has gotten to him. If he doesn't win, people are going to say he "choked." The tortoise, on the other hand, as the clear underdog, has nothing to lose—no one expects him to win. When the race starts the hare is really stressed. He starts off too fast! He gains a pretty commanding lead, but by the time he gets half-way to the finish he's got so much lactic acid buildup in his muscles that he's got to take a break. (What causes him to drift off to sleep? Aesop wants you to think its overconfidence. I think someone might have slipped a "mickey" into his Gatorade—probably some bookie who had a lot of money riding on the tortoise.) By the time he wakes up the tortoise is almost across the finish line. The hare makes a heroic attempt to beat the tortoise. He runs the last quarter-mile in world-record time. Bearing down on the tortoise, the hare closes the gap in big chunks, and for a breathless moment it looks like he might win after all! Then, seemingly out of nowhere, a rock hits the hare in the foot, he stumbles, and the tortoise beats him by half a meter.

Along with pondering this tragic story of how money can taint the purity of amateur athletics, let's try some simple kinematics calculations to their match race. Let's suppose that hares and tortoises traditionally have raced distances of about 800 meters. The running speed of the average hare is about 10 m/s while a well-conditioned tortoise can only manage about 0.1 m/s for any length of time. If the tortoise beats the hare by only half a meter, how long was the hare unconscious before he resumed his heroic attempt to recapture the title?

To solve this we need only write equations of motion for both the hare and tortoise determine the total running time for each. We'll assume they both start at position $x = 0$.

For the tortoise:

$$x_t = v_t t_t$$

This tells us how long it took the tortoise to run the race.

$$t_t = \frac{x_t}{v_t} = \frac{800 \text{ m}}{0.10 \text{ m/s}} = 8{,}000 \text{ s or } 2.22 \text{ hr}$$

For the hare:

$$x_h = v_h t_h.$$

Because we know that they both start at the same place, the hare is 0.5 m behind when the tortoise crosses the line, and the hare's actual running time is equal to that of the tortoise minus the time he has been unconscious we get:

$$x_h = 799.5 \text{ m} = v_h (t_t - t_{unconscious}).$$

Solving for the time the hare was out cold, we get:

$$t_{unconscious} = 7{,}920 \text{sec}.$$

This means that the hare was asleep for over two hours. It's hard to believe that was a normal nap. Consider that his time of 80 seconds for the 800 meter run would have broken the world record by over 20 seconds if he hadn't stopped in the middle, and *then* tell me that you're not convinced that there was foul play!

THE REAL STORY OF ICARUS

Greek mythology is a treasure trove of action-packed story telling. Just like the film industry, it is full of pathos, violence, drama, and desperation. It fulminates with the capriciousness of fate and the vicissitudes of injustice, all within the context of a make-believe world full of whim and fancy.

Remember the story of Daedalus and his son, the ill-fated Icarus? Daedalus was a master artisan and inventor imprisoned with his son on the Isle of Crete by the tyrannical King Minos. In a desperate attempt to escape, Daedalus fashions some artificial wings out of feathers and wax, and when no one is looking he and Icarus lift off from the walls of the city to apparent freedom. Daedalus tells Icarus not to fly too high because then he will be too close to the sun, which could melt the wax, resulting in equipment malfunction. At first Icarus heeds his father's advice but he finds flying to be so much fun that after a while he can't help himself. He flies higher and higher, his wings melt, and he plummets to his death in the Mediterranean Sea far below.

Our purpose here is to debunk the entire premise of this story. First of all anyone who has ever climbed a mountain knows that the air temperature drops the higher you climb. At elevations of a few miles above sea level it is both figuratively and literally freezing.

Let's look at the forces you would have to exert to fly with a pair of artificial wings.

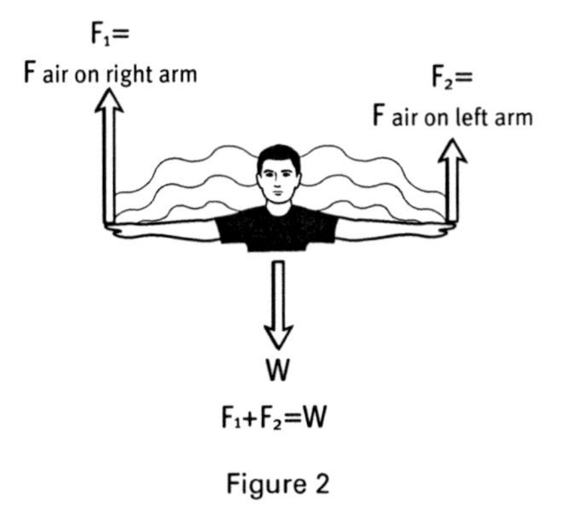

Figure 2

Each of Icarus's outstretched arms would have to be capable of supporting a force at least equal to half his weight (and more than that if he actually wants to accelerate in an upward direction). To simulate Icarus's flight (and you really probably shouldn't try this at home), imagine lying on your back on a bench with a dumbbell equal to half your weight strapped to each arm. Imagine flapping your outstretched arms up and down for a few seconds or even just holding them there. Now you know why they came up with hang gliders.

In fact, what Daedalus actually must have invented was the prototype for a hang glider. He probably suggested that Icarus learn to use "thermals" which are rising masses of hot air to stay aloft, and thereby the confusion in the story with the "thermal energy" of the sun. Icarus must have hit a downdraft and gotten tangled up in some telephone cables, and the story changed afterwards for dramatic effect.

THE LITTLE PRINCE IN ZERO GRAVITY

The Little Prince, written by Antoine de Saint Exupery in 1943, while not technically a fairy tale, still counts in our study of fairy-tale physics because of the title character "the prince." In this compact little story we see firsthand that the author has a complete ignorance of principles of gravitation, centripetal forces, and interstellar space travel. The little prince lives on asteroid B612 about the size of a medium wood shed, and yet he walks around and tends to his roses and his volcano just as if an asteroid that size had any significant gravitation.

Let's look at the amount of the gravity force acting on *Le Petit Prince*. We'll make the following assumptions: The asteroid has a radius of about 1.5 m and is made mostly of iron with a density of 13,000 kg/m^3. The little prince, being just a wee lad, has a mass of 40 kg. Next we'll find the approximate mass of the asteroid assuming a roughly spherical shape.

$$M = DV = D(\frac{4}{3}\pi r^3) = (13{,}000\ \text{kg/m}^3)(\frac{4}{3}\pi[1.5\ \text{m}]^3) \approx 180{,}000\ \text{kg}.$$

Using Newton's universal law of gravitation we get the force of gravity between the asteroid B612 and the little prince to be

$$F = \frac{GMm}{r^2} = \frac{6.67\times 10^{-11}\ \text{Nm}^2/\text{kg}(180{,}000\ \text{kg})(40\ \text{kg})}{(1.5\ \text{m})^2} = 2\times 10^{-4}\ \text{N}.$$

That's not a very big force. It tells us that the escape velocity from the asteroid would be only:

$$v_{esc} = \sqrt{\frac{2GM}{r}} = \sqrt{\frac{2(6.67\times 10^{-11}\ \text{Nm}^2/\text{kg}^2)(180{,}000\ \text{kg})}{1.5\ \text{m}}} \approx 3\times 10^{-3}\ \text{m/s}.$$

This is a very small velocity. (At this speed it would take over five minutes to move one meter.) Therefore, this means that the little prince must be incredibly careful with every step he takes. The smallest upward motion and he will be flung into space never to return!

At one point in the book (apparently after being flung off of B612 to a neighboring rock) the little prince meets "The Lamplighter," whose asteroid has been gradually increasing its rotation rate over the years to its present rate of once a minute. Based on the above data it is clear that anyone landing on this asteroid (except very close to the poles) will be instantly flung off into space. Assuming an asteroid of the same size as B612 the tangential speed of a point on the surface near the equator will be:

$$v = \frac{2\pi r}{60\ \text{s}} = \frac{2\pi(1.5\ \text{m})}{60\ \text{s}} = 0.16\ \text{m/s}.$$

Because this is about 50 times the escape velocity it would be impossible to set foot anywhere near the equator. For the little prince to stay put on the asteroid surface at the equatorial latitudes, it would require a centripetal force of magnitude

$$F_c = \frac{mv^2}{r} = \frac{(40 \text{ kg})(0.16 \text{ m/s})^2}{1.5 \text{ m}} = 0.68 \text{ N},$$

which (assuming the asteroid is the same size and mass as B612) is over 200 times the centripetal force (gravity) actually acting on him. Even at a spot less than a foot from the poles we can show that he will exceed the escape velocity by at least ten times. We conclude that this asteroid should be completely uninhabitable, and the lamplighter would have been flung into space decades earlier.

In addition to problems with staying anchored to the surface, it would be impossible to keep an atmosphere anchored to the asteroid. With no atmosphere not only would they all suffocate, but their blood would boil as well! The boiling temperature of a liquid depends on pressure. As pressure drops so does the boiling temperature, and with an atmospheric pressure of zero blood, which is mostly water, would boil at room temperature.

We can only conclude from this that the author of *The Little Prince,* in an attempt to present his idealistic philosophical views, is unwilling, or too squeamish, to face the harsh realities and the brutal conditions of life (or lack thereof) on inhospitable chunks of space jetsam.

Appendix C: The Laws of Cartoon Physics

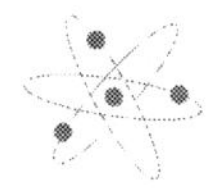

THIS BRIEF SECTION SUMMARIZES SOME of the principles of physics from the cartoon universe. While some of the laws resemble those from our own world, there are some interesting differences. In researching the topic we spent many hours poring over old *Looney Tunes* videos, Coyote and Roadrunner reruns, and a couple of animated films including the pivotal *Who Framed Roger Rabbit.*

Because the study of cartoon physics is a relatively new field in the realm of physics research there are still many fascinating unanswered questions. We have tried here to point out some of the more well-understood phenomena, but if you watch cartoons religiously you will undoubtedly uncover many other intriguing cartoon physics mysteries.

THE FIRST LAW OF CARTOON PHYSICS (THE FLCP): *GRAVITY WORKS ONLY WHEN YOU LOOK DOWN*

This principle is perhaps the most well known in cartoon physics. You can find evidence in almost any Bugs Bunny or Coyote/Roadrunner episode, and many others as well. The most consistent example occurs when an unsuspecting "toon" (cartoon character) is fooled into running off the edge of a cliff. For the first second the unfortunate toon moves in a straight line several feet off of the edge and hangs there suspended for a moment or two. Normally gravity then kicks in—in two stages. The first is called the latent, or "realization," phase. Once the character realizes he is hanging suspended in thin air, the gravitational force can then begin to work. Sometimes this is all it takes, but usually this phase is followed by the "look downward phase." It is usually a moment after the downward look that the acceleration due to gravity becomes inevitable.

THE SECOND LAW OF CARTOON PHYSICS: *HORIZONTALLY PROJECTED BODIES CONTINUE TO MOVE IN A STRAIGHT LINE UNAFFECTED BY THE FORCE OF GRAVITY*

This is really a corollary to the first law. Because most horizontally projected objects in cartoons are inanimate (no pun intended) they are unable to look down, and therefore cannot be affected by gravity. Characters also may move in a straight line as long as they feel no imminent danger from falling. Therefore, these cases almost always occur close to the ground.

THE SECOND COROLLARY TO THE FIRST LAW OF CARTOON PHYSICS: *IT IS SOMETIMES POSSIBLE TO WALK ON AIR AS LONG AS THE DISTANCE IS NOT TOO GREAT*

Occasionally you will see a toon accidentally find himself in the air a few feet from a precipice. Upon becoming aware of this he will then accelerate towards the ground far below in accordance with the first law. Sometimes however, if the toon is highly motivated, he is able to run a few steps on the air to safety. (This only works for very short distances but it does suggest that the viscosity of air may be highly variable depending on the emotional state of the observer (toon).

THE LAW OF CARTOON FRICTION: *FRICTION MUST BUILD UP BEFORE IT CAN ACTUALLY EXERT A FORCE*

Consider the following scenario. The Roadrunner is being pursued by the hapless Coyote who has a rocket engine strapped to his back (Newton's third law clearly works in cartoons). The Roadrunner needs to get away in a hurry but first he must build up friction. We therefore see his legs move in a circular path for at least four or five full strides before his feet "catch" and he is able to accelerate away. Notice also that the acceleration is *extremely* rapid, and the Roadrunner seems to increase his velocity from zero to a very high value almost instantly. Large accelerations require large forces giving further evidence that friction is building up until it can be unleashed with great force. It's no wonder the Coyote never catches the Roadrunner!

THE LAW OF CARTOON ELASTICITY: *ALL CHARACTERS IN CARTOONS (AND MANY SOLID OBJECTS AS WELL) HAVE EXTREMELY HIGH ELASTICITIES*

This principle is extremely well illustrated in *Roger Rabbit*. Everyone thinks Judge Doom (Christopher Lloyd) is an actual person, but after being flattened by a steamroller he is able to pop back into shape. He is in fact an insane toon! We see the elastic nature of toons demonstrated clearly during collisions with heavy objects like anvils all of the time. In addition, notice what happens to cartoon characters' eyes when surprised—they pop or stretch out of their heads like large elastic bands. This super-elasticity is present in a toon's whole body, or single body parts.

THE ANVIL PRINCIPLE: *THE HARDEST SUBSTANCE KNOWN IN THE CARTOON UNIVERSE IS THE ANVIL*

This is self evident.

THE GEOLOGY OF CARTOON ROCK HYPOTHESIS: *THE STRUCTURAL INTEGRITY OF ROCK IS VARIABLE*

There are many examples of head-on collisions between cartoon characters and solid or semisolid objects. How many times have we seen the Coyote fooled into a collision by a fake tunnel (usually painted on to a mountainside) or some other ruse perpetrated by the more sophisticated Roadrunner? In many cases the wall is extremely unforgiving and remains structurally coherent. However, the Coyote is usually flattened (literally) until he regains his shape due to the elastic principles that we previously mentioned. Nevertheless, in other situations, particularly in collisions with the ground after a long fall, the rock or soil is highly compressible, allowing the Coyote to create a large trench or hole out of which he must climb after the impact. Again this seeming inconsistency in physical principles is probably a result of the emotional state of the toon. With more time to contemplate the impact, the toon seems to be able to make the impact slightly less rigid, although obviously still painful.

THE HEISENBERG CARTOON PRINCIPLE: *THE OBSERVER (TOON) AFFECTS THE PHYSICAL OUTCOME OF ANY EXPERIMENT*

This phenomenon is encapsulated in many of the principles just discussed. In the cases of gravity, projectile motion, and collisions, clearly, the *perception* of the situation by a toon affects the outcome. (If only the makers of *What the Bleep* were attempting to describe the laws of *cartoon* physics we might have given them a much more favorable review.) As we discussed in the last chapter the equations of quantum mechanics in the real world can only be meaningfully applied at the atomic scale, but in the cartoon universe apparently quantum mechanical principles, as demonstrated by the Heisenberg cartoon principle, *do* apply to macroscopic systems.

THE MOVABLE HOLE PHENOMENON (THE LAW OF CONSERVATION OF HOLES): *THE TOTAL NUMBER OF HOLES IN THE UNIVERSE CAN NEITHER BE CREATED NOR DESTROYED, BUT THEY CAN EXIST OR CEASE TO EXIST FOR FINITE PERIODS OF TIME*

In layman's terms: A hole can be moved from one place to another but cannot be destroyed. Consider the following example. The Coyote is chasing the Roadrunner who paints a tunnel on the side of a mountain. The Roadrunner disappears into the tunnel, and yet the closely pursuing Coyote ends up smashing into a rigid wall. What exactly is happening here? This phenomenon is in fact both a conservation law and a quantum mechanical artifact. According to the uncertainty principle particles must be flashing in and out of existence over periods of time too small to detect the particle directly. These are called virtual particles, and their effects are detected indirectly in various experiments. In the cartoon universe the same phenomenon can sometimes be observed with holes. If the Road Runner paints a hole or tunnel in the rock, there must be a finite probability that the hole actually exists. However, in the time it takes the Coyote to get there he will not be able to detect it. If he did, it would be a violation of conservation of energy, and therefore it is no surprise to us that he inevitably collides with solid rock!

THE LAST-SECOND JUMP

Another thought-provoking cartoon physics phenomenon involves what might be called *the last second jump*—also known as *instantaneous frame of reference transfer*. The classic example involves

Bugs Bunny on an airplane heading vertically downward at a high velocity about to crash into the ground. We know, if we have seen enough Saturday morning *Looney Tunes,* that if Bugs jumps upward out of the plane at the last second before the impact, he'll be all right. This is because, unlike the world in which *we* live, as soon as Bugs jumps he instantly transfers his frame of reference. For example let's say that the airplane is traveling at a downward speed of 100 miles per hour relative to the ground. If Bugs can jump upward with an initial upward speed of 10 miles per hour (relative to the plane), which represents quite an impressive vertical leap potential (over a meter and a half) he should still be moving at 90 miles per hour relative to the ground upon impact. However, in the cartoon universe Bugs instantly transfers into the ground frame of reference. The downward airplane velocity vector vanishes and Bugs touches down without a scratch! The explanation for this phenomenon is still being researched but scientists at the Center for Advanced Study of Cartoon Physics have suggested that it has something to do with the silly way in which space and time are curved in the cartoon universe.

Index

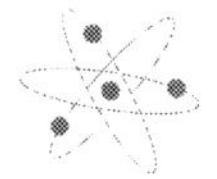